ISBN 978-1-330-02364-8
PIBN 10006551

PRACTICAL MATHEMATICS FOR BEGINNERS

PRACTICAL MATHEMATICS
FOR BEGINNERS

BY

FRANK CASTLE, M.I.M.E.

MECHANICAL LABORATORY, ROYAL COLLEGE OF SCIENCE, SOUTH KENSINGTON;
LECTURER IN MATHEMATICS, PRACTICAL GEOMETRY, MECHANICS, ETC.,
AT THE MORLEY COLLEGE, LONDON

London

MACMILLAN AND CO., Limited

NEW YORK : THE MACMILLAN COMPANY

1905

First Edition 1901.
Reprinted 1902. New Edition 1903, 1904, 1905 (twice).

CAJORI

GLASGOW : PRINTED AT THE UNIVERSITY PRESS
BY ROBERT MACLEHOSE AND CO. LTD.

PREFACE.

THE view that engineers and skilled artizans can be given a mathematical training through the agency of the calculations they are actually called upon to make at their work, steadily gains in popularity. · The ordinary method of spending many years upon the formal study of algebra, geometry, trigonometry, and the calculus may be of value in the development of the logical faculty, but it is unsuitable for the practical man, because he has neither the time nor the inclination to study along academic lines.

But though Practical Mathematics secures more and more adherents, the subject is still in a tentative stage. The recent revision of the syllabus issued by the Board of Education only two years after its first appearance, is evidence of this.

The present volume is designed to help students in classes where the new course of work issued from South Kensington forms the basis of the lessons of the winter session. Such students are supposed to be familiar with the simple rules of arithmetic, including vulgar fractions, hence the present volume commences with the decimal system of notation. The modern contracted methods of calculation, which are so useful in practical problems, are not taught in many schools and they are therefore introduced at an early stage.

In the extensive range of subjects included in the present volume care has been taken to avoid all work that partakes of the mere puzzle order, and only those processes of constant practical value have been introduced. Since, in mathematical teaching especially, "example is better than precept," a prominent place is given to typical worked out examples. In nearly all cases these are such as occur very frequently in the workshop or drawing office.

911250

The order in which the subjects are presented here merely represents that which has been found suitable for ordinary students. Teachers will have no difficulty in taking the different chapters in any order they prefer. Any student working without the aid of a teacher is recommended to skip judiciously during the first reading any part which presents exceptional difficulty to him.

So many practical examples of a technical kind, not usually to be found in mathematical books, have been included in this volume that some errors may have crept into the answers, but in view of the careful method of checking results which has been adopted, these will in all probability prove to be small in number.

I desire again strongly to emphasize what I have already said in another volume of somewhat similar scope. "Readers familiar with the published works of Prof. Perry, and those who have attended his lectures, will at once perceive how much of the plan of the book is due to his inspiration. But while claiming little originality, the writer has certainly endeavoured to give teachers of the subject the results of a long experience in instructing practical men how to apply the methods of the mathematician to their everyday work."

Mr. A. Hall, A.R.C.S., has read through some of the proof sheets, and I am indebted to him for this kindness. I also gratefully acknowledge my obligations to Prof. R. A. Gregory and to Mr. A. T. Simmons, B.Sc., not only for many useful suggestions in the preparation of my MSS., but also for their care and attention in reading through the whole of the proof sheets.

F. CASTLE.

LONDON, *August*, 1901.

PREFACE TO NEW EDITION.

Several important additions have been made in this edition. Sections dealing with Square Root, Quadratic Equations, and Problems leading to Quadratic Equations, have been added, and, where possible, more exercises have been introduced. Some corrections in the Answers have been made, and I am indebted to many teachers for calling my attention to the necessity for them; as it is too much to hope that there are no more mistakes in so large a number of figures, I shall be grateful to anyone who may call my attention to other inaccuracies.

In its present form the book is not only suitable for students of classes in connection with the Board of Education, but for candidates for the Matriculation examination of the London University under the new regulations; it will also assist students preparing for the Army and Navy Entrance examinations to answer the new type of questions recently introduced into the mathematical papers at these examinations.

F. C.

LONDON, *November*, 1902.

CONTENTS.

CHAPTER I.

In a similar manner, 8·073 would be read as *eight, point, nought, seven, three.*

The relative values of the digits to the left and right of the decimal point can be easily understood by tabulating the number 4321·2345 as follows :—

Thousands	Hundreds	Tens	Units		Tenths	Hundredths	Thousandths	Ten Thousandths
4	3	2	1	·	2	3	4	5

Also it will be obvious that in multiplying a decimal by 10 it is only necessary to move the decimal point *one* place to the *right*; in multiplying by 100 to move it *two* places to the right, and so on.

Similarly, the decimal point is moved *one* place to the *left* when dividing by 10, and two places when dividing by 100.

Addition and subtraction of decimal fractions.—When decimal fractions are to be added or subtracted, the rules of simple Arithmetic can be applied. The addition and subtraction of decimal fractions are performed exactly as in the ordinary addition and subtraction of whole numbers; the only precaution necessary to prevent mistakes is· to keep the decimal points under each other. For instance :

Ex. 1. Add together 36·053. ·0079, ·00095, 417·0, 85·5803, and ·00005.

```
 36·053
   ·0079
   ·00095
417·0
 85·5803
   ·00005
─────────
538·64220
```

Ex. 2. Subtract 578·9345 from 702·387.

```
702·387
578·9345
────────
123·4525
```

The decimal points are placed under each other, and the addition and subtraction are carried out as in the familiar methods for whole numbers.

EXERCISES. I.

Add together

1. 47·001, 2·110116, ·0401, and 75·81983.
2. 23·018706, 1·907, ·07831, and 1·006785.
3. 471·5132, 17·927, 8·00704, and 208·98.
4. 32·98764, 5·0946, ·087259, and ·56273·
5. 65·095, ·63874, 214 89, and ·0568.
6. 372·0647, 41·62835, ·964738, and 876.
7. ·7055, 324·88, 7·08213, and ·0621·

Subtract

8. 15·01853 from 47·06.
9. 708·960403 from 816·021.
10. 28·306703 from 501·28601.
11. 39·765496 from 140·3762.
12. 27·9876543 from 126·0123.
13. 13·9463 from 15·10485.
14. 23·872592 from 35·07316.
15. 22 94756 from 23·002.
16. 11·72013 from 113·408.

Multiplication of decimal fractions.—The process of the multiplication of decimal fractions is carried out in the same manner as in that of whole numbers. When the product has been obtained, then : **The decimal point is inserted in a position such that as many digits are to the right of it as there are digits following the decimal points in the multiplier and the multiplicand added together.**

Ex. 1. $36·42 \times 4·7$.

Multiplying 3642 by 47, we obtain the product 171174.   As there are two digits following the decimal point in the multiplicand and one digit following the decimal point in the multiplier, we point off three digits from the right of the product, giving as a result 171·174.

Ex. 2.   ·000025 × ·005.

Here $25 \times 5 = 125$.

In the multiplicand there are six digits following the decimal point, and in the multiplier three.   Hence the product is ·000000125.   The positions to the right of the decimal point, occupied by the six digits and the three digits referred to, are often spoken of as "decimal places"; thus ·000025 would be said to consist of six decimal places.

A similar method is used when three or more quantities have

to be multiplied together, as the following example will make clear :

Ex. 3. $2.75 \times .275 \times 27.5$.

The continued product of $275 \times 275 \times 275$ will be found to be 20796875. Now, there are two decimal places in the first multiplier, three in the second, and one in the last. This gives a total of six decimal places to be marked off from the right of the product. Hence, the required product is 20.796875.

In addition to applying this rule for determining the number of decimal places in the way shown, the student should mentally verify the work wherever possible. Thus, by inspection, it is seen that $.275$ is nearly $\frac{1}{4}$, and $\frac{1}{3}$ of 27 is 9. This result multiplied by 2 shows that the final product will contain two figures, followed by decimal places.

Ex. 4. $73.0214 \times .05031$.

The product obtained as in previous cases is 3.673706634.

In practice, instead of using the nine decimal places in such an answer as this, an approximate result is, as a rule, more valuable than the accurate one. The approximation consists in leaving out, or, as it is called, *rejecting* decimals, and the result is then said to be true to one, two, three, or more significant figures, depending upon the number of figures which are retained in the result.

The rule adopted is as follows :—*If the rejected figure is greater than 5, or, five followed by other figures, add one to the preceding figure on the left ; if the rejected figure is less than 5, the preceding figure remains unaltered.* When only one figure is to be rejected and that figure is 5, it is doubtful whether to increase the last figure or to leave it unaltered. An excellent rule is in such a case to leave the last figure as an even number, thus using this rule we should express 35.15 and 36.85 as 35.2 and 36.8 respectively.

In this manner a result may be stated to two, three, four, or more significant figures ; the last figure, although it may not be the actual one obtained in the working, is assumed to be the nearest to the true result.

Thus in Example 4, above, the result true to one decimal place is 3.7 ; the rejected figure 7 being greater than 5, the preceding figure 6 is increased by unity. The result, true to two places, is 3·67 ; the rejected figure 3 is less than 5, and the preceding figure is therefore unaltered. The result true to three and four decimal places would be 3·674 and 3·6737 respectively. Applying a rough check, in the way previously mentioned, it is easily seen, that as the multiplier lies between $\frac{5}{100}$ and $\frac{6}{100}$, the result lies between $73 \times \frac{5}{100}$ and $73 \times \frac{6}{100}$. In other words the result lies between 3·65 and 4·38.

In simple examples of this kind it may at first sight seem to be unnecessary to use a check, but, if in all cases the result is verified, the common mistakes of sending up, in examinations or on other occasions, results 10, 100, or more times, too great or too small (which the exercise of a little common sense would show to be inaccurate) would be avoided.

Significant figures.—When, as in Ex. 2 (p. 3), the result is a decimal fraction in which the point is followed by a number of cyphers, the result must include a sufficient number of *significant figures* to ensure that the result is sufficiently accurate. The term significant figure indicates the first figure to the right of the decimal point which is not a cypher. Thus, if the result of a calculation be ·0000026 this includes seven decimal figures ; but an error of 1 in the last figure would mean an error of 1 in 26, or nearly 4 per cent. (p. 21). If the result were 78·6726, then an error of 1 in the last figure would simply denote an error of 1 in 780,000, or, ·00013 per cent.

Again, in Ex. 2 (p. 3), the result ·000000125 must include the three significant figures 125, for an error of 1 in the last figure would mean an error of 1 in 125 or ·8 per cent.

Some common values.—There are many decimal fractions of such frequent occurrence in practice that it may be advisable to commit them and their equivalent vulgar fractions to memory.

Thus $·125 = \frac{125}{1000} = \frac{1}{8}$; $·25 = \frac{25}{100} = \frac{1}{4}$; $·375 = \frac{375}{1000} = \frac{3}{8}$; $·5 = \frac{5}{10} = \frac{1}{2}$; $·75 = \frac{75}{100} = \frac{3}{4}$.

It will be noticed that by remembering the first of the above results the other fractions can be obtained by multiplying it by 2, 3, etc., or in each case the result is obtained by mentally dividing the numerator by the denominator.

Conversion of a vulgar to a decimal fraction.—To convert a vulgar fraction to a decimal fraction, reduce the vulgar fraction to its lowest terms and then divide its numerator by its denominator.

Ex. 1. $\frac{9}{24} = \frac{3}{8} = 3 \div 8 = \cdot375$; $\frac{7}{8} = 7 \div 8 = \cdot875$.

Ex. 2. $\frac{1}{160} = \cdot00625$.   *Ex.* 3. $\frac{54}{125} = \cdot432$.

In many cases it will be found simpler and easier to reduce a fraction to its equivalent decimal if the numerator and denominator are first multiplied by some suitable number.

Ex. 4.   Reduce $\frac{7}{250}$ to a decimal.

Multiplying by 4 we get $\frac{28}{1000} = \cdot028$.

In a similar manner $\frac{1}{25} = \frac{4}{100} = \cdot04$.

Other examples can be worked in like manner.

In some cases the figures in the quotient do not stop, and we obtain what are called **recurring** (they are also called **repeating**, and sometimes **circulating**) decimals.

Ex. 5. $\frac{1}{3} = \cdot333\ldots$.

The result of the division is shown by as many threes as we care to write.   The notation $\cdot\dot{3}$ is used to denote this unending row.

Ex. 6.   Again $\frac{2}{3} = \cdot666 = \cdot\dot{6}$.

In each of these, and in similar cases, the equivalent vulgar fractions are obtained by writing 9 instead of 10 in the denominator, thus $\cdot\dot{3} = \frac{3}{9} = \frac{1}{3}$, etc.   In a similar manner $\frac{1}{7} = \cdot142857$, and these figures again recur over and over again as the division proceeds, hence $\frac{1}{7} = \cdot\dot{1}4285\dot{7}$.

When it is necessary to add or subtract recurring decimals, as many of the recurring figures as are necessary for the purpose in hand are written, and the addition or subtraction performed in the usual manner.   With a little practice the student soon becomes familiar with the more common recurring decimals.

Any decimal fraction, such as $\cdot\dot{3}$, $\cdot\dot{1}4285\dot{7}$ in which all the figures recur is called a **pure recurring decimal** ; the equivalent vulgar fraction is obtained by writing *for a numerator the figures that recur, and for the denominator as many nines as there are figures in the recurring decimal.*

When the decimal point is followed by some figures which do not recur and also by some which do recur, the fraction is called a **mixed recurring decimal**, and the equivalent fraction is obtained by *subtracting the non-recurring figures from all the figures to obtain the numerator, and by writing as many nines as there are recurring figures, followed by as many cyphers as there are non-recurring figures for the denominator.*

Ex. 7   Express as a vulgar fraction the recurring decimal $\cdot 1\dot{2}\dot{3}$.
Here there are two recurring figures and one not recurring,

$$\therefore \cdot 1\dot{2}\dot{3} = \frac{123-1}{990} = \frac{122}{990} = \frac{61}{495}.$$

Ex. 8. $\cdot 3\dot{2}65\dot{7} = \frac{32657-32}{99900} = \frac{32625}{99900} = \frac{145}{444}.$

Decimals of concrete quantities.—It is often necessary to express a given quantity as a fraction of another given quantity of the same kind.   Thus, in the case of £1. 15s., it is obvious that $15s. = \frac{15}{20}$ of 20 shillings, and £1. 15s. may be written £$1\frac{3}{4}$; or, $\frac{3}{4} = \cdot 75$, we may also write £1. 15s. as £1·75.

Ex. 1.   To reduce 10d. to the decimal of a pound.
As there are 240 pence in £1,

$$\therefore \text{ required fraction is } \tfrac{10}{240} = \tfrac{1}{24} = £\cdot 04167 \dots .$$

Ex. 2.   Express 7s. 6½d. as the decimal of a pound.
Here 7s. 6½d. $= 90\cdot 5$d.

$$\therefore \frac{90\cdot 5}{240} = \cdot 377.$$

And £1. 7s. 6½d. may be written £1·377.

Ex. 3.   Express 6 days 8 hours as the decimal of a week.
As there are 24 hours in a day,

$$6 \text{ days } 8 \text{ hours} = 6\tfrac{8}{24} = 6\tfrac{1}{3} \text{ days, } .$$

$$\therefore 6 \text{ days } 8 \text{ hours} = \frac{6\cdot\dot{3}}{7} = \cdot 90476\dot{1} \text{ week.}$$

Ex. 4.   Reduce 5d. to the decimal of 1s.

$$\tfrac{5}{12} = \cdot 41\dot{6}s.$$

Ex. 5.   Express in furlongs and poles the value of ·325 miles.
Here, multiplying by 8, the number of furlongs in a mile, we obtain 2·6, and multiplying the decimal ·6 by 40 (the number of poles in a furlong) we get 24 poles.
Hence ·325 mile = 2 fur. 24 po.

$$\begin{array}{r} \cdot 325 \\ 8 \\ \hline 2\cdot 600 \\ 40 \\ \hline 24\cdot 0 \end{array}$$

Ex. 6.   Reduce 9 inches to the decimal of a foot.

There are 12 in. in a foot.   Hence the question is to reduce $\frac{9}{12}$ to a decimal.

$$\therefore \ 9 \text{ in.} = \cdot 75 \text{ ft.}$$

Given a decimal of a quantity, its value can be obtained by the converse operation to that described.

Ex. 7.   Find the value of ·329 of £1.

The process is as follows : First multiplying by 20 we obtain the product 6580, and marking off three decimals we get the value 6·580 shillings.   In a similar manner multiplying by 12 and 4 as shown, we obtain the value of ·329 of £1, which is read as 6 shillings 6 pence 3 farthings and ·84 of a farthing.

$$
\begin{array}{r}
\cdot329 \\
20 \\
\hline
6\cdot580 \\
12 \\
\hline
6\cdot960 \\
4 \\
\hline
3\cdot840
\end{array}
$$

The result could be obtained also by multiplying ·329 by 240, the number of pence in £1, giving 78·96d. and afterwards expressing in shillings, etc.

Ex. 8.   Find the number of feet and inches in ·75 yard.

Here $\cdot75 \times 3 = 2\cdot25 \text{ feet,}$

and $\cdot25 \text{ ft.} = \cdot25 \times 12 \text{ in.}$

$$= 3 \text{ in.}$$

$$\therefore \ \cdot75 \text{ yard} = 2 \text{ ft. } 3 \text{ in.}$$

Contracted methods.—The results of all measurements are at best only an approximation to the truth.   Their accuracy depends upon the mode of measurement, and also, to some extent, on the quantity measured.   All that is requisite is to be sure that the *magnitude of the error is small compared with the quantity measured.*

It is clear that in a dimension involving several feet and inches, an error of a fraction of an inch would probably be quite unimportant.   But such an error would obviously not be allowable in a small dimension not itself exceeding a fraction of an inch.

By means of instruments such as *verniers, screw gauges,* etc., measurements may be made with some approach to accuracy. But these, or any scientific appliances, rarely give data correct beyond three or four decimal places.   Thus, if the diameter of a circle has been measured to ·001 inch, then, since *no result can be more exact than the data,* there is no gain in calculating the circumference of such a circle to more than three decimal places. Hence 3·1416 is a better value to use for the ratio of the diameter of a circle to its circumference than 3·14159. In such cases, too, the practical contracted methods of calculation are the best.

In a similar manner when areas and volumes are obtained by the multiplication of linear measured distances the arithmetical accuracy to any desired extent may be ensured by extending the number of significant figures in the result, but it should be remembered that the accuracy of any result does not depend on the number of significant figures to which the result is calculated, but on the accuracy with which the measurements or observations are made.

In any result obtained the last significant figure may not be accurate, but the figure preceding should be as accurate as possible. It is therefore advisable to carry the result to one place more than is required in the result.

It is evident that loss of time will be experienced if we multiply together two numbers in each of which several decimal figures occur, and after the product is obtained reject several decimals. Especially is this the case in practical questions in which the result is only required to be true to two or more significant figures. In all such cases what is known as Contracted Multiplication may be used.

Contracted multiplication.—In this method the multiplication by the highest figure of the multiplier is first performed. By this means the first partial product obtained is the most important one.

The method can be shown, and best understood by an example.

Ex. 1. Multiply ·006914 by 8·652.

The product of the two numbers can of course be found by the ordinary methods; and to compare the two methods, "ordinary" and "contracted," the product is obtained by both processes :

Ordinary Method.	*Contracted Method.*
6914	6914
8652	2568
13828	55312
34570	41484
41484	34670
55312	14828
·059819928	·059820

The ordinary method will be easily made out. In the contracted method the figures in the multiplier may be reversed, and the process continued as follows : Multiply first by 8, so obtaining 55312 ; next by 6—this step we will follow in detail —6 × 4 = 24, the 4 need not be written down (but if written it is cancelled as indicated), and the 2 is carried on. Continuing, 6 × 1 = 6, and adding on 2 gives 8. Next, 6 × 9 = 54, the 4 is entered ; and 6 × 6 gives 36, this with the 5 from the preceding figure gives 41, hence the four figures are 4148.

In the next line, multiplying by 5, we can obtain the two figures 0 and 7, but as these are not required unless there is some number to be carried, it is only necessary to obtain 69 × 5, and write down the product 345, add 1 for the figure rejected (because it is greater than 5) thus making 346. Finally, as 2 × 9 will give 18, we have to carry 1, and therefore we obtain 2 × 6 = 12, together with the one carried from the preceding figure which gives 13, add 1 for the figure (8) rejected, which gives 14. Adding all these partial products together we obtain the final product required.

Thus, in the second row one figure is rejected, in the next row two figures, and in the last row three figures are left out.

It may be noticed again, with advantage, that when the rejected figure is 5 or greater, the preceding figure is increased by 1, also that the last figure of the product is not trustworthy. Having noted (or cancelled) the rejected figures, as will be seen from the example, the decimal point is inserted as in the ordinary method, i.e. marking off in the product as many decimal places as there are in the multiplier and multiplicand together.

Though the multiplier is very often reversed, this is not necessary, except to avoid mistakes. The multiplier may be written in the usual way, and the work will then proceed from the left hand figure of the multiplier, i.e. the work is commenced by multiplying by 8 and not by 2.

Ex. 2. The circumference of a circle is obtained by multiplying the diameter of the circle by 3·1416. Find the circumference of a circle 13·25 inches diameter.

Here, we require the product of 13·25 and 3·1416.

$$\therefore \quad \begin{array}{r} 13\cdot25 \\ 61413 \\ \hline 3975 \\ 132\cancel{5} \\ 53\cancel{00} \\ 1\cancel{325} \\ \cancel{7950} \\ \hline 41\cdot63 \end{array}$$

Hence the required circumference is 41·63.

EXERCISES. II.

1. Multiply 6·234 by ·05473, leaving out all unnecessary figures in the work.

2. 4·326 by ·003457.   **3.** 8·09325 by 62·0091.

4. ·72465 by ·04306.   **5.** 5·80446 by ·10765·

6. 21·0021 by ·0098765.   **7.** 24·9735 by 30·307.

8. ·73001 by 7·30121.   **9.** ·053076 by 98·0035.

10. 3·12105 by ·905008.   **11.** ·0435075 by 3·40604.

12. 76·035 by ·0580079.   **13.** 5·61023 by ·597001.

14. 59·6159 by 3·0807.   **15.** ·020476 by 2·406.

16. 43·7246 by ·24805·   **17.** ·01785 by 87·29.

18. 40·637 by ·028403.   **19.** 2·030758 by 36·409·

20. 82·5604 by ·08425.   **21.** 6·04 by ·35.

22. 8·0327 by ·00698.   **23.** 390·086 by ·00598.

24. 4·327615 by ·003248.

25. Add together five-sevenths, three-sixteenths, and eleven-fourteenths of a cwt., and express the sum in lbs.

26. Express 9s. 4½d. as the decimal of £1. 7s.

27. Subtract ·035 of a guinea from 1·427 of a shilling.

28. Subtract 3·062 of an hour from 1·5347 of a day.

29. Add together ·0029 of a ton and ·273 cwts.

30. Reduce ·87525 of a mile to feet.

31. Find the sum of 2·35 of 2s. 1d. and 0·0̇3̇ of £6. 3s. 9d.

32. Add together $\frac{5}{9}$ of a guinea, $\frac{17}{20}$ of a half-crown, I$\frac{7}{36}$ shilling, and $\frac{1}{6}$ of a penny, and reduce the whole to the decimal fraction of a pound.

33. Express 3s. 3d. as the decimal of 10s.

34. Add together $\frac{5}{9}$ of 7s. 6d., 2·07 of £1. 8s. 2d., and $\frac{3}{4}$ of ·0671 of 16s. 8d. Express the answer in pence.

Division of Decimal Fractions.—The division of one quantity by another when decimals enter into the operation, is performed exactly as in the case of whole numbers. The process can be best explained by an example as follows :

Ex. 1.   Divide ·7 by ·176·

This may be described as finding a number, which, when multiplied by ·176, gives a product equal to ·7.

Though decimals may be divided as in the case of whole numbers, care is necessary in marking off the decimal point.   In the present, and in all simple cases, the position of the decimal point is evident on inspection.   Practically, it is often convenient to multiply both terms by 10, or some multiple of 10—100, etc.—and so obtain at once, without error, the position of the unit's figure, and hence of the decimal point.

Thus, in the above example, multiplying both terms by 10, we have to divide 7 by 1·76, and it is evident that the number required lies between 3 and 4.   This determines the position of the unit's figure.   As 7·0 is unaltered by adding any number of ciphers to the right, we add two for the purpose of the division.   Multiplying 1·76 by

$$1·76 \overline{)\,7·00\,(\,3·97}$$
$$\underline{5\ 28}$$
$$1\ 720$$
$$\underline{1\ 584}$$
$$1360$$
$$\underline{1232}$$
$$1280$$

3 we obtain 5·28, which, subtracted from 7·00, gives a remainder 1·72 ; to this we affix a cipher and carry on the division as far as necessary ; when this is done, we find ·7 ÷ ·176 = 3·9772727·

It will be seen that the ordinary method of performing division necessarily requires considerable space, especially when there are several figures in the quotient.

Italian Method.—Another method, referred to as the *Italian* method, in which only the results of the several subtractions are written down, is often used ; the method of procedure is as follows :
Note, as before, that 1·76 will divide into 7 ;
then, since 3 × 6 = 18, the 8 is not written down but is instead mentally subtracted from 10, leaving 2.   Next 3 × 7 = 21 and 1 carried makes 22 ; the 2 is again not written down, but instead, after the addition of unity (from the multiplication of 6 by 3), we say 3 from

$$1·76 \overline{)\,7·00\,(\,3·97}$$
$$1\ 720$$
$$\overline{}$$
$$1360$$
$$\overline{}$$
$$1280$$

10 = 7.   In a similar manner the remaining figure is obtained ; the next row of figures is arrived at by a like method and so on.

Comparing the two examples it will be seen, that as at each step of the work one line of figures is dispensed with, the working takes up far less room than is the case in the ordinary method.

It is obviously bad in principle to use more figures than are essential for the work in hand ; these are not only unnecessary, but give additional trouble, and also increase the risk of making mistakes. In many cases, students are found to work with ten or more decimal figures, when, owing to errors of observation, or measurement, or to slightly incorrect data, even the first decimal place may not be trustworthy. It is, of course, in-advisable to add an error of arithmetic to an uncertainty of measurement or data, but even a slight error is preferable to working out ten, or fifteen, places of decimals to a practical question, and when the result is arrived at, to proceed to reject the greater part of the figures obtained, leaving only two or three decimal places. To avoid this, what is known as con-tracted division is often adopted.

Contracted Division.—It is assumed that the student is familiar with the ordinary method of obtaining the quotient in the case of division. The long process of division can, how-ever, also be advantageously contracted. The method of doing this will be clear from the following worked example.

Ex. 1. Divide ·03168 by 4·208.

We shall work this example by the contracted method alone.

To begin with, the number 7 is obtained by the usual process of division. By multiplying the divisor by 7 the product 29456 is arrived at. When this is subtracted from 31680 the remainder 2224 is left. It is seen that if we drop or cancel the 8 from the divisor 4208, thus obtaining 420, it can be divided into the remainder 2224, five times. In multiplying by five we take account of the 8, thus, as 5×8 is 40, the 0 is not entered but the 4 is carried. Proceeding we have $0 \times 5 = 0$, and adding 4, we see this is the figure to be entered. Now proceed to the next and the following figures, obtaining in the usual way 2104 ; subtract this from 2224, and the remainder 120 is obtained.

$$4208 \,)\, 31680 \,(\, 7529$$

$$
\begin{array}{r}
29456 \\
\hline
2224 \\
2104 \\
\hline
120 \\
84 \\
\hline
36 \\
36 \\
\hline
\end{array}
$$

Proceeding in like manner with the multiplier 2, we obtain 84, which, subtracted from 120 leaves 36, and our last figure in the quotient is 9. By the method described on p. 12 the answer is written ·007529.

As the product of the divisor and quotient, when there is no remainder, is equal to the dividend, it follows that the dividend may be multiplied by any number if the quotient is divided by the same number. Thus, in the last example, if ·03168 is multiplied by 1000, then, 31·68 divided by 4·208 gives the result 7·259. Dividing this by 1000 we obtain the answer ·007259. This process of multiplying and dividing by 1000 simply means shifting the decimal point three places to the right in the divisor, and three places to the left in the quotient.

The above example shows that the method of contracted division consists in leaving out or, as it is called, rejecting a figure at each operation. Any number which would be added on to the next figure by the multiplication of the rejected figure is carried forward in the usual way. To avoid mistakes it may be convenient either to draw a line through each rejected figure of the divisor, or to place a dot under it.

Ex. 2.   When the circumference of a circle is given, the diameter is obtained by dividing the circumference by 3·1416.

The circumference of a circle is 41·63 inches; find the diameter of the circle.

$$
\begin{array}{r}
3\cdot1416\,)\,41\cdot630\,(\,13\cdot25 \\
31\cdot416 \\
\hline
10\cdot214 \\
9\cdot424 \\
\hline
790 \\
628 \\
\hline
162 \\
157 \\
\hline
5
\end{array}
$$

EXERCISES.   III.

Divide the following numbers, leaving out all unnecessary figures in the work.

1. ·43524 by 219·7962.
2. ·00729 by ·2735.
3. 24·495 by ·0426.
4. 131·95 by 4·375.
5. 33·511 by ·0713.
6. ·414 by 34·5·

7. 32·121 by 498.　　　　　　**8.** 166·648 by ·000563.

9. 1·6023 by 294.　　　　　　**10.** 7·3 by 584.

11. ·292262 by 32·7648.

12. Find the value of ·09735 ÷ 5·617 to four significant figures.

13. How many lengths of ·0375 of a foot are contained in 31·7297 feet ?

14. If sound travels at the rate of 1125 feet per second, in what time would the report of a gun be heard when fired at a distance of 1·375 miles ?

15. Find the value to four significant figures of 6·234 × ·05473, also ·09735 ÷ 5·617.

Divide

16. 19·305 by ·65.　　**17.** 325·46 by ·0187.　　**18.** 172·9 by 0·142.

Find the value of

19. $\dfrac{\frac{4}{5} \text{ of } 8\cdot236}{\frac{9}{10} \text{ of } \cdot138}$.　　　　　　**20.** $\dfrac{12\cdot4 + \cdot064 - \cdot066}{\cdot022}$.

21. Compute by contracted methods 23·07 × 0·1354, 2307 ÷ 1·354.

22. Compute 4·326 × 0·003457 and 0·01584 ÷ 2·104 each to four significant figures, leaving out all unnecessary figures in the work.

CHAPTER II.

Ratio.—The relation between two quantities of the same kind with respect to their relative magnitude is called Ratio.

In comparing the relative sizes of two objects it is a matter of common experience to refer to one as a multiple—two or three times, etc., the other ; or a sub-multiple—one-half, or one-third, etc., the other. This relation between two quantities of the same kind in respect of their relative magnitude, and in which the comparison may be made without reference to the exact size of either, is called Ratio.

Ratio may be written in three ways ; thus, if one quantity be 12 units and another 6 units the ratio may be expressed as $\frac{12}{6}$, $12 \div 6$, or omitting the line $12 : 6$. If there are two quantities in the ratio of 12 to 6, then the statement $12 : 6$ or $\frac{12}{6}$ indicates that the first number is twice the second ; or, the second quantity is one-half the first. Again, if two quantities are in the ratio $5 : 7$, then the first is $\frac{5}{7}$ of the second, or the second is $\frac{7}{5}$ of the first.

Quantities of the same kind are those which may be expressed in terms of the same unit.

The ratio of 12 things to 6 similar things is definite, and indicates that the number of one kind is twice that of the other ; but the ratio of 12 tables to 6 chairs conveys no meaning. Also it will be obvious that it is impossible to compare a length with an area, or an area with a volume, as, for example, the ratio of 3 inches to 4 square inches, or 4 square inches to 20 cubic inches, although in comparing two quantities of the same kind we can assert that one is twice, three times, or some multiple or sub-multiple of the other, without defining what the unit implies.

In making the comparison the magnitudes may be either abstract or concrete numbers, but the ratio between them must always be abstract, that is, merely a number.

Hence, it is necessary, in comparing magnitudes, that the quantities be written in terms of a common unit. For example, the ratio of 3 tons to 14 lbs., or the ratio of 10 feet to 4 inches is obtained by considering that as there are 2240 lbs. in a ton, the first named ratio would be $3 \times 2240 : 14$; the second, since 12 inches make 1 foot, would be $10 \times 12 : 4$.

When it is required to divide a number in a given ratio, it is only necessary to add together the two terms of the ratio for a common denominator, and take each in turn for a numerator.

Ex. 1. Divide £35 in the ratio of 2 : 5. The denominator becomes $2 + 5$, and the required amounts are $\frac{2}{7}$ of 35 and $\frac{5}{7}$ of $35 = £10$ and £25 respectively.

Beginners are often confused when required to divide a given number in the proportion of two or more fractions, and begin by taking the given fractions, instead of proceeding to reduce them to a common denominator. The way to proceed may be shown by an example :

Ex. 2. Divide £70 in the ratio of $\frac{1}{3}$ and $\frac{1}{4}$. This does *not* mean $\frac{1}{3}$ and $\frac{1}{4}$ of 70 ; but, as fractions with the same denominators are in the same proportion as their numerators, it is necessary to write $\frac{1}{3}$ as $\frac{4}{12}$ and $\frac{1}{4}$ as $\frac{3}{12}$. Then the question is to divide £70 in the ratio 3 : 4, and the required amounts are $\frac{3}{7}$ of $70 = £30$, and $\frac{4}{7}$ of $70 = £40$.

Ex. 3. Find the ratio of 1 ft. 3 in. to 6 ft 3 in.

Here as 1 ft. 3 in. = 15 in. and 6 ft. 3 in. = 75 in. the required ratio is $\frac{15}{75} = \frac{1}{5}$; or, the quantities are in the ratio of 1 to 5.

Ex. 4. A yard is 36 in. and a metre 39·37 in. Find the ratio of the length of a yard to that of a metre.

The ratio is $\dfrac{36}{39 \cdot 37} = \cdot 9144$.

Ratios of very small quantities.—In finding the ratio of one quantity to another, it is only the relative magnitudes of the two quantities which are of importance. The quantities themselves may be as small as possible, but the ratio of two very small quantities may be a comparatively large number.

Thus $\frac{1}{1000} = \cdot001$ is a small quantity, and so is $\cdot00001$, but the ratio of $\cdot001$ to $\cdot00001$ is $\dfrac{\cdot001}{\cdot00001} = 100$. Again $\cdot0000063$ is a very small number, and so is $\cdot0000081$, but the ratio of $\dfrac{\cdot0000081}{\cdot0000063}$ is simply $\frac{81}{63}$ or $\frac{9}{7} = 1\frac{2}{7}$. This very important fact concerning ratio is often lost sight of by beginners, and it must be carefully noted in making calculations.

Proportion.—The two ratios $2 : 4$ and $8 : 16$ are obviously equal, and their equality is expressed either by $2 : 4 = 8 : 16$, or by $2 : 4 : : 8 : 16$. The former is the better method of the two. When as in the given example, the two ratios are equal, the four terms are said to be in *proportion*, hence :

Four quantities are proportional, when the ratio of the first to the second is equal to the ratio of the third to the fourth. That is, when the first is the same multiple or sub-multiple of the second, which the third is of the fourth, the quantities are proportional.

We may thus state that the numbers 6, 8, 15, and 20 form a proportion. The proportion is written as $6 : 8 = 15 : 20$, and should be read " that the ratio of 6 to 8 is equal to the ratio of 15 to 20."

The first and last terms of a proportion are called the *extremes*, and the second and third terms the *means* ; in the last example 6 and 20 are the extremes, and 8 and 15 are the means in the proportion.

When four quantities are proportional, the product of the extremes is equal to the product of the means.

Thus $6 \times 20 = 8 \times 15$, or $\frac{6}{8} = \frac{15}{20}$, in which the proportion is written as the equality of two ratios.

Since the product of two of the terms of a proportion is equal to the product of the other two, it follows at once that if three terms of a proportion are given, the remaining one can be calculated.

Ex. 1. Find the second term of a proportion in which 14, 12 and 15 are respectively the 1st, 3rd and 4th terms.

$$14 : \text{required term} = 12 : 15 ;$$
$$\therefore \quad \text{required term} = \frac{15 \times 14}{12} = 17\frac{1}{2}.$$

To find the fourth proportional to three given quantities.—
When the first three terms of a proportion are given to obtain the
fourth we proceed as follows : Multiply the second by the third term
and divide the product by the first term.

Ex. 2. Find the fourth proportional to 2·5, 7·5 and 4·25.

$$\text{Fourth proportional} = \frac{7\cdot5 \times 4\cdot25}{2\cdot5} = 12\cdot75.$$

Hence 2·5 : 7·5 = 4·25 : 12·75.

Mean proportional to two given numbers.—This may be
taken to be a particular case of the last problem, in which the
second and third terms are alike. Thence, we have the rule :

Multiply the two given numbers together and find the square root
of the product. This will be the mean proportional required.

Ex. 3. Find the mean proportional to 10 and 40.

Here, $10 \times 40 = 400$;

∴ Mean proportional $= \sqrt{400} = 20.$

Hence, 20 is the mean proportional required.

[The rules for square root are explained on p. 27.]

Unitary method.—By the previous methods of simple proportion, we may proceed to find the remaining one when three
out of the four terms are known. In practice this plan of procedure may often be replaced by a convenient modification of it
called the *Unitary Method*, in which, given the cost, or value, of
a definite number of articles, or units, we may, by division, find
the value of one unit, and finally, the value of any number of
similar units by multiplication.

The method may be shown by the following simple example :

Ex. 1. If the cost of 112 articles be 10s., what will be the cost of
212 at the same rate ?

Using the three given terms, we may write the following proportion :

$$112 : 10 = 212 : \text{required term},$$

∴ required term $= \dfrac{10 \times 212}{112} = 18$s. 11½d.

By the unitary method we should proceed as follows :

If the cost of 112 articles be 10s., then the cost of one article
at the same rate is $\frac{10}{112}$s.,

therefore the cost of 212 articles is $\frac{10}{112}$ by 212s. = 18s. 11½d.

EXERCISES.   IV.

1. If a train travel 215 miles in 10 hrs. 45 min., what distance will it travel in $24\frac{1}{4}$ hrs. at the same rate?

2. In what time will 25 men do a piece of work which 12 men can do in 15 days?

3. Divide £814 among 3 persons in the ratios $\frac{2}{5} : \frac{3}{7} : \frac{5}{8}$.

4. If the carriage of 8 cwt. for 120 miles be 24s., what weight can be carried 32 miles, at the same rate, for 18s.?

5. Find a fourth proportional to ·45, ·8 and ·367.

6. Divide £56 between A, B, C and D in the ratio of the numbers 3, 5, 7 and 9.

7. Divide 204 into three parts proportional to the numbers 7, 8, 9.

8. Find the number that is to $7\frac{2}{3}$ in the ratio of £3. 1s. 3d. to £4. 13s. 11d.

9. A sum of £32818 is to be divided among four persons in the proportion of the fractions $\frac{2}{3}$, $\frac{3}{4}$, $\frac{4}{5}$ and $\frac{5}{6}$. Find the share of each.

10. Define ratio. Does it follow from your definition that it would be wrong to speak of the ratio of 5 tons to 3 miles, and if so how does it follow?

11. What should be the price of 194 dozen articles, if 391 such articles cost £21. 3s. 7d.?

12. If a parish pays £2165. 12s. 6d. for the repairing of $7\frac{1}{8}$ miles of road, what length of road would £1500 pay for at the same rate?

13. If the carriage of $3\frac{1}{2}$ tons for a distance of 39 miles cost 14s. 7d., what will be the carriage of 20 tons for a distance of 156 miles at half the former rate?

Percentages.—The ratio of two quantities, or the rate of increase or diminution of one quantity as compared with another of the same kind, is often expressed in the form of a percentage. The word "cent" simply denotes a "hundred," hence a percentage is simply a fraction with a denominator of a 100. This fact enables a comparison to be made at once, without the preparatory trouble of reducing the fractions to like denominators. Examples on percentages occur so frequently, and are so varied, that it is difficult to select typical illustrations. The following, however, may make the matter clear.

Suppose that two classes, of 20 and 50 students respectively, are expected to attend an examination. In the first named, 18 students, and in the second, 47 students, present themselves. Then, we may say that 2 in 20 and 3 in 50 were away from the examination ; but the comparison is most easily made by finding the percentage in each case. Thus, in the first case we

have absent $\dfrac{2}{20} \times 100 = 10$ per cent. ;

in the second case $\dfrac{3}{50} \times 100 = 6$ per cent.

These results would be written as 10% and 6%.

Ex. 1. Suppose the population of a town in 1885 was 15,990, and in 1890 was 20,550. The actual increase is $20550 - 15990 = 4560$; but although the actual increase is useful, it is much better to be able to state the rate at which the population is increasing for each 100 of its inhabitants. The increase for each 100 of its population is found by simple proportion as follows :

$$15990 : 100 :: 4560 : \text{increase required.}$$
$$\therefore \text{Increase for each } 100 = \frac{4560 \times 100}{15990} = 28 \cdot 5.$$

Thus, the increase for each 100 of its population, is 28·5. This number is called 28·5 per cent., and is written 28·5%. The rate per cent. permits of a ready reference to an increase or diminution of any kind.

Ex. 2. The population of another town in 1885 was 20,400, and in 1890 was 24,960. The actual increase (as before) is 4560, but it does not follow from this that the two towns are increasing at the same rate. In this case the rate of increase is obtained from :

$$20400 : 4560 :: 100 : \text{rate of increase} ;$$
$$\therefore \text{Rate of increase} = \frac{4560 \times 100}{20400} = 22 \cdot 3.$$

From these examples it is clear that the population of the latter town is not increasing as fast as the former by 6 per hundred, or, as usually written, by 6%.

In like manner, percentages are often used to compare the proportions of lunatics, paupers, criminals, etc., in the population of different towns.

Rate or debt collectors and others in many cases are paid at the rate of so much per cent. If a rate collector is paid at the rate of 2 per cent., for example, this would mean that for every £100 collected he is allowed £2 ; for every £50, £1, etc.

Ex. 3. If in a machine it is found that a quarter of the energy expended is wasted in frictional and other resistances, we should say that 25 per cent. is wasted, meaning that $\frac{25}{100}$ is useless.   This does not tell us the actual numerical amount of the loss ; all that we can infer is that for every 100 units of work expended on the machine 25 units disappear.   Such a percentage also enables a comparison to be made, and is a convenient method of expressing the efficiency of machines.   If one machine has an efficiency of 75 per cent. and another of 80 per cent., we know that the second is 5 per cent. more efficient than the first.

If, in addition, we know that 25 per cent. is the total loss due to all resistances, but 10 per cent. of this is due to the resistance of a particular part of the mechanism, this gives a *percentage of a percentage* and its numerical value is

$$\frac{25}{100} \text{ of } \frac{10}{100} = \frac{25}{100} \times \frac{10}{100} = \frac{250}{10000} = 2\cdot5, \text{ or } 2\tfrac{1}{2} \text{ per cent. } (\%).$$

Ex. 4.   A reef of quartz contains ·0044 per cent. of gold.   If the quartz produces £5. 12s. per ton, find the weight of a sovereign in grains.

$$£5. \quad 12s. = £5\tfrac{6}{10} = 5\cdot6 \text{ sovereigns,}$$
$$1 \text{ ton} = 2240 \times 7000 \text{ grains,}$$

and $\qquad$ ·0044 per cent. $= \dfrac{\cdot0044}{100} = \cdot000044.$

$$\therefore \text{ weight of } 5\cdot6 \text{ sovereigns} = \cdot000044 \times 2240 \times 7000$$
$$= \cdot44 \times 224 \times 7.$$
$$\therefore \text{ weight of 1 sovereign} = \frac{\cdot44 \times 224 \times 7}{5\cdot6}$$
$$= 123\cdot2 \text{ grains.}$$

Percentages and averages.—The data for practical calculations are in many cases either the result of measured quantities, or experimental observations—in each case liable to error.   To obtain a trustworthy resulct a omparatively large number of observations may be taken and the average or mean result calculated.   Such an average may be obtained by adding all the results together and dividing by the number of them.   When the average is thus arrived at it may generally be accepted as the best approximation to the truth.   Accepting it as correct, then the difference between it and any single value got by observation can be ascertained, and the error expressed most conveniently as a percentage.

EXERCISES. V.

1. The composition of bronze or gun metal is to be 91 % copper and 9 % tin. Find the weight of a cubic foot of the material. Also find the amount of copper and tin required to make 1000 lbs. of the alloy. (See Table I., p. 123.)

2. The composition of white or Babbit's metal is to be 4 parts copper, 8 antimony and 96 tin. Express these as percentages, find the weight of a cubic foot of the alloy, also the amount of each material required to make 200 lbs. of the metal.

3. If the cost of travelling by rail for 42 miles is 5s. 3d., what is the cost of travelling 35 miles at a price per mile 20 per cent. higher?

4. A collector receives 5 per cent. commission on all debts collected, and this commission amounts to £4. Find the amount collected.

5. In a class of 80 boys, $12\frac{1}{2}$ % failed to pass an examination. How many passed?

6. If the annual increase in the population of a state is 25 per thousand, and the present number of inhabitants is 2,624,000; what will the population be in three years' time? and what was it a year ago?

7. By selling coal at 15s. a ton a merchant lost 12 per cent. What would he have gained or lost per cent. if he had raised the price to 18s. 9d. per ton?

8. A man sold a horse for £36, losing 4 per cent. of the cost price. How much did he pay for the horse?

9. A 56-gallon cask is filled with a mixture of beer and water, in which there is 84 per cent. of beer. After 8 gallons are drawn off, the cask is filled up again with water. What is the percentage of beer in the new mixture?

10. If a dozen eggs are bought for 1s. 8d., for how much must they be sold singly to make a profit of 20 per cent.?

11. In an examination 60 per cent. of the candidates pass in each year. In 5 successive years the numbers examined are 1000, 840, 900, 1260, 1400; what is the average number of candidates per annum, and the average number of failures?

12. A farmer purchased 120 lambs at 30s. a head in the autumn. During the winter 12 died, but he sold the rest in May at 45s. each. What was his gain per cent.?

13. If 3500 baskets are purchased at $1\frac{3}{4}$d. each and sold at $2\frac{1}{2}$d. apiece, what will be the total gain and the gain per cent.?

14. For what amount should goods worth £1,900 be insured at 5 per cent., so that, in case of total loss, the premium and the value of the goods may be recovered?

CHAPTER III.

POWERS AND ROOTS.

Squares and cubes. Powers.—When a number is multiplied by itself the result is called the *square* or the *second power* of the number. Thus, the square of 3, or 3×3, is 9 ; and the square, or second power, of 4 is 4×4, that is, 16.

When three numbers of the same value, are multiplied together the result is called the *cube* or *third power* of the number ; thus, the cube of 3 is $3 \times 3 \times 3$, that is, 27. The cube of 4 is $4 \times 4 \times 4$, or 64.

In the same manner, the result of multiplying four numbers of the same value together is called the *fourth power* of the number. Hence, the fourth power of 4 is $4 \times 4 \times 4 \times 4$, that is 256.

By multiplying five of the same numbers together we should obtain the fifth power of the number. The sixth, seventh, or any other power, of a number is determined in a similar manner.

Index.—A convenient method of indicating the power of a number is by means of a small figure placed near the top and on the right-hand side of a figure ; thus the square or second power of 2 may be written 2×2, but more conveniently as 2^2, and the fourth power of 2 either as $2 \times 2 \times 2 \times 2$, or 2^4. Similarly the square, cube, fourth or fifth power of 4 would be written 4^2, 4^3, 4^4, and 4^5.

This method is also adopted when a number containing several digits is used. Thus 2574^2 signifies 2574×2574, and 63^3 means $63 \times 63 \times 63$. The small figure used to show how many times a given number is supposed to be written is called the *index* or exponent of the number.

Involution.—The process by which the powers of a number are obtained is called Involution.—The number itself is called the first power or the *root*, and the products are called the *powers* of the number.

The powers of 10 itself are easily remembered, and are as follows :

$$10^2 = 100, \quad 10^3 = 1000, \quad 10^6 = 1,000,000, \text{ etc.}$$

This method of indicating large numbers is very convenient in physical science, in which such numbers as 5 or 10 millions, etc., are of frequent occurrence ; for, in place of writing 5,000,000, for instance, we may write it more shortly, as 5×10^6.

Again, 5,830,000 could be written as $5 \cdot 83 \times 10^6$, and 5,830 as $5 \cdot 83 \times 10^3$.

The squares of all numbers from 1 to 10 are easy to remember ; they are as follows :

$1^2 = 1,$	$4^2 = 16,$	$7^2 = 49,$
$2^2 = 4,$	$5^2 = 25,$	$8^2 = 64,$
$3^2 = 9,$	$6^2 = 36,$	$9^2 = 81.$

The squares of all numbers from 10 to 20 might with advantage also be written down.

When a number consists of three or more figures its square or any higher power can be obtained by multiplication, or in many cases better by logarithms, which will be described later.

Evolution.—The process which is the reverse of Involution is that of extracting, or finding, the roots of any given numbers.

The root of a number is a number which, multiplied by itself a certain number of times, will produce the number.

Thus, the *square root* of a given number is that number which, when multiplied by itself, is equal to the given number.

The root of a given number may be denoted by the symbol $\sqrt{\ }$ placed before it, with a small figure indicating the nature of the root placed in the angle.

In this manner the cube root of 27 is denoted by $\sqrt[3]{27}$, the fourth root of 64 by $\sqrt[4]{64}$, and so on.

The square root would be denoted by $\sqrt[2]{9}$, but the 2 is usually omitted, and it is written more simply as $\sqrt{9}$.

Another, and for many purposes a better method, is to indicate the root by a fraction placed as an index, and referred to as a **fractional index**; thus, for example, the square root of 9 is written $9^{\frac{1}{2}}$, and is read as nine to the power one-half. Similarly, the cube root of 27 is written as $27^{\frac{1}{3}}$, meaning 27 to the power one-third.

Square root.—*Method I.*—To extract the square root of any quantity in cases where it is not possible to ascertain such root by inspection, we have to adopt a rule. The following example will illustrate the method of extracting a square root.

Ex. 1.   *Method I.*—Find the square root of 155236.

$$
\begin{array}{r}
15\dot{5}2\dot{3}\dot{6}\ \left(300+90+4\right. \\
90000 \\
\hline
(2\times300)+90=690\ )\ 65236 \\
62100 \\
\hline
2\times390+4=784\ )\ 3136 \\
3136 \\
\hline
\end{array}
$$

The process is as follows :

Divide the given number into periods of two figures each, by putting a point over the *unit's figure*, another on the figure 2 which is in the second place to the left of the 6, and a third also on the 5, as shown. The given number consists of six figures; the required square root contains three.

As $300^2 = 90,000$ and $400^2 = 160,000$, the required square root lies between 300 and 400; hence, we put 300 to the right of the given number, and subtract its square 90,000 from the number the quare root of which is required; this gives a remainder of 65236.

Put twice 300 to the left of the remainder 65236; this 600 divides into 65236 a little over 90 times; place 90 with the 300 in the place occupied by the square root and add 90 to 2×300, and thus obtain 690; this result multiplied by 90 gives a product of 62100; subtract this product from 65236, and the remainder 3136 is obtained.

Next, set down to the left of the remainder 3136, $2 \times 390 = 780$; this will divide into 3136, 4 times. Place 4 with the answer as shown.

Add 4 to 780, obtaining 784; multiply by 4 and obtain 3136; this subtracted from 3136 leaves no remainder; or 394 is the square root required.

Ex. 1.   *Method II.*—The ordinary practical method is as follows: Point as before, and find the largest number the square of which is less than 15 ; 3 is such a number. Set the figure 3 to the right of the given number and its square 9 under the first pair of figures 15 ; subtract 9 from 15, obtaining a remainder 6.

Bring down the next two figures, making the number 652.

Now put the double of 3, that is 6, on the left of the number 652, and by trial find that 6 will divide into 65 nine times. Put the 9 with the first figure of the square root on the right, and also on the left with the 6, and multiply 69 by 9 obtaining 621, which when subtracted from 652 gives a remainder of 31.

$$155236 \,(\, 394$$
$$\underline{9}$$
$$69 \,)\, 652$$
$$\underline{621}$$
$$784 \,)\, 3136$$
$$\underline{3136}$$

Bring down the next two figures, thus obtaining 3136. Double the number 39, the part of the root already found, and put the result 78 on the left, as shown.

By trial, find that 78 will divide into 313 four times. Put the 4 on the right with the other numbers, 39, of the square root which is being obtained, and also with the 78, making the number 784 on the left ; this last number multiplied by 4, the figure just added, gives 3136, which subtracted, leaves no remainder. Hence 394 is the square root required. If we proceed to extract the square root of 394 we obtain 19·85, and this is the *fourth root* of 155236 ;

$$\therefore\ 155236^{\frac{1}{4}} = \sqrt[4]{155236} = 19 \cdot 85.$$

The student should always begin to point at the unit's place, whether the given number consists of integers, or decimals, or both.

Ex. 2.   Find the square root of 1481·4801.

The pointing begins at the unit's place, and every alternate figure to the right and left of the unit's place is marked as indicated in the adjoining example.   As there are two dots to the left of the unit's place, the square root consists of the whole number 38 and the decimal ; the working is exactly the same as in the previous example.

$$1481 \cdot 4801 \,(\, 38 \cdot 49$$
$$\underline{9}$$
$$68 \,)\, 581$$
$$\underline{544}$$
$$764 \,)\, 3748$$
$$\underline{3056}$$
$$7689 \,)\, 69201$$
$$\underline{69201}$$

It should be observed that, to obtain the square root of a decimal fraction, the pointing should commence from the second figure of the decimal place.

Ex. 3.   Find the square root of ·9216.

$$·9\overset{.}{2}1\overset{.}{6} (·96$$
$$\underline{81}$$
$$186) \overline{1116}$$
$$\underline{1116}$$

The method adopted will be evident from the working shown.

As examples, obtain the square roots of the following frequently occurring numbers ; these should be worked out carefully, and the first two at least committed to memory.

$$\sqrt{2}=1·414\ldots,$$
$$\sqrt{3}=1·732\ldots,$$
$$\sqrt{5}=2·236\ldots,$$
$$\sqrt{6}=2·449\ldots.$$

The square root of each of these numbers is an unending decimal.   Thus, the square root of 3 can be carried to any number of decimal places, but the operation will not terminate. Such a square root is often called a **surd**, or an **incommensurable** number.

In any practical calculation in which *surds* occur, the value is usually not required to more than two or three decimal places.

If a number can be easily separated into factors, the square root can be obtained more readily.   The method adopted is to try in succession if the number is divisible by 4, 9, 16, and other numbers of which the square roots are known.

Ex. 4.   To find the square root of 1296.

$$1296 = 4 \times 324 = 4 \times 4 \times 81,$$
$$\therefore \ \sqrt{1296} = \sqrt{16 \times 81}$$
$$= 4 \times 9 = 36.$$

A similar method may be employed in the case of numbers the roots of which cannot be expressed as whole numbers.

Ex. 5. $\qquad\qquad \sqrt{128} = \sqrt{64 \times 2}$
$$= 8\sqrt{2} ;$$

and remembering that the $\sqrt{2}$ is 1·414 approximately, the value $8 \times 1·414 = 11·312$ can be found.

Ex. 6. $\qquad\qquad \sqrt{243} = \sqrt{81 \times 3}$
$$= 9\sqrt{3}.$$

In many cases where a surd quantity occurs in the denominator of a fraction, it will be found advisable, before proceeding to find the numerical value of the fraction, to transfer the surd from the denominator to the numerator. This is readily effected by multiplication.

Thus, if as a result to a given question we obtain the fraction $\frac{100}{\sqrt{3}}$, we may proceed to divide the numerator by $\sqrt{3}$ or $1\cdot732\ldots$ in order to obtain the numerical value of the fraction; but it is better, and simpler, to multiply both numerator and denominator by $\sqrt{3}$. This gives $\frac{100\sqrt{3}}{\sqrt{3}\times\sqrt{3}}=\frac{100\sqrt{3}}{3}$; and in this form, knowing that $\sqrt{3}=1\cdot732\ldots$, it is only necessary to move the decimal point two places to the right and divide by 3.

Square root of a vulgar fraction.—In finding the square root of such a fraction, it is necessary to obtain the square root of numerator and denominator.

Ex. 7. Find the square root of $2\frac{7}{9}$.

Here $\sqrt{2\frac{7}{9}}=\sqrt{\frac{25}{9}}=\frac{5}{3}$.

In a similar manner the square root of $20\frac{1}{4}$ is $4\cdot5$.

When the denominator is not a perfect square, we may proceed in some cases to first multiply both the numerator and denominator by the number which will make the denominator a perfect square. Or, we may multiply both numerator and denominator by the denominator.

Thus, to find the square root of $\frac{3}{8}$, we might find the square root of 3 and of 8 and divide one long number by another; but it is better to multiply thus

$$\frac{\sqrt{3}}{\sqrt{8}}=\frac{\sqrt{3}\times\sqrt{8}}{8}=\frac{\sqrt{24}}{8}=\frac{4\cdot898}{8},$$

when we only require to find one root instead of two; or, convert the given fraction to a decimal fraction, and find the root in the usual manner.

Ex. 8. Find the square root of $\frac{7}{8}$.

Here if the numerator and denominator be multiplied by 2, the fraction becomes $\frac{14}{16}$ and its square root is $\frac{\sqrt{14}}{4}$, which leaves only one root to be extracted.

Contracted method.—In practical calculations the square root of any quantity is never required to more than a few significant figures, and when more than half the required number of digits have been found, the remainder may be found by contracted division.

Ex. 9.   Obtain the square root of 13 to five places of decimals.

Here, proceeding as in the preceding examples, the square root of $13 = 3 \cdot 605$ is obtained together with a remainder 3975. The remaining figures of the square root may now be obtained by contracted division (see p. 13), viz., by dividing 3975 by 7205, giving 55, which is placed with the number already obtained.

Hence the required root is $3 \cdot 60555$.

$$
\begin{array}{r}
13\,(\,3 \cdot 60555 \\
9 \\
\hline
66\,)\ \ 400 \\
396 \\
\hline
7205\,)\ \ 40000 \\
36025 \\
\hline
7205\,)\ \ \ 3975\,(\,55 \\
3602 \\
\hline
373 \\
360 \\
\hline
13
\end{array}
$$

Cube root.—The arithmetical method of ascertaining the cube root of a number in all except the simplest cases is too tedious and unwieldy to be of any practical use. Indeed, it is not worth the time necessary to learn it, and it will be better to leave a consideration of cube roots until the student is familiar with the use of logarithms or the slide rule, by the help of which cube roots can be found easily and readily in any case.

EXERCISES. VI.

Find the square root of :

1. 37249. **2.** 4·9284. **3.** 1006009. **4.** 18671041. **5.** 122·1025.

6. 65 and 50 in each case to four decimal places, also of 8 to six decimal places.

7. 3263·8369· **8.** 450643·69. **9.** (i) $39\frac{1}{16}$; (ii) 40008·0004.

10. 90018·0009. **11.** 6877219041. **12.** 99·8001· **13.** 42436.

14. ·00501264. **15.** 18671041. **16.** 1085·0436.

17. Add together $\sqrt{53111 \cdot 8116}$, $\sqrt{20\frac{1}{4}}$ and $\sqrt{9}$.

18. Divide the square root of ·04 by $\sqrt{2\frac{7}{9}}$.

19. Find the value of

(i) $1 \cdot 2\sqrt{\frac{3}{8}}$,   (ii) $1 \cdot 2\sqrt{\frac{1}{2}}$,   (iii) $1 \cdot 2\sqrt{\frac{9}{16}}$,   (iv) $1 \cdot 2\sqrt{\frac{7}{8}}$,

in each case to two significant figures.

CHAPTER IV.

PLANE GEOMETRY.

Use of instruments.—Graphic methods are applicable to the majority of the problems which a practical man is called upon to solve. By means of a few mathematical instruments results may often be obtained which could only be arrived at

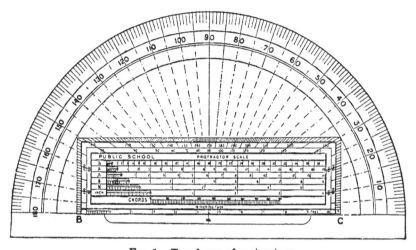

Fig. 1.—Two forms of protractors.

by mathematical methods after the solution of many difficulties. Even in problems in which sufficiently accurate results are not obtainable by graphical methods mathematical instruments may be used with advantage to check roughly the conclusions reached by calculation. To take a simple example ; if the lengths of the three sides of a triangle are given, then, by means of a simple formula, the area can be obtained to any degree of accuracy necessary. But it is also advisable to draw the triangle to a fairly large scale, for, since the area is one-half the product of the base and the perpendicular drawn from the base to the opposite vertex, the length of this perpendicular and the base can be measured and the product obtained, then half the product gives the area, and furnishes a ready method of checking the calculated result.

Accuracy and neatness are absolutely necessary in graphic work of any description. Distances should be measured and circles drawn as accurately as the instruments will permit. The instruments should be of fairly good quality ; the following are necessary, but others may be added if it is thought desirable.

(*a*) *Pair of pencil compasses*, (*b*) *pair of dividers*, (*c*) *protractor.* The last may be rectangular, as shown at *BC* (Fig. 1), or, better,

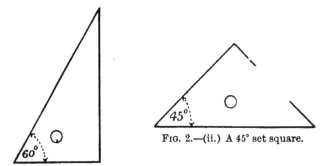

FIG. 2.—(ii.) A 45° set square.

FIG. 2.—(i.) A 60° set square.

a semi-circular one about 6″ diameter.   (*d*) *A* 60° *set square* about 9″ long, (*e*) 45° *set square* about 6″ long, (i), (ii) (Fig. 2), (*f*) a good *boxwood scale* (Fig. 10), (*g*) *drawing board*, imperial size (30″ × 22″), or if preferred half·imperial, and (*h*) a *tee-square* to suit the board.

The best point for the compass-lead is the flat or chisel point, and the lead used should be of medium hardness, for if it is too soft the point requires constant sharpening, and it is difficult to draw a good firm circle or arc ; if it is too hard scratches are made and the surface of the paper is spoilt ; when closed the point of the compass and lead should be on the same level.

Pencils.—Two pencils are requisite ; one an H., H.H., or H.H.H. sharpened to a flat chisel point (Fig. 3) ; the other an H.B., sharpened to a fine round point. The chisel-point should be sharpened so that when looked at from the end of the pencil the edge is invisible. The edge is made and maintained by using a small rectangular slip of fine glass paper.

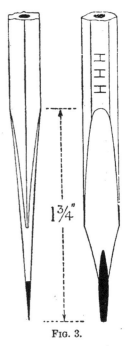

Measurement of angles.—If a straight line *OA* (Fig. 4) initially coincident with a fixed line *OC*, rotate about a centre *O*, and in the opposite direction to that in which the hands of a clock turn, then the number of **degrees** in the angle *COA* is the numerical measure of the angle.

Draw a circle of any convenient radius, and divide its circumference into 360 equal parts, then if two consecutive divisions be joined to the centre, the lines so drawn contain a length of arc equal to $\frac{1}{360}$th part of the circumference of the circle, and the angle between them is known as an angle of **one degree.** If, in the circle, two radii are drawn perpendicular to each

Fig. 3.

other, they enclose a quarter of the circle, and hence a right angle consists of 90 degrees, written 90°. Each degree is divided into 60 equal parts, or **minutes** ; and each minute is again subdivided into 60 equal parts called **seconds.**

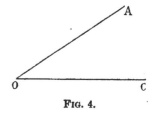

Abbreviations are used for these denominations. 52° 14′ 20″·5 denotes 52 degrees 14 minutes 20·5 seconds.

Fig. 4.

Fractional parts of a degree may be expressed either as decimals of a degree, or in minutes, seconds, and decimal parts of a second.

Thus, an angle may be expressed as, say, 54°·563, or, multiplying the decimal part by 60, we get 33·78 minutes ; again, multiplying ·78 by 60 we get 46·8 seconds. Therefore the given angle may be written either as 54°·563 or 54° 33′ 46″·8.

The length of the lines forming the two sides of the angle have no connection with the magnitude of the angle. Hence with centre O and any convenient radius describe a circle $CBDE$ as in Fig. 5 ; the line OA may be assumed to be a movable radius of the circle free to move about a centre O. When at A if an arc one-sixth of the circumference has been described, then the angle COA is an angle of 60 degrees, written as 60°. When coincident with OB the angle traced out will be an angle of 90°. When in a position OA' the angle COA' is greater than a right angle and is called an *obtuse angle*. The angle COA is less than 90° and is called an *acute angle*.

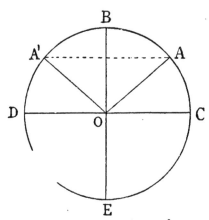

FIG. 5.—Measurement of angles.

Comparison of the magnitudes of angles.—A comparison of the magnitudes of two angles ABC and DEF (Fig. 6) may

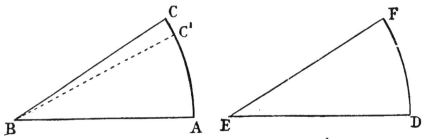

FIG. 6.—Comparison of the magnitudes of two angles.

be made by placing the angle DEF on the angle ABC, so that the point E exactly falls upon the point B, and the line DE coincides with the line AB. Then if the line EF falls on the line BC the angles are said to be equal. The angle DEF is less than the angle ABC if EF falls within BC, as shown by the dotted line BC'. It is larger if it falls outside BC.

This method of *superposition* is readily performed in the following way : Draw from centres B and E two equal arcs AC and DF, so that DE and EF in the one case are equal to AB and BC, respectively, in the other. If the point A be joined to the point C, and D to F, then, if AC is equal to DF, the angles ABC and DEF are obviously equal. Or, using a piece of tracing paper, make a tracing of DEF, and by placing the tracing on ABC, the comparison is readily made.

When, as in the angle DEF, there is only one angle at E it is usually written simply as the angle E.

To copy a given angle ABC.—This is obviously only a modification of the preceding construction. Thus, with centre B (Fig. 6) and any radius describe an arc cutting the two sides of the triangle in points A and C. With centre E, and same radius, describe an arc cutting ED in D, mark off a length DF equal to AC, join E to F. DEF is the required angle.

Ex. 1. Set out by a protractor an angle of 53°.

At the point P (Fig. 7) place the mark * shown on the pro-

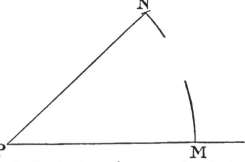

Fig. 7.—To set out an angle by means of a protractor.

tractor in Fig. 1 coincident with P, and the edge of the protractor BC with the line PM.

Make a mark opposite the division indicating 53° on the protractor. Remove the protractor and join the mark to P. An angle MPN containing 53° will have been made with the given line PM.

How to use a protractor to measure an angle.—When used to measure a given angle, the edge BC of the protractor is placed coincident with one of the lines forming the angle. The mark * on the protractor is made to coincide with the vertex of the angle, and the point where the other line crosses the division on the scale is noted ; this shows, in degrees, the angle required.

Scale of chords.—The protractors in general use, even when expensive instruments, are often inaccurate. It is not, moreover, an easy matter to read off an angle nearer than half a degree, and for many purposes this is not sufficiently accurate. A scale of chords is consequently often used. Such a scale is shown below.

Thus, to set out an angle of 53°·7. From the table of chords the chord of 53° is found to be ·892, difference for ·7° to be added is ·011 ∴ chord of 53°·7 = ·903.

With centre P (Fig. 7), radius 1 or 10 units, describe the arc MN. With centre M, and a distance ·903 if the radius is 1, or 9·03 if radius is 10, cut the arc at N; join P to N. Then MPN is an angle of 53°·7.

Conversely given an angle MPN (Fig. 7), with a radius 1 (or 10), describe arc MN, measure MN and refer to the scale of chords.

CHORDS OF ANGLES.

	0°	1°	2°	3°	4°	5°	6°	7°	8°	9°	·1° ·2° ·3°	·4° ·5° ·6°	·7° ·8°
0°	·000	·017	·035	·052	·070	·087	·105	·122	·140	·157	2 3 5	7 9 10	12 14
10°	·174	·191	·209	·226	·243	·261	·278	·296	·313	·330	2 3 5	7 9 10	12 14
20°	·347	·364	·382	·399	·416	·433	·450	·467	·484	·501	2 3 5	7 9 10	12 14
30°	·518	·534	·551	·568	·585	·601	·618	·635	·651	·667	2 3 5	7 8 10	12 13
40°	·684	·700	·717	·733	·749	·765	·781	·797	·813	·829	2 3 5	6 8 10	11 13
50°	·845	·861	·867	·892	·908	·923	·939	·954	·970	·985	2 3 5	6 8 9	11 12
60°	1·000	1·015	1·030	1·045	1·060	1·075	1·089	1·104	1·118	1·133	1 3 4	6 7 9	10 12
70°	1·147	1·161	1·175	1·190	1·203	1·218	1·231	1·245	1·259	1·272	1 3 4	6 7 8	10 11
80°	1·286	1·209	1·312	1·325	1·338	1·351	1·364	1·377	1·389	1·402	1 3 4	5 6 8	9 10

To bisect a given angle EBF.—With centre B (Fig. 8), and any convenient radius, draw an arc of a circle, cutting EB and BF in the points A and C. With A and C as centres and any equal radii, draw arcs intersecting at D. Join D to B, then BD bisects the given angle, EBF.

In this construction it is desirable to make the distances BA and BC as large as convenient, and also to arrange that the two arcs cutting each other shall cross as nearly as possible at right angles, for the point of intersection is then easily seen. If the given lines do not meet, then we may either produce them until they meet, or draw *CB* and *AB* two lines meeting at *B* parallel to and at equal distances from the two given lines.

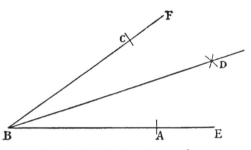

Fig. 8.—To bisect a given angle.

Angles of 60° and 30°.—*To set out OE at an angle of 60° to OC* (Fig. 9). With *O* as centre and any radius, draw an arc cutting *OC* at *B*. From *B* as centre with the *same* radius draw another arc cutting the former at *E*. Join *O* to *E*. Then *BOE* is an angle of 60°. Bisecting the angle we obtain an angle of 30°, by again bisecting an angle of 15°, and so on.

The angles referred to may be set out also by using the 60° set-square, or by the protractor.

Scales and their use.—The majority of problems considered are supposed to be solved by the process known as drawing to

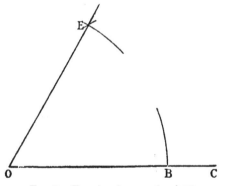

Fig. 9.—To set out an angle of 60°.

scale. In making a drawing of any large object, such as a building, it would be inconvenient, if not impossible, to make it as large as the object itself. In other words, to draw it *full size* is out of the question, but if a drawing were made in which every foot length of the building were represented on the drawing by a length of half an inch, the drawing would be said to be drawn to a scale of $\frac{1}{2}$ inch to a foot, or to a scale of $\frac{1}{24}$.

By means of a suitable scale any required dimension could be obtained as readily as if the drawing were made full size. In a

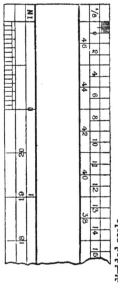

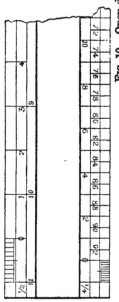

FIG. 10.—Open divided scale.

similar manner, if the drawing were made so that every length of 3 inches on the drawing represented an actual length of 12 inches, the scale would be said to be $\frac{1}{4}$. The fraction of $\frac{1}{24}$, or $\frac{1}{4}$, etc., is called the **representative fraction of the scale.**

Hence, *Representative fraction of a scale*

$$= \frac{\text{number of units in any line on the drawing}}{\text{number of units the line represents}}.$$

The term representative fraction is not always used, but, more shortly, the drawing is said to be made to a scale of $\frac{1}{4}$ or $\frac{1}{24}$.

When dimensions are inserted on a drawing a convenient notation is to use one dash $'$ to denote feet and two dashes $''$ for inches, thus a dimension of 1 ft. 3 in. could be written 1$'$ 3$''$.

Scales of boxwood or ivory are readily obtainable ; the former are cheaper than the latter, and the student should possess at least one good boxwood scale about 12 inches long. What is called an *open* divided scale will be found most useful. These can be obtained with the following scales : 1$''$, $\frac{1}{2}''$, $\frac{1}{4}''$, and $\frac{1}{8}''$ on one side, all divided in eighths. The same scales in tenths are found on the obverse side. Such a scale is shown in Fig. 10. These scales are divided up to the edge, which is made thin, as shown in the sections a and b, and so allows dimensions to be marked off direct from the scale with a fine-pointed pencil or pricker.

It is not advisable to use compasses or dividers, if it can be avoided, when transferring dimensions from scales. The frequent use of dividers soon wears away the divisions on the scale, and renders them useless for accurate measurements.

Division of a line into equal parts—Given any line AB (Fig. 11), to divide it into a number of equal parts is comparatively an easy task when an even number of parts are given, such as 2, 4, 8, etc. In such a case the line would be bisected by using the dividers, each part so obtained again bisected, etc.

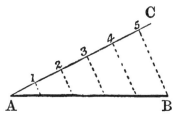

FIG. 11.—Division of a line into five equal parts.

When an odd number of parts are required, such as 5, etc., a length may be taken representing about one-fifth of AB (Fig. 11). This, on trial, may prove to be slightly longer or shorter than the necessary length. By alteration of the dimension in the required direction, and by repeated trials, a length is ultimately found which is exactly one-fifth. Much unnecessary time and labour may be spent in this way.

A better method is to set off a line AC (Fig. 11) at any convenient angle to AB, and to mark off *any* five equal lengths along AC from A to 5, and join 5 to B. If lines are drawn through the successive points 1, 2, 3, 4, parallel to the line $5B$, then AB will be divided into the required number of equal parts.

It will be obvious that the process of marking off a given number of equal distances along the line AC may be carried out by using the edge of a *strip of squared paper*, or a *piece of tracing paper* or *celluloid* on which a number of parallel lines have been drawn.

Conversely, given a line denoting a number of units, then the length of the unit adopted can be found.

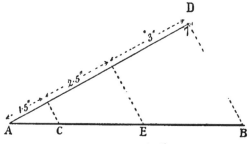

FIG. 12.—Division of a line into three segments in a given proportion.

Division of a line into three segments in a given proportion (say 1·5, 2·5, 3). Draw a line AD making any convenient acute angle with AB. Set off 7 units (equal to the sum of three segments) along AD. Join point 7 to B. Through points 1·5 and 2·5 draw lines parallel to $7B$. AB is thus divided at C and E in the required proportion.

Ex. 1. *To cut off a fraction of a given line.*—To cut off a fraction say $\frac{4}{7}$ of AB (Fig. 12). Draw AD at an acute angle to AB, and along AD mark off 7 equal divisions. Join 7 to B. Through 4 draw $4E$ parallel to $7B$; then $AE = \frac{4}{7}AB$.

Construction of scales.—The cheaper kinds of scales are often very inaccurate, those which are machine divided of the type shown in Fig. 10 are expensive, and it sometimes becomes necessary to substitute some simple form which can be readily made for oneself. For this purpose good cartridge paper, thin cardboard, or thin celluloid may be used. If the latter is employed the lines may be scratched on the surface by using a small needle mounted in the end of a penholder and projecting about a $\frac{1}{4}$ in. or $\frac{1}{2}$ in.

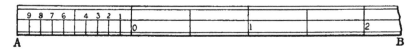

FIG. 13.—Construction of a simple scale.

To make a scale, two lines are drawn about a $\frac{1}{4}$ in. or $\frac{1}{2}$ in. apart. A number of divisions are then marked off along AB, Fig. 13, each one inch in length. The end division is subdivided into 10 equal parts. The lines denoting inches are made slightly longer than those indicating half inches, and these in turn longer than the remaining divisions. Finally, numbers are inserted as shown, the larger divisions being numbered from left to right, the smaller from right to left. When this notation is adopted any dimension such as 1·7″ can be estimated without risk of error by counting.

Other similar scales may be made as required.

In the preceding scale, although a dimension such as 1·7″ involving only one decimal place can be made accurately, yet to obtain a dimension such as 1·78 it would be necessary to further divide mentally the space between the 7th and 8th division into 10 equal parts, and to estimate as nearly as possible a length

equal to 8 of such parts. Such a method is a mere approximation. When distances involving two or more decimal places
have to be estimated other measuring instruments, such as
diagonal scales, verniers, screw-gauges, etc., are used.

Diagonal scale.—A diagonal scale of boxwood or ivory is
usually supplied with sets of mathematical instruments. They
can be purchased separately at a small outlay. To make such a

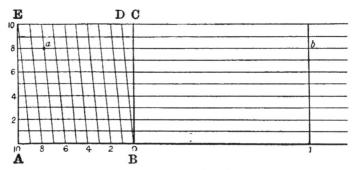

FIG. 14.—Diagonal scale.

scale, set off AB (Fig. 14) equal to 1 inch, draw BC perpendicular
to AB, and divide AB and BC each into ten equal parts; join
the point B to the first division on CE and draw the remaining
lines parallel to it as in the figure. A dimension 1·78 is the
distance from the point b to a, the point of intersection of a
sloping line through 7 and the horizontal line through 8.

Proportion. — It has
been shown (p. 19) that
when four quantities are
proportional we may
write them as $A:B=C:D$.
Given A, B, and C,
we proceed to find the
fourth proportional geo-
metrically as follows :

Draw two lines at any
convenient angle to each
other. In Fig. 15 the
lines are at right angles. Set off a distance $oa=A$ along
the vertical line, and a distance $ob=B$ along the horizontal

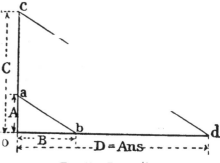

FIG. 15.—Proportion.

line. Join a to b. Set off the third quantity C along the vertical line, making $oc = C$; draw a line cd parallel to ab, and meeting ob produced at d. Then $od = D$ is the fourth proportional, or answer required.

Denoting the fourth proportional by x, $A : B = C : x$; or, multiplying extremes and means,

$$A \times x = B \times C ;$$

$$\therefore \ x = \frac{B \times C}{A}.$$

In many cases the value of a complicated fraction can be found by proportion by a similar geometrical method.

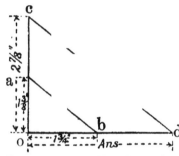

FIG. 16.—Simplification of a fraction.

Ex 1. Find the value of $\dfrac{1\frac{3}{4} \times 2\frac{7}{8}}{1\frac{3}{8}}$.

In Fig. 16, on a convenient scale, ob is made $= 1\frac{3}{4}$, $oa = 1\frac{3}{8}$, and $oc = 2\frac{7}{8}$.

Join a to b and through c draw cd parallel to ab, meeting ob produced at d.

Then od is the required result.

When measured, od will be found to be 3·7 units.

Mean proportional.—*To find a mean proportional to two given lines AB and AC.* Draw the two lines, as in Fig. 17, so

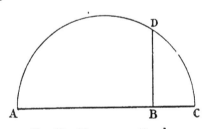

FIG. 17.—Mean proportional.

that they form together one line AC. On AC describe a semicircle, and at B draw a perpendicular BD meeting the semicircle in D. Then BD is the mean proportional required.

If the line AB to a given scale represent a certain number of units and BC one unit on the same scale, then BD is the square root of AB.

Square root.—The square root of a number is often required in practical calculations, and may be calculated as already explained on p. 27, or obtained by means of the slide rule (p. 149), or by graphical construction, as follows :

Ex. 1. *Find the square root of* $4\frac{3}{4}$.

Using any convenient scale, mark off $ab = 4\frac{3}{4}$, and $bc =$ unity on same scale (Fig. 18).

On ac describe a semicircle, and at b draw bd perpendicular to ac, and meeting the semicircle in d.

Then bd is the *square root* required.

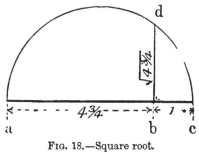

FIG. 18.—Square root.

The construction shown in Fig. 18 is the same as that of finding a Geometrical Mean or *the mean proportional of the two numbers* $4\frac{3}{4}$ *and unity*.

Ex. 2. *To obtain the fourth root of* $4\frac{3}{4}$ *or* $\sqrt[4]{4\frac{3}{4}}$.

Having obtained, as in the previous example, the square root bd, make be (Fig. 19) equal to bd. This is effected by using b as centre, bd as radius, and describing the arc be, meeting ac in e.

On ec describe a semicircle. Let f be the point of intersection of bd with the semicircle.

Then bf is the fourth root required.

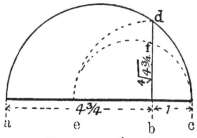

FIG. 19.—Fourth root.

In a similar manner the 8th, 16th, etc., roots, can be obtained.

Plane figures.—A triangle is a figure enclosed by *three lines* as AB, BC, and CA.

These lines form at their points of intersection three angles. The three *lines* are called the *sides* of the triangle. The angle formed at the point of intersection of the sides AB and BC may be called the angle ABC, but more simply the angle B. The two remaining angles are called A and C. Any one of its three angular points A, B, or C (Fig. 20) may be looked upon as the *vertex* and the opposite side is then called the *base* of

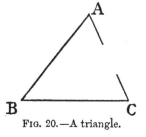

FIG. 20.—A triangle.

the triangle.   The *altitude* of a triangle is the perpendicular distance of the vertex from the base.

Equilateral triangle.—When the three sides of a triangle are equal, the triangle is an equilateral triangle ; the angles of the triangle are equal, each being 60°.

Isosceles triangle.—When two sides of a triangle are equal, the triangle is an isosceles triangle.

A right-angled triangle (Fig. 21) is a triangle one angle· (*C*) of which is a right angle ; the side (*AB*) opposite the right angle is called the *hypotenuse.*

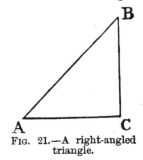

Fig. 21.—A right-angled triangle.

A parallelogram is a four-sided figure, the opposite sides of which are equal and parallel.

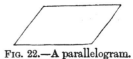

Fig. 22.—A parallelogram.

A **rectangle** is a parallelogram having each of its angles a right angle, or, in other words, each side is not only equal in length to the opposite side, but is also perpendicular to the two adjacent sides.

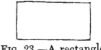

Fig. 23.—A rectangle.

A **square** is a parallelogram which has all its sides equal, and all its angles right angles.

Fig. 24.—A square.

Rhombus.—A rhombus is a parallelogram in which all the sides are equal but the angles are not right angles.

Fig. 25.—A rhombus.

The altitude of a parallelogram is the perpendicular distance between one of the sides assumed as a base and the opposite side.

The circle.—The curved line *ABED* (Fig. 26) which encloses a circle is called the *circumference.*   Any straight line such as *OA*, *OB*, etc., drawn from the centre to the circumference is a **radius,** and a line such as *AD* passing through the centre and terminated by the circumference is a **diameter** of the circle. A portion of a circle as *OBEC*, cut off by two radii, is a se**ctor** of a circle.

A line such as *BC* which does not pass through the centre is a chord, and the portion of the circle *BEC* cut off by it is called a segment of a circle. A line touching the circle is a ta**ngent,** the line joining the point of contact to the centre is at right angles to the tangent.

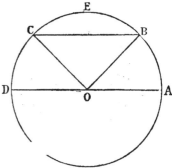

FIG. 26.—A circle.

Perimeter.—The term **perimeter** of a figure is used to denote *the sum of the lengths of all its sides*, thus the perimeter of a parallelogram is the sum of the lengths of its four sides.

Geometrical truths illustrated by means of instruments. —The following important geometrical truths may be verified by means of drawing instruments. Lengths are measured by a scale ; angles by a protractor or a scale of chords ; any necessary calculations are made arithmetically, and *tracing paper* may be used to show the equality of angles.

Parallel lines.—*When two parallel straight lines are crossed by a third straight line, the alternate angles are equal.*

Draw any two parallel lines (Fig. 27), and a third line *EF* crossing them. Show, by tracing the angles on a sheet of paper, or by measuring the angles, that the alternate pairs of ·angles marked × and *O* are in each case equal to each other.

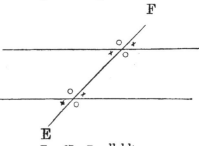

FIG. 27.—Parallel lines.

Also show by measurement that the four angles formed by the intersection of the third line with each of the parallel lines are equal to 360°.

Parallelogram.—Draw a parallelogram *ABCD* (Fig. 28), the longer sides being 3″, and the shorter 2″ long. Verify by measurement that the angle at B is equal to the angle at D, and the angle at A equal to the angle at C.   Join the points A and C and B and D.   The lines AC and BD are called the *diagonals* of the parallelogram. Verify that the two diagonals are bisected at O, their point of intersection.

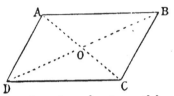

FIG. 28.—Opposite angles of a parallelo-gram are equal.

$\therefore AO = OC$, and $DO = OB$.

If a triangle and a parallelogram are on the same or equal bases and the altitude of each is the same, the area of the parallelogram is double that of the triangle.   Draw any parallelogram $ABCD$; join A to C (Fig. 29).   Cut the paper along AC and make the triangle ABC coincide with the triangle ADC. Or, using a piece of tracing paper, trace carefully the triangle ABC, then place it on ADC with B at D.   Note that the lines forming the triangles are coincident.

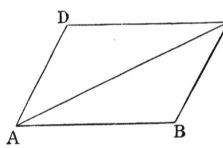

FIG. 29.—The area of a parallelogram is double that of a triangle on the same base and the same altitude.

Triangles.—The angles at the base of an isoceles triangle are equal.—Draw to scale an isoceles triangle ABC, that is, make the side $AB =$ side AC (Fig. 30). Show by measuring that the angle at C is equal to the angle at B.

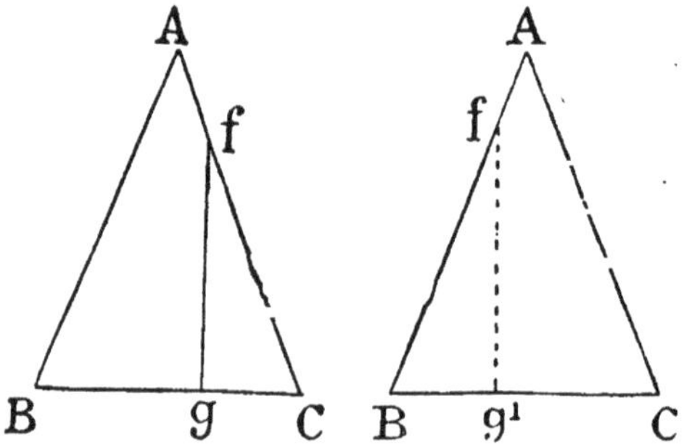

Also prove the equality by cutting out the angle at C and placing it on B. This may be effected by marking off a length

FIG. 30.—The angles at the base of an isosceles triangle are equal.

Cg not greater than half BC, and drawing gf perpendicular to

BC, meeting *AC* in *f* If *Cfg* be placed as shown in Fig. 30 with the angle *C* on *B* and *Cg* coinciding with *Bg'*, then the line *Cf* will be found to coincide with the line *BA*.

Or, using a piece of tracing paper, trace the triangle, fold the paper and see that the angles are equal.

If a line be drawn at right angles to the base of a triangle, and passing through the vertex it will bisect the base. Draw a triangle *ABC* with *AD* at right angles to the base. Make a tracing of the triangle *ACD*, then place it on the triangle *ABD*, with the point C coincident with *B*, all the other lines of the triangles can be made to coincide. Hence verify that the triangles *ACD* and *ABD* are equal, and *D* is the middle point of *BC*.

Equilateral triangle.—Make a triangle having all its three sides equal. (*a*) Measure by means of a protractor any one of the three angles and write down its magnitude; (*b*) carefully trace one of the angles on a piece of tracing paper, and placing the paper on each of the other two angles, verify that all the angles are equal and that the sum of the three angles is 180°.

Each side may be made equal to 3″; draw a line perpendicular to the base and passing through the vertex of the triangle. Verify by measurement that the line so drawn bisects the base, and also the vertical angle at *A*.

The three angles of a triangle are together equal to 180°—Draw the triangle *ABC* (Fig. 31), making the two sides *AC* and *BC* respectively equal to 4 and 3 units of length. Join *AB*, measure by the protractor the angles at *A*, *B*, and *C*; add the

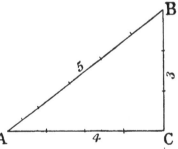

FIG. 31.—The three angles of any triangle are together equal to 180°.

values of the angles measured in degrees together, and ascertain if the angles *A*, *B* and C make 180°. It will be found that
.*B*=53° 7′, *A*=36° 53′, and C=90°.

Complement of an angle.—When the sum of two angles *A*, *B*, is equal to 90°, one angle is said to be the *complement* of the other. That is, *A* is the complement of *B*, or, *B* is the complement of *A*.

Right angled triangle.—The side opposite the right angle is called the *hypotenuse* and in a right-angled triangle the following relation always holds :

The square on the hypotenuse is equal to the sum of the squares on the other two sides.

Thus in Fig. 31, $3^2 + 4^2 = 25 = 5^2$.

As the two lines BC and CA in each case represent a certain number of units of length we can write the above statement simply as

$$AB = \sqrt{BC^2 + CA^2}.$$

Various values for BC and CA should be taken, and in each case it will be found that the relation holds good.

This property of a triangle, that when the three sides are proportional to the numbers 3, 4, and 5, the angle opposite the greater side is a right angle, is largely used by builders and others to obtain one line at right angles to another ; instead of 3, 4, and 5, any multiples of these numbers such as 6, 8, 10, etc., may be used ; also it is obvious that the unit used is not necessarily either an inch or foot, it may if necessary be a yard, or a chain, etc.

The greater angle of every triangle is subtended by the greater side.—Draw a triangle having its sides 9, 7, and 3 units, measure the angles with a protractor, verify that the sum of the three angles is equal to 180°, and carefully observe that the greatest angle is opposite the greatest side 9, and the smallest angle is opposite the smallest side 3.

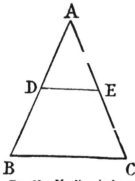

FIG. 32.—If a li ie is drawn parallel to the rbase of a triangle, the segments of the . sides are proportional.

If a line be drawn parallel to the base of a triangle, cutting the side or sides produced, the segments of the sides are proportional.

Draw a triangle ABC, and a line DE parallel to the base (Fig. 32) ; show by measurement that

$$\frac{BC}{AB} = \frac{DE}{AD} ;$$

∴ if DE is one-half BC, then AD is one-half of AB.

Similar figures.—Similar figures may be defined as exactly alike in form or shape although of different size; or, perhaps better, as figures having the same shape but drawn to different scales.

Two triangles are similar when the three angles of one are respectively equal to the three angles of the other. The student

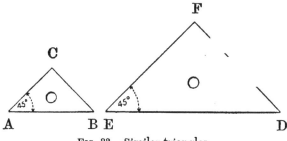

FIG. 33.—Similar triangles.

is already familiar with similar triangles in the form of set squares which may be obtained of various sizes. Obviously all the three angles of a 45° or 60° set square are the same whatever be the lengths of the sides.

As a simple illustration suppose that one side of a 45° set square be twice that of a corresponding side in another 45° set square, then the remaining sides of the second square are each twice those of the former. Thus if EF (Fig. 33) be twice AC, then it follows that ED is twice AB and DF is twice BC. It also follows that if one or more lines be drawn parallel to one side of a triangle the two sides are divided in the same proportion. Thus if in Fig. 34 BC be drawn parallel to the base DF, then $AB : AD = AC : AF = BC : DF$. This is sufficiently clear from Fig. 34, in which AD and AF are each divided into a number of equal parts and the ratio of AB to AD

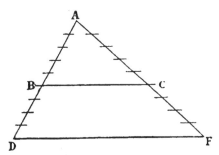

FIG. 34.—When two or more lines are drawn parallel to one side of a triangle, the two sides are divided in the same proportion.

is seen to be equal to the ratio of AC to AF or BC to DF. These important relations may be verified by drawing various triangles to scale.

D

Similar figures as in Fig. 35 may be divided into the same number of similar triangles. If each side of the figure *ABCDE* is three times the corresponding side of the other, then the line *AC* is three times *A'C'* and *AD* is three times *A'D'*.

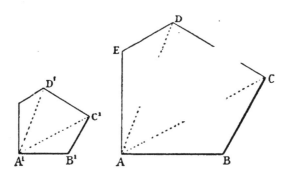

Fig. 35.—Similar figures.

Similar figures are not necessarily bounded by straight lines, the boundaries may consist of curved lines, as for example two maps of the same country may be drawn one to a scale of 1 inch to a mile, the other to a scale of $\frac{1}{4}$ inch to a mile, and any straight or curved line on the one will be four times the corresponding line on the other.

Circles.—The angle in a semicircle is a right angle. Draw a line *AB* to any convenient length, say 3 inches. On *AB* describe a semicircle, Fig. 36 and from any point *P* on the semicircle draw lines to *A* and *B*. Measure the angle *APB*, or test it by inserting the right angle of a set square. It will be found by taking several positions, and in each case joining to *A* and *B*, that the angle at *P* is always a right angle.

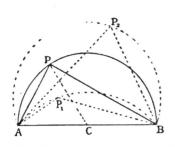

Fig. 36.—Angle in a segment of a circle.

In a similar manner it may be proved that in a segment less than a semicircle such as *AP₁B* Fig. 36, the angle formed is greater than a right angle, and when as at *AP₂B* the segment is greater than a semicircle the angle is less than a right angle.

Another important result is shown, where a line is drawn from P to the centre of the circle C, then as CA, CP, and CB are all radii of the same circle, they are obviously equal. Hence the line joining the middle point of the hypotenuse of a right angled triangle to the opposite angle is equal in length to half the hypotenuse.

Ex. 1. Construct a right-angled triangle in which the hypotenuse is $3.75''$ and one side is $1.97''$.

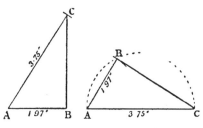

Draw two lines at right angles intersecting at a point B (Fig. 37); measure off $AB = 1.97''$, with A as centre and radius $3.75''$ describe an arc cutting BC at C.

Or, make AC equal to $3.75''$, and describe a semicircle on it.

FIG. 37.—Construction of a right-angled triangle.

Then with A as centre and with a radius $1.97''$ describe an arc intersecting the semicircle at B. Join B to A and C, then ABC is the triangle required.

If two chords in a circle cut one another, the rectangle on the segments of one of them is equal to the rectangle on the segments of the other.

Thus, if AC and BD be two chords in a circle cutting each other at a point E, the rectangle $AE . EC =$ rectangle $BE . ED$.

If, as shown in Fig. 38, the lines are perpendicular to each other, and one passes through the centre, then $BE = ED$;

$$\therefore\ AE . EC = ED^2.$$

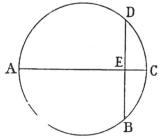

FIG. 38.—If two chords in a circle cut one another, the rectangle on the segments of one is equal to the rectangle on the segments of the other.

From this the graphic method of obtaining square root as shown on p. 43 is obtained.

If a quadrilateral ABCD be inscribed in a circle the sum of the opposite angles equals 180°.

Thus angle $ABC +$ angle $ADC = 180°$ (Fig. 39) and similarly angle $A +$ angle $C = 180°$.

Ex. 1.   Draw a circle 4″ diameter, take any four points on the circumference and join by straight lines as in Fig. 39.   Verify that the sum of the opposite angles in each case is 180°.

In any circle the angle at the centre is double the angle at the circumference, on the same or on equal arcs as bases.

Draw a circle of 3 or 4 inches diameter, select any two points, *A* and *C* (Fig. 40) on the circumference, and join to centre *O*.

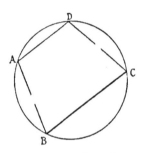
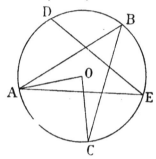

FIG. 39.—Quadrilateral in a circle.

FIG. 40.—On the same or on equal arcs the angle at the centre is double the angle at the circumference.

Also join *A* and *C* to any point *B* on the circumference. Measure the angles *AOC* and *ABC* by the protractor and prove the theorem.

Make $AD = AC$; join *A* and *D* to any point *E* on the circumference ; show that the angle

$$AED = ABC = \tfrac{1}{2}AOC.$$

Angles on the same, or on equal arcs, are equal to one another. Prove this by joining points *D*, *A* and *C* to different points on the circumference.

The products of parts of chords of a circle cutting one another are equal.—Draw a circle of 3 or 4 inches diameter and draw any diameter such as *DC* (Fig. 41) from any convenient point in *DC* produced, draw a line *PAB* cutting the circle in points *A* and *B*.   Measure the lengths, *PC*, *PD*, *PA*, and *PB*.   Show that $PD \times PC = PB \times PA$.

If from a given point P (Fig. 41) **a line PT be drawn touching the circle, and a line PAB cutting the circle in points A and B, then the rectangle contained by PB . PA, is equal to the square on PT.** Using the previous construction draw from *P* a tangent to the circle, measure the line PT, and verify that $PT^2 = PB . PA$, and therefore from previous result we have also

$$PT^2 = PC \times PD.$$

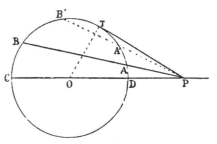

It will be noticed that as the angle DPB is increased the points A and B approach each other, thus taking some position as $PA'B'$, then the chord $A'B'$ is less than AB. When the two points of intersection are coincident we obtain the tangent PT.

Fig. 41.—$PT^2 = PB . PA.$

Hence the tangent may be taken to be a special case of a chord in which the two points of intersection are coincident.

To construct a triangle having given the length **of its three sides.** Draw a line AB (Fig. 42) equal in length to one of the given sides ; with one of the remaining lengths as radius, and A as centre, describe an arc ; with B as centre and remaining length obtain an arc intersecting the former in C. Join C to A and to B. ACB is the required triangle.

. *Ex.* 1. Three sides of a triangle measure 2·5, 1·83, and 2·24 inches respectively. Construct the triangle.

Draw AB (Fig. 42) equal to 2·5 inches.

With B as centre, radius 2·24 inches, describe an arc, and with A as centre and a radius of 1·83 describe an arc intersecting the former in C.

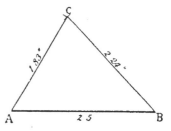

Fig. 42.—To construct a triangle having given the length of its three sides.

Join CA and CB. ABC is the required triangle.

The construction is obviously impossible when the sum of two sides is less than the third.

The angles may be measured by using the scale of chords. Do this and show the angles are as follows :—A is 60°, B is 45°, and C is 75°.

To construct a triangle **having given two sides and the** angle **included between the two** sides.

Ex. 2. Two sides of a triangle each measure 5·4 in. and the angle included between these sides is 40°. Construct the triangle and find the length of the remaining side of the triangle.

By means of the table of chords (p. 36), or by a protractor, set out at C (Fig. 43) an angle of 40°. Make CB and CA each equal to 5·4 in., join B to A. Then BCA is the triangle required. The length

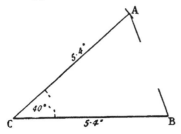

Fig. 43.—To construct a triangle given two sides and included angle.

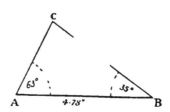

Fig. 44.—To construct a triangle given one side and two adjacent angles.

of BA will be found to be 3·69, and this is the length of the third side. If the angles be measured it will be found that A and B are each 70°.

To construct a triangle given one side and two adjacent angles.

Ex. 3. Construct a triangle having one side equal to 4·78 in. and the two adjacent angles equal to 35° and 63° respectively. Make the base AB (Fig. 44) 4·78 in. in length. At A and B set out, by the scale of chords or protractor, the angles $BAC=63°$ and the angle $ABC=35°$. If C is the point of intersection of the two lines, then ACB is the required triangle. Measuring the sides we find AC to be 2·77 in., and BC to be 4·3 in.

Given two sides and the angle opposite one of them to construct the triangle.

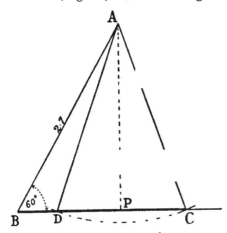

Fig. 45.—To construct a triangle given two sides and the angle opposite one of them.

Ex. 4. Construct a triangle having two sides 2·7 in. and 2·5 in. respectively and the angle opposite the latter equal to 60°. At any convenient point B (Fig. 45) set out an angle of 60°. Make $BA = 2·7$ in. With A as centre and a radius equal to 2·5 in. describe an arc. The arc so drawn may intersect the base in two points D and C. Join D and C to A. Then each of the triangles DBA or CBA satisfies the required conditions, and the remaining side may be either BD or BC. This is usually called the *ambiguous case.* If the arc just touches the line BC, one triangle only is possible; if the radius is less than AP the problem is impossible.

EXERCISES VII.

1. The side of an equilateral triangle is 10 inches; find the length of the perpendicular from the vertex to the base.

2. Two sides of a triangle are 12 feet and 20 feet respectively, and include an angle of 120°; find the length of the third side.

3. (i) Construct a right-angled triangle, base 1·75″, hypothenuse 3·25.

(ii) One side of a right-angled triangle is 29 ft. 6 in. and the adjacent acute angle 27°; find the hypothenuse.

4. Two sides of a triangle are 5 and 7, base 4 feet; find the length of the perpendicular drawn to the base from the opposite vertex, also find the area of the triangle.

5. Two sides of a triangle are 10·47 and 9·8 miles respectively, the included angle is 30°; find the third side.

6. Measure as accurately as you can the given angle BOA, also the length OA (Fig. 46). From A draw AM perpendicular to OB, measure OM and AM, divide OM by OA and AM by OA, and in each case give the quotient.

7. Draw a circle of 2·25″ radius. In this circle inscribe a quadrilateral $ABCD$ having given:

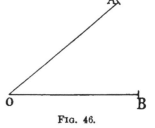

FIG. 46.

Sides $AB = 2·87″$, $DC = 2·5″$,
Angle $BCD = 76·5°$.

Measure the angle BAD. Draw the tangent to the circle at A. Join AC, and measure the angles which AC makes with the tangent. Also measure the angles ABC and ADC.

8. (i) Prove that the sides of a quadrilateral figure are together greater than the sum of its diagonals.

(ii) A quadrilateral $ABCD$ is made from the following measurements: The diagonals AC and BD cut in O at right angles, $OA = 3$ in., $OB = 4$ in., $OC = 8$ in., $OD = 6$ in. Show that a circle may be described about the quadrilateral.

9. In a triangle one side is 11·14 ft. and the two adjacent angles are 38° and 45°. Find the length of the side opposite the former angle.

10. A quadrilateral framework is made of rods loosely jointed together, and has its opposite sides equal. Show that when one side is held fast, all positions of the opposite side are parallel to one another.

11. Divide a line 6 in. long into nine equal parts. Construct a triangle with sides equal to 2, 3, and 4 of these parts respectively. Measure the angles of the triangle. Circumscribe the triangle with a circle and measure its radius.

12. Construct a triangle having its sides in the ratio $5 : 4 : 2$, the longest side being $2\frac{3}{4}''$ long.

13. In a triangle given that $a = 1\cdot56'$, $B = 57°$, $C = 63°$, find the two remaining sides.

14. The angles of a triangle being 150°, 18°, and 12°, and the longest side 10 ft. long, find the length of the shortest.

15. Find the least angle of the triangle whose sides are $7\cdot22$, 7, and $9\cdot3$.

16. Two sides of a triangle are $1\cdot75$ ft. and $1\cdot03$ ft., included angle 37° ; find the remaining parts of the triangle.

17. The perimeter of a right-angled triangle is 24 feet and its base is 8 feet ; find the other sides.

18. Find the least angle of the triangle whose sides are 5, 9, and 10 ft. respectively.

19. Determine the least angle and the area of the triangle whose sides are $72\cdot7$ feet, 129 feet, and $113\cdot7$ feet.

20. The two sides AB and BC of a triangle are $44\cdot7$ ft. and $96\cdot8$ ft. respectively, the angle ABC being 32°. Find (i) the length of the perpendicular drawn from A to BC ; (ii) the area of the triangle ABC ; (iii) the angles A and C.

21. In a triangle ABC, AD is the perpendicular on BC ; AB is $3\cdot25$ feet ; the angle B is 55°. Find the length of AD. If BC is $4\cdot67$ feet, what is the area of the triangle?
Find also BD and DC and AC. Your answers must be right to three significant figures.

22. The sides of a triangle are $3\cdot5$, $4\cdot81$ and $5\cdot95$ respectively ; find the three angles.

23. Construct a triangle, two sides 5 and 6 inches respectively, and having an angle of 30° opposite the former side ; find the remaining side.

24. Construct a triangle, one side $2\cdot5$ inches long and adjacent angles 68° and 45° respectively. What are the lengths of the remaining sides? Also find the area of the triangle.

25. Two parallel chords of a circle 12 in. diameter lie on the same side of the centre, and subtend 72° and 144° at the centre. Show that the distance between the chords is $3''$.

SECTION II. ALGEBRA.

CHAPTER V.

EVALUATION. ADDITION. SUBTRACTION.

Explanation of symbols.—In dealing with numbers, or digits, as the numerals 1, 2, 3 ... are called, accurate results can be obtained whatever be the unit employed. Thus the digit 7 may refer to 7 shillings, ounces, yards, or other units. In adding two digits, such as 7 and 5, together, we obtain the sum 12, whatever the unit employed may be.

The signs already made use of in Arithmetic are also employed in Algebra, but in Algebra representations of quantities

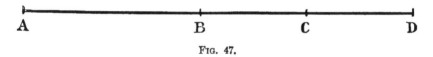

A   B   C   D

Fig. 47.

are utilised which have a further generality. Both letters and figures are used as symbols for numbers, or quantities. These numbers may be *known* numbers, and are then usually represented by the first letters of the alphabet, a, b, c, etc. ; or, they may be numbers which have to be found, called *unknown* numbers, and these are often denoted by x, y, z.

A more general meaning is given to the signs + and − in algebraical expressions than in arithmetic.

If a distance AC measured along a line AD (Fig. 47) from left to right is said to be positive, the distance CA measured in the opposite direction from right to left would be negative.

The result of the first measurement AC could be indicated by **+a**, while the same distance CA, but *measured in the opposite direction*, would be indicated by **−a**.

Again, if a length CD be measured in the same direction from left to right and be denoted by **+ b**, the length DC measured from right to left would be indicated by **−b**.

Hence $+a+b$ would mean the sum, or addition, of the two lines AC, CD, and so a line of length equal to AD is obtained, where

$$AD = AC + CD.$$

Similarly, $+a-b$ would denote the length AB obtained by measuring a length a in the positive, and a length b in the negative direction.

The beginner will probably experience some difficulty in the consideration of these positive and negative quantities. In Arithmetic the difficulty is avoided, for the only use that is made of the sign − (minus) is to denote the operation of subtraction, and in this idea the assumption is made that a quantity cannot be subtracted from another smaller than itself. Moreover in Arithmetic we are apt to assume that no quantity is less than 0.

In Algebra, on the contrary, we must get the conception of a quantity less than 0, in other words, of a *negative quantity*. Thus, as an illustration, consider the case of a person who neither owes nor possesses anything; his wealth may be represented by 0. Another person, who not only possesses nothing but owes £10, is worse off than the first, in fact he is worse off to the extent of £10 compared with the first person. His wealth may, therefore, be denoted by − £10.

Again, in what is called the Centigrade Thermometer the temperature at which water freezes is marked 0°, and that at which water boils 100°, and any temperature between these may be at once written down. But it is often necessary to refer to temperatures below the freezing point. To do this we represent the reading in degrees, but prefix a negative sign to indicate that we measure downwards instead of upwards. Thus +5° C. or, as it is usually written, 5° C. indicates five degrees above freezing point, whereas −5° C. indicates five degrees below zero.

If AC denote a distance of 4 miles in an easterly direction, and AB a distance of $2\frac{1}{2}$ miles (Fig. 47), then a person starting from A and walking 4 miles in an easterly direction will arrive at C. If when he arrives at C he then proceeds due west a distance equal to $1\frac{1}{2}$ miles he will arrive at B, and his distance from A will be $2\frac{1}{2}$ miles ; or, if as before, a denote the distance AC, and c the distance AB, then if BC be denoted by b, the distance from A would be expressed by $+a-b=+c$.

Algebraical sum.—When writing down an expression it is usual, where possible, to place the positive quantity first and to dispense with the $+$ sign. The above expressions would, therefore, always be written as $a+b$ and $a-b$. The signs placed between the numbers indicate in the first case the sum of two positive quantities, and in the second case the subtraction of one positive quantity from another. In the latter case the quantity $a-b$ could also be described as the addition of a minus quantity b to a positive quantity a, by which what is called the **algebraical sum** of the two quantities is obtained. **The algebraical sum of two or more quantities is, therefore, the result after carrying out the operations indicated by the signs before the quantities.**

The algebraical sum of $+10$ and $-18=-8$.

In the quantity $a-b$, if a represents a sum of money received, then $-b$ may represent a sum of money paid away. The algebraic sum is represented by the balance $a-b$.

It will be seen that in Algebra the word sum is used in a different and wider sense than in Arithmetic. Thus, in Arithmetic $7-3$ indicates that 3 is to be subtracted from 7, but in Algebra a similar expression means the sum of the two quantities whether the expression is in numbers, as $7-3$, or in letters, as in $a-b$.

How a product is expressed.—The arithmetical symbols of operation, $+$, $-$, $\times$, and $\div$, are used in Algebra, but are varied according to circumstances. The general sign for the multiplication of quantities is $\times$; but the product of single letters may be expressed by placing the letters one after another ; thus the product of a and b may be written $a \times b$ or $a \cdot b$, but is usually written as ab.

In a similar manner the product of 4, a, x, and y is expressed by $4axy$. It will be obvious that this method is not applicable in Arithmetic. Thus 5×7 cannot be written as 57.

The product of two quantities such as $a+b$ and $c+d$ may be expressed as $(a+b) \times (c+d)$, or, and usually, as $(a+b)(c+d)$.

Expression.   Quantity.—When, as in $a+b-c$, or $4axy$, several terms are joined together with or without signs, they together form what is called an *algebraic expression or quantity*.

Other names used in Algebra.—Any quantity, such as $4a$, indicates that a quantity a must be taken four times ; the multiplier, 4, of the letter is called a **coefficient** ; and $4a$, containing a coefficient and a letter, is called a **term**.

Multiples of the quantities a, b, c, etc., may be expressed by placing numbers before them as, $2a$, $3b$, $5x$; the numbers 2, 3, and 5 thus prefixed are called the *coefficients* of the *letters a, b, and x*.

When no coefficient is used the coefficient must be taken to be 1, thus x means $1 \times x$, bc indicates $1 \times b \times c$, etc.

The product of a quantity multiplied by itself any number of times is called a **power** of that quantity, and is indicated by writing the number of factors on the right of the quantity and above it. Thus :

$a \times a$ is called the square of a and is written a^2 ;

$b \times b \times b$ is called the cube of b and is written b^3.

Similarly, the multiplication of any number of the same letters together may be represented; thus c^n indicates c to the power n, where n represents a given number.

The number denoting the power of a given quantity is called its **index** (plural, indices) or **exponent**.

It is very important that the distinction between coefficient and index be clearly understood. Thus $4a$ and a^4 are quite different terms.

Let $a=2$, then $4a=8$; but $a^4=2^4=16$.

The use of signs may be exemplified in the following manner ·

Ex. 1.   In the expression a^2+b-c,
Let $a=4$, $b=7$, and $c=3$.
Then $a^2+b-c=4^2+7-3=23-3=20$.

Ex. 2.   Find the value of d from the equation $d=1\cdot2\sqrt{t}$, (i) when $t=\frac{9}{16}$; (ii) when $t=\frac{3}{8}$.

(i) $d=1\cdot2\sqrt{\frac{9}{16}}=1\cdot2 \times \frac{3}{4}=\cdot9$.

(ii) $d=1\cdot2\sqrt{\frac{3}{8}}=1\cdot2\frac{\sqrt{24}}{8}=\frac{1\cdot2 \times 4\cdot898}{8}=\cdot7347\cdot$

Ex. 3. Find the value of
$$\frac{ax^2+b^2}{bx-a^2-c}, \text{ when } a=3,\ b=5,\ c=2,\ x=6.$$

Here $\qquad ax^2+b^2=3\times 6^2+5^2\quad =133,$

Also $\qquad bx-a^2-c=5\times 6-3^2-2=\ 19,$

$$\therefore\ \frac{ax^2+b^2}{bx-a^2-c}=\frac{133}{19}=7.$$

Ex. 4. Find the value of $c\sqrt{10ab}+b\sqrt{8ac}+a\sqrt{45bc}$, when $a=8$, $b=5$, $c=1$.

Substituting the given values in the expression given we obtain
$$\sqrt{10\times 8\times 5}+5\sqrt{8\times 8\times 1}+8\sqrt{45\times 5\times 1}$$
$$=\sqrt{400}+5\times 8+8\times 15$$
$$=20+40+120=180.$$

Ex. 5. Find the value of
$$(ac-bd)\sqrt{a^2bc+b^2cd+c^2ad-2},$$
when $a=1$, $b=2$, $c=3$, $d=0$.

Substituting these values in the given expression we obtain
$$(3-0)\sqrt{1\times 2\times 3+4\times 3\times 0+9\times 1\times 0-2}$$
$$=3\sqrt{6-2}=6.$$

In Ex. 5 it should be carefully noticed that, as one of the given terms d is equal to 0, any *term* containing that letter must be 0. Hence, we may either omit all the terms containing that letter or, by writing them as in the above example, the terms in which the letter occurs are each seen to be equal to zero.

EXERCISES. VIII.

If $a=4$, $b=3$, $c=2$, $d=0$, find the value of

1. $3a+2b+c.$
2. $2a-2b-c.$
3. $ab^2c^3d^4.$
4. $a-b^2+c^3-d^4.$

Find the value of

5. $a^3+3a^2b+3ab^2+b^3$ when $a=1$, $b=2$.

6. The resistance of a wire is given by $R=\dfrac{l}{a}k.$ Given $l=100$, $a=\cdot002$, and $k=\cdot00002117$ find the value of R.

7. The heating effect of a current is given by $H=\cdot24C^2Rt.$ Find H given $C=20$, $R=30$, $t=60$.

8. $HP=\dfrac{CV}{746}.$ This is the relation between horse power, current in amperes, and volts. Given $C=30$, $V=100$. Find HP.

If $a=1$, $b=-1$, $c=2$, $d=0$, find the value of

9. $\dfrac{a+b}{a-b}+\dfrac{c+d}{c-d}+\dfrac{ad-bc}{bd+ac}-\dfrac{c^2-d^2}{a^2+b^2}$

When $a=10$, $b=3$, $c=7$ of,

10. $\dfrac{b+c}{2c-3b}$.

11. $\dfrac{3c}{4a+2}+\dfrac{4bc}{10a-16}$.

12. $\sqrt{\dfrac{3b+c}{c-b}}-\sqrt{\dfrac{9b+7c-12}{c-b}}$.

When $a=5$, $b=4$, $c=3$ of,

13. $\dfrac{7b+c}{3a+4b}$.

14. $\dfrac{7a}{11b-3c}+\dfrac{11b}{8b-7c}-\dfrac{10c}{7a-5b}$.

Addition.—The addition of algebraical quantities denotes the expression in one sum of all like quantities, regard being had to their signs.

When like quantities have the same sign, their sum is found by adding the coefficients of similar terms and annexing the common letters. Thus $7a+4a=11a$. Also,
$$7a+3a+3b+5a=15a+3b.$$

Ex. 1.   Add together : $7a+5b$, $-5a+4b$, $3a-2b$.

When several such quantities have to be added together, they may be arranged so that all terms having the same letter, or letters, and the same powers of the letter, or letters, are in columns as shown ; the positive and negative coefficients in each column are then added separately, and the sign of the greater value is prefixed to the common letters. The operation would proceed as follows. Arrange in columns, placing the letters in alphabetical order.

$$\begin{aligned} 7a+5b \\ -5a+4b \\ \underline{3a-2b} \\ 5a+7b \end{aligned}$$

Commencing with the row on the left-hand side, we have $7a+3a=10a$. Now add to this $-5a$; or, in other words, from $10a$ subtract $5a$, and the result is $5a$. Again $5b+4b=9b$; and $9b-2b=7b$. Hence the sum required is $5a+7b$.

Ex. 2.   Add   together : $7x^2-9y^2+20xy$, $-11x^2+10y^2-6xy$, $8x^2-7y^2-4xy$.

First adding together the coefficients of the terms in x^2 we get $7-11+8=4$.

In a similar manner the coefficients of y^2 are $-9+10-7=-6$; and of xy are $20-6-4=10$.

Hence, the required sum is $4x^2-6y^2+10xy$.

It is not necessary to separate the coefficients and to write them down.   It is much better to perform the addition mentally.

EXERCISES. IX.

Add together

1. $b + \frac{1}{2}c - \frac{1}{3}a$, $c + \frac{1}{2}a - \frac{1}{3}b$, $a + \frac{1}{2}b - \frac{1}{3}c$.

2. $ax^2 - bx^2 + cy^2 - ab^2c$, $3\frac{1}{3}ax^2 - 3\frac{1}{3}bx^2 + cy^2 + 4ab^2c$,
$\qquad 5\frac{1}{2}ax^2 - 5\frac{1}{2}bx^2 - 3cy^2 - 2ab^2c$, $-7ax^2 + 7bx^2 + cy^2 - ab^2c$.

3. $6m - 13n + 5p$, $8m - 9p + n$, $-p + m$, $n + 5p + m$.

4. $7a + 5b - 13c$, $-b + 4c + a$, $3b + 3c - 3a$, and find the value of the result when $a = 1$, $b = 2$, $c = 3$.

5. $2x + 4y - z$, $2z - 3y - 2x$, $3x - z$, $2y - x$.

6. $a - 2b + 2c$, $b - 3c + 3a$, $c - 4a + 4b$.

7. $ax^2 - 2dx^2 - 2x + 2c - 3f$, $ax^2 + 2dx^2 - bx + cx - 1$,
$\qquad ax^2 - dx^2 - bx - cx - c + 1$, $-x^2 + 3bx - c + 2f$.

Find the sum of

8. $2(a + b + c)$, $3(a + b - c)$, $4(a + c - b)$, $b + c - a$; and obtain its numerical value if $a = \frac{1}{4}$, $b = \frac{1}{6}$, $c = \frac{1}{12}$.

9. $\frac{1}{4}x^4 + \frac{2}{7}y^4 + \frac{1}{3}x^3y - \frac{1}{5}x^2y^2$ and $\frac{3}{4}x^4 + \frac{5}{7}y^4 - \frac{1}{3}x^3y - \frac{4}{5}x^2y^2$.

Simplify

10. $4x^2 + 8xy + 4y^2 - 9x^2 + 18xy - 9y^2$.

11. $x^6 + 3x^5 + 7x^4 + 15x^3 + 10x^2 - 3x^5 - 9x^4 - 21x^3 - 45x^2 - 30x$
$\qquad + 2x^4 + 6x^3 + 14x^2 + 30x + 20$.

12. Simplify and find the value of $x^2 + xy + x - x^2 + xy - x^2$
$\qquad + y^2 + 66$, when $x = 100$, $y = 50$.

13. $4x - 16x^2 + 16y^2 + 4x + 16x^2 - 16y^2 - 4y$ when $x = 1$, $y = 0$.

Subtraction.—In Algebra, **to perform the operation of subtraction, arrange like terms together as in addition, change the signs of all the terms to be subtracted, and then add to the other expression.** Thus, to subtract $7a$ from $13a$, we reverse the sign of $7a$ and so make it minus; for $13a - 7a$ is only another way of expressing that $7a$ is to be subtracted from $13a$. Thus $13a - 7a = 6a$.

Ex. 1. From $5a + 3x - 2b$ subtract $2c - 4y$. The quantity to be subtracted, when its signs are changed, is $-2c + 4y$,
$\qquad \therefore$ the remainder is $5a + 3x - 2b - 2c + 4y$.

Ex. 2. Subtract $a^2 - 2b - 2c$ from $3a^2 - 4b + 6c$.

Here, after arranging as in addition and changing the signs, we proceed as in addition, thus:
$3a^2 - a^2 = 2a^2$; $2b - 4b = -2b$; and, finally,
$\qquad 6c + 2c = 8c$.

$$\begin{array}{r} 3a^2 - 4b + 6c \\ -\ a^2 + 2b + 2c \\ \hline 2a^2 - 2b + 8c \end{array}$$

Hence, the result is $2a^2 - 2b + 8c$.

It is not necessary to perform on paper the actual operation of changing the signs of all the terms in the expression to be subtracted. The operation should be carried out mentally.

Ex. 3.   From $7x^2 - 2x + 5$ subtract $3x^2 + 5x - 1$.

Here we may, as in Ex. 2, write the terms under each other. Then, after mentally altering the sign of $3x^2$, we obtain by addition $4x^2$. Again mentally altering the sign of $5x$ and adding to $-2x$ we obtain $-7x$; and, finally, repeating the operation for the last figure we get the number 6. Hence the result is $4x^2 - 7x + 6$.

$$\begin{array}{r} 7x^2 - 2x + 5 \\ 3x^2 + 5x - 1 \\ \hline 4x^2 - 7x + 6 \end{array}$$

It will be noticed that the process of subtraction in the last Example, where we subtract $3x^2$ from $7x^2$, simply means to find a quantity such that when added to $3x^2$ will give $7x^2$. So that in (Ex. 3) *adding the second and third rows together the sum is equal to the first row.* This affords a ready check and should always be used.

The subtraction of a negative quantity is equivalent to adding a corresponding positive quantity.—If a length AB,

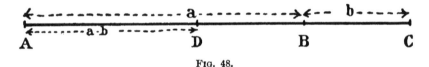

FIG. 48.

Fig. 48, be denoted by a, and another length BC by b, then $a + b$ would be represented by AC, a line equal in length to $AB + BC$, both being measured in the positive direction (from left to right).

Also $a - b$ would be a quantity obtained by subtracting b from a, and could be obtained by measuring off a length BD in a negative direction from B, so that $a - b$ is appropriately represented by AD.

As BC is positive, the reversal of direction indicated by CB is negative, and would be indicated by $-b$. Thus, a *minus* sign before a quantity reverses the direction in which the quantity is measured. Now, to subtract b from a, we reversed the direction of b and added it on to a. If, then, we have to subtract a negative quantity, $-b$ or CB, from a positive quantity a, by reversing the direction we obtain BC or $+b$, and adding on to a we get AC or $a + b$. We could indicate this by $a - (-b)$, the negative sign outside the bracket indicating that the quantity inside the bracket has to be subtracted from a. The change in sign is true whether the quantity subtracted be positive or negative. Hence, the diagram illustrates the rule already given, namely, to subtract one quantity from another—change the sign of the quantity to be subtracted and proceed to add the two together.

EXERCISES. X.

1. From $6a - 2b + 2c$ take $3a - 3b + 3c$.
2. $8a + x - 6b$ take $5a + x + 4b$.
3. $9x^2 - 3x + 5$ take $6x^2 + 5x - 3$.
4. Subtract $2a - 2b - 3b + 3c$ from $2a + 2b + 3b + 3c$.
5. $ax - bx - yd + yc$ from $2ax - bx + 2yc - yd$.
6. $xc - xd + ya + yb$ from $xa - xd - yc + yb$.
7. $3p - 3m + 2m - 2n$ from $2m - 2p - 3n + 3p$.
8. $3yz - 3xz - 4xy$ from $3yz - 3xy - 4xz$.
9. $a - b - d - c - e$ from $2d + 2b + 2a + 3e - 2c$.
10. $x^2 - 3y^2 + 6xz - 3xy$ from $x^2 + 4xy - 5xz + z^2$.
11. Add together
$a - 2b - c$, $4a - 3b + c - 2d$, $4d - 3c + 2b$, $c - 5a - b - d$, $a + 6b + 5c + 3d$; and subtract $c - a + 2d$ from the sum.
12. What must be added to $a + b - c$ to make $c + d$, and what must be taken away from $x^2(1 + 2y) - y^2(1 + 2x)$ to give as remainder $2xy(x - y)$?
13. Add together $4a^3 + 3bc^2 - 6a^2b$, $5a^2b - c^3 - 3bc^2$, $c^3 - 2a^3 - 2a^2b$.
14. From
$6x^3y + 10x^2y^2 + 13xy^3 + 19y^4$ take $5x^3y - 2x^2y^2 + 3xy^3 - 2y^4$.
15. Subtract
$2a^4 + 3a^3b + 5a^2b^2 + 8ab^3 + 11b^4$ from $4a^4 + 6a^3b + 8a^2b^2 + 10ab^3 + 12b^4$.
16. Find the sum of $4x^3 - 5ax^2 + 6a^2x - 5a^3$, $3x^3 + 4ax^2 + 2a^2x + 6a^3$, $-17x^3 + 19ax^2 - 15a^2x + 8a^3$, $13ax^2 - 27a^2x + 18a^3$, $12x^3 + 3a^2x - 20a^3$.
17. Find the sum of $ab + 4ax + 3cy + 2ez$, $14ax + 20ez + 19ab + 8cy$, $13cy + 21ez + 15ax + 24ab$.

CHAPTER VI.

MULTIPLICATION. DIVISION. USE OF BRACKETS.

Multiplication.—As in Arithmetic, multiplication may be considered as a concise method of finding the sum of any quantity when repeated any number of times. The sum thus obtained is called the *product*.

In multiplying, what is called the **Rule of Signs** must be observed, *i.e.* **The product of two terms with like signs is positive; the product of two terms with unlike signs is negative.**

If a is to be multiplied by b, it means that a has to be added to itself as often as there are units in b ; hence, the product is ab.

If $-a$ is to be multiplied by $-b$, it means that $-a$ is to be subtracted as often as there are units in b, and the product is ab.

Again, if $-a$ is to be multiplied by b, it means that $-a$ is to be added to itself as often as there are units in b, hence the product is $-ab$. The same result would be obtained by multiplying a by $-b$.

At the present stage it is necessary to be able to apply the rules of multiplication readily and accurately. The proofs will come later.

Rule.—**To multiply two simple expressions together, first multiply the coefficients, then add the indices of like letters. Remember that like signs produce ✚ (plus), unlike signs produce ━ (minus).**

Ex. 1. Multiply $4ab^3$ by $3a^2b^2$.

Here the product of the coefficients is $4 \times 3 = 12$. Adding the indices of like letters we have $a \times a^2 = a^{1+2} = a^3$, and $b^3 \times b^2 = b^{3+2} = b^5$.

Hence the product is $12a^3b^5$:
$$\therefore \quad 4ab^3 \times 3a^2b^2 = 12a^3b^5.$$

Ex. 2. $\quad 4a^3b^2c^5e^4 \times 6a^4b^3c^4e^3 = 24a^7b^5c^9e^7.$

When the expressions each consist of two terms, the process of multiplication may be arranged as follows :

Ex. 3.   Multiply $x+5$ by $x+6$.

Write down the two expressions as shown, one under the other ; multiply each term of the first expression by each term of the second, and arrange the results as here indicated ; begin at the left-hand side, thus, $x \times x = x^2$. Write the x^2, and after it the product of x and 5 or $5x$. As the signs are alike, the sign of each of these products is $+$. Next multiplying by the second term 6 we get $6x+30$; the term $6x$ is placed immediately below the corresponding term $5x$, and the term 30 on the extreme right-hand side ; finally,

$$\begin{array}{r} x + 5 \\ x + 6 \\ \hline x^2 + 5x \\ 6x + 30 \\ \hline x^2 + 11x + 30 \end{array}$$

add the terms together to obtain the product.   By arranging the terms one under the other, and multiplying, the result can always be obtained.   But this is not enough ; in such a simple expression the student should be able to at once write down the product by inspection.

This is effected by noting that the first term x^2 of the product is obtained by multiplying together the two first terms in the given expressions ; the last term is the product of the two second terms 6 and 5, and the middle term is the product of x and the sum of the two second terms.

$$\therefore \quad (x+5)(x+6) = x^2 + 11x + 30.$$

In a similar manner,

$$(a+b)^2, \text{ or } (a+b)(a+b) = a^2 + 2ab + b^2,$$
$$(a-b)^2, \text{ or } (a-b)(a-b) = a^2 - 2ab + b^2,$$
$$(x-5)(x-6) = x^2 - 11x + 30,$$
$$(x-5)(x+6) = x^2 + x - 30,$$
$$(x+5)(x-6) = x^2 - x - 30.$$

When the product of two expressions containing more than two terms is required, it is usually convenient to arrange the terms one under the other, and to proceed as in the following example :

Ex. 4.   Multiply together $14ac - 3ab + 2$ and $ac - ab + 1$.

$$\begin{array}{l} 14ac - 3ab + 2 \\ ac - \ ab + 1 \\ \hline 14a^2c^2 - \ 3a^2bc + \ 2ac \\ \quad - 14a^2bc \qquad\quad + 3a^2b^2 - 2ab \\ \quad\qquad\quad + 14ac \qquad\quad - 3ab + 2 \\ \hline 14a^2c^2 - 17a^2bc + 16ac + 3a^2b^2 - 5ab + 2 \end{array}$$

Proceeding as in Ex. 3 we multiply each term of the top line by ac, by $-ab$, and finally by 1, thus we have
$$14ac \times ac = 14a^2c^2.$$
Next $-3ab \times ac = -3a^2bc,$
and finally $2 \times ac = 2ac.$

By multiplying by the second term $-ab$ the second line is obtained. After writing down the third line, the terms are added and the product is thus found.

Continued Product.—When several quantities are multiplied together the product obtained is called the *continued product* of the quantities.

Ex. 1.   The continued product of $3b$, $7c$, and $2a$ is $42abc$.

Ex. 2.   Obtain the continued product of $x+2$, $x+3$, x, and $x+1$.
The product of $x+2$ and $x+3$ is x^2+5x+6.
Also the product of x and $x+1$ is x^2+x.
Hence

$$x^2+5x+6$$
$$x^2+x$$

$$\overline{x^4+5x^3+6x^2}$$
$$x^3+5x^2+6x$$

$$\overline{x^4+6x^3+11x^2+6x}$$

EXERCISES.   XI.

Multiply
1. x^2-ax+a^2 by x^2+ax+a^2.
2. $2a^3-a^2b+3ab^2-b^3$ by $2a^3+a^2b+3ab^2+b^3$.
3. $x^6+x^5y-x^3y^3+xy^5+y^6$ by x^2-xy+y^2.
4. $x^4+3x^3+7x^2+15x+10$ by x^2-3x+2.
5. $x^2+4xy+4y^2$ by $x^2-4xy+4y^2$.
6. $x^3-12x-16$ by $x^3-12x+16$.
7. $a^6-3a^4b^2+6b^6$ by $a^6-2a^2b^4+b^6$.
8. $x^4+x^2y^2+y^4$ by x^2-y^2.
9. $4a^3b-6a^2b^2-2ab^3$ by $2a^2+3ab-b^2$.
10. $a^4-b^4+c^4$ by $a^2-b^2-c^2$.
11. $1-y^2-y^3$ by $1-y^4-y^5$.
12. $3x^4-x^2-1$ by $2x^4-3x^2+7$.
13   x^3+6x^2+8x-8 by x^2-2x+4.
14. $4a^2+12ab+9b^2$ by $4a^2-12ab+9b^2$.
15. $13a^2-17a-45$ by $-a-3$.

Division.—In Algebra, as in Arithmetic, the terms **divisor**, **dividend**, and **quotient** are used. From a given dividend and divisor, we can by the process of division proceed to find the quotient of two or more algebraical expressions. When the divisor is exactly contained in the dividend, then the product of the divisor and the quotient is equal to the dividend. When the divisor is not exactly contained in the dividend, and there is a remainder, the remainder must be added to the product of the quotient and the divisor in order to give the dividend.

Ex. 1.   Divide $70abx^3y$ by $2ax^2y$;

$$\therefore \quad \frac{70abx^3y}{2ax^2y}.$$

As in Arithmetic the work may be done by cancelling, thus,

$$70 \div 2 \text{ gives } 35,$$

and $\qquad\qquad abx^3y \div ax^2y$ gives bx;

hence the required quotient is $35bx$.

Ex. 2.   Divide $15a^2b^2$ by $-5a$;

$$\therefore \quad \frac{15a^2b^2}{-5a} = -3ab^2.$$

When the dividend and divisor both consist of several terms, we arrange both dividend and divisor according to the powers of the same letter, beginning with the highest. The following example worked out in full will show the method adopted :

$$a^2+2ax+x^2 \,\big)\, a^5+5a^4x+10a^3x^2+10a^2x^3+5ax^4+x^5 \,\big(\, a^3+3a^2x+3ax^2+x^3$$

$$\begin{array}{l}
a^5+2a^4x+a^3x^2 \\
\hline
\quad 3a^4x+9a^3x^2+10a^2x^3 \\
\quad 3a^4x+6a^3x^2+3a^2x^3 \\
\hline
\qquad\qquad 3a^3x^2+7a^2x^3+5ax^4 \\
\qquad\qquad 3a^3x^2+6a^2x^3+3ax^4 \\
\hline
\qquad\qquad\qquad\qquad a^2x^3+2ax^4+x^5 \\
\qquad\qquad\qquad\qquad a^2x^3+2ax^4+x^5 \\
\hline
\end{array}$$

Divide the first, or left-hand, term of the divisor into the dividend. Thus a^2 divided into a^5 gives a^3 ; write this quantity on the right-hand side as shown, and put the term a^5 under the first term of the dividend. In a similar manner by multiplying the remaining two terms $2ax$ and x^2 by a^3, and subtracting, we obtain $3a^4x+9a^3x^2$. Now bring down the next term $10a^2x^3$, and proceed as before.

EXERCISES. XII.

Divide

1. $4a^5 - a^3 + 4a$ by $2a^2 + 3a + 2$.
2. $x^2 + 4yz - 4y^2 - z^2$ by $x - 2y + z$.
3. $a^3 + a^2b + a^2c - abc - b^2c - bc^2$ by $a^2 - bc$.
4. $2x^4 + 27xy^3 - 81y^4$ by $x + 3y$.
5. $\dfrac{x^3 - 3x^2y + 3xy^2 - y^3}{x^3 + y^3}$ by $\dfrac{x^2 - 2xy + y^2}{x^2 - xy + y^2}$.
6. $9a^2 - 4b^2 - c^2 + 4bc$ by $3a - 2b + c$.
7. $25a^2 - b^2 - 4c^2 + 4bc$ by $5a - b + 2c$.
8. $-207a^5b^7c^4$ by $23a^4b^3c^2$.
9. $4x^2y - 8xy^2 - 4y^2 + 3x^2 + 4x + 1$ by $x - 2y + 1$.
10. $12x^4 - 17x^3 - 9x^2 + 13x - 63$ by $4x^2 - 3x + 7$.
11. $2a^3 + 6a^2b - 4a^2c - 2ab^2 + 3ac^2 - 6b^3 + 4b^2c + 9bc^2 - 6c^3$ by $a + 3b - 2c$.
12. $a^5 - a^4 + a^3 - a^2 + a - 1$ by $a^3 - 1$.
13. $(a^2 - bc)^3 + 8b^3c^3$ by $a^2 + bc$.
14. (i) Express in algebraical symbols: The difference of the squares of two numbers is exactly divisible by the sum of the numbers.

(ii) The sum of the cubes of three numbers diminished by three times their product is exactly divisible by the sum of the numbers.

Use of Brackets.—In Algebra it is frequently necessary to group parts of an expression, and the use of brackets for this purpose is very important. There are several forms of brackets in general use; for example, ( ), { }, [ ]. Sometimes a line is placed over two numbers, and such a line has the same meaning as enclosing in brackets. Thus, if a quantity $b + c$ has to be multiplied by $d + f$ the terms may be written as $\overline{b + c} \times \overline{d + f}$, or as $(b + c)(d + f)$. In this way the use of brackets gives a short method of indicating multiplication. The use of the different forms of brackets can be shown by the following examples:

Ex. 1. $3a - (4b - 7c)$.

Here, the brackets indicate that the expression $4b - 7c$ is to be subtracted from $3a$. We have already found (p. 63) that in the process of subtraction we change the sign of each term and then add; hence it is obvious that the result obtained will be the same whether we first subtract $7c$ from $4b$ and afterwards subtract the remainder from $3a$, or first add $7c$ to $3a$ and subtract $4b$ from the sum.

If a positive sign occur before a bracket the signs of all the terms remain unaltered when the brackets are removed ; if a minus sign is placed before the bracket the signs of each term inside the brackets must be changed when the brackets are removed.

Ex. 2.　$3a + (4b - 7c + 3d) = 3a + 4b - 7c + 3d.$
$3a - (4b - 7c + 3d) = 3a - 4b + 7c - 3d.$

The other forms of brackets which are used are [ ] and { }. In each case they denote that whatever is in one pair of them is to be regarded as one quantity to be added, subtracted, multiplied, or divided, as a whole, in the manner which the signs and quantities outside the brackets indicate.

Ex. 3.　Express the product of $2a + 3b$ and $4c + 5d$.
The quantities may be written as $(2a + 3b)(4c + 5d)$.

Further, to indicate that $3f$ is to be subtracted from the product and the result multiplied by $7e$, we use another pair of brackets, thus, $7e\{(2a + 3b)(4c + 5d) - 3f\}$; and to express that when $3x$ is subtracted from the last obtained product the whole must be multiplied by 8, we have to use another bracket, thus,

$$8[7e\{(2a + 3b)(4c + 5d) - 3f\} - 3x].$$

In removing the brackets the work may commence either from the inside pair, removing one form of bracket at each operation until the outside pair is reached. Or, the work may with advantage commence at the outside pair, repeating until the inside pair is reached. Moreover, to prevent mistakes, it is advisable only to remove one pair of brackets at each step.

Ex. 4.　Simplify
$$8(a^2 + x^2) - 7(a^2 - x^2) + 6(a^2 - 2x^2)$$
$$= 8a^2 + 8x^2 - 7a^2 + 7x^2 + 6a^2 - 12x^2$$
$$= 7a^2 + 3x^2.$$

Ex. 5.　Simplify
$$4x - [(4x - 4y)(4x + 4y) - \{4x + (4x + 4y)(4x - 4y)\} + 4y].$$
Multiplying the terms in the brackets we get
$$4x - [16x^2 - 16y^2 - \{4x + 16x^2 - 16y^2\} + 4y]$$
or　　$4x - [16x^2 - 16y^2 - 4x - 16x^2 + 16y^2 + 4y]$
$$= 4x - 16x^2 + 16y^2 + 4x + 16x^2 - 16y^2 - 4y$$
$$= 8x - 4y.$$

Ex. 6.   Simplify

$$3a - [a+b - \{a+b+c - (a+b+c+d)\}],$$
$$\therefore \quad 3a - [a+b - \{a+b+c - a - b - c - d\}]$$
$$= 3a - [a+b+d] = 2a - b - d.$$

EXERCISES.   XIII.

Simplify

1. $\dfrac{(a-b)^2}{a+b} + \dfrac{(b-c)}{b+c} + \dfrac{(a-c)^2}{a+c}$, and find the value when
$$a = 5, \ b = 3, \ c = 1.$$

2. When $a = 2$, $b = 5$, $x = 4$, of
$$\frac{2a^2 - b}{b - x} - \frac{a^2 - 2b}{x+b} + \frac{3a^2(b-x)}{25 - x^2}.$$

3. $a - [2b + 3c + \{2a - 4b - (a - 2b + 4c)\}]$.

4. $6x + y - [3x + 2y - \{7x - 2z - (6y - 18z)\}]$, and subtract the result
from $13x - 7(x-y) - 16(z+x)$.

5. Remove the brackets from the expression
$$4(a+2b) - [3a - 4\{a - (2b - 3a)\}].$$

6. Find the value of
$$\frac{a+b+c}{a-b-c} - \frac{abc}{d} \quad \text{when } a = \tfrac{3}{4}, \ b = -\tfrac{1}{2}, \ c = \tfrac{2}{3}, \ d = \tfrac{3}{5}.$$

Simplify the following, find the value in each case when
$$x = 1, \ y = -\tfrac{1}{2}, \ z = -2.$$

7. $3x - 4y - (2x - 3y + z) - (5x + 2y - 3z)$.

8. $x^2 + y[x - (y+z)] - [x^2 + 5xy - y(x+z)]$.

9. $11x + 2y - [4x - \{7y - (8x + \overline{9y - 3z})\}]$.

Simplify

10. $12c - [\{3b - (2b - a)\} - 4a + \{2c - \overline{3b - a - 2b}\}]$.

11. $\dfrac{1 + a^{\frac{1}{2}} - a - a^{\frac{3}{2}}}{1 - a}$.

12. $(a+b+c)^2 - (a-b+c)^2 + (a+b-c)^2 - (-a+b+c)^2$.

13. Simplify $7(a-b)(b-2a) - (2a-b)(b-7a) - 6b^2$.

14. From $9(a-b)^2$ take the sum of $(3a-b)^2$ and $(a-3b)^2$.

CHAPTER VII.

FACTORS. FRACTIONS. SURDS.

FACTORS.

A KNOWLEDGE of factors must be obtained before the student can hope to deal successfully with algebraic expressions and their simplification.

Factors.—When an algebraic expression is the product of two or more quantities, each of these quantities is called a factor of it.

Thus, if $x+5$ be multiplied by $x+6$, the product is

$$x^2+11x+30,$$

and the two quantities $x+5$ and $x+6$ are said to be the *factors* of

$$x^2+11x+30.$$

The determination of the factors of a given expression, or, as it is called, the **resolution** of the expression into its factors, may be regarded as the inverse process of multiplication.

The following results, easily obtained by multiplication, occur so frequently, and are of such importance, that they should be carefully remembered :

$$(a+b)(a+b) \text{ or } (a+b)^2 = a^2+2ab+b^2 \dots\dots\dots\dots(1)$$
$$(a-b)^2 = a^2-2ab+b^2 \dots\dots\dots\dots\dots(2)$$

The results are equally true when any other letters are used instead of a and b. We can write with equal correctness

$$(x+y)^2 = x^2+2xy+y^2.$$

Or, **The square of the sum of two quantities is equal to the sum of the squares of the quantities increased by twice their product.**

Similarly, **The square of the difference of two quantities is equal to the sum of the squares of the quantities diminished by twice their product.**

By multiplying $(x+y)(x-y)$ we obtain x^2-y^2. Conversely given x^2-y^2 we can at once write down the factors as $x+y$ and $x-y$.

The first of these relations may be expressed as : **The product of the sum and the difference of two numbers is equal to the difference of their squares.**

Ex. 1. $40^2-39^2=(40+39)(40-39)=79\times1=79.$

Ex. 2. $1000^2-998^2=(1000+998)(1000-998)=1998\times2=3996.$

Ex. 3.   To obtain the factors of $(a^2+b^2-c^2)^2-4a^2b^2$. [This may be written $(a^2+b^2-c^2)^2-(2ab)^2.$]

Using the last rule given above we get
$$(a^2+b^2-c^2+2ab)(a^2+b^2-c^2-2ab),$$
or
$$\{(a+b)^2-c^2\}\{(a-b)^2-c^2\}.$$

Again using the rule we get
$$(a+b+c)(a+b-c)(a-b+c)(a-b-c).$$

Ex. 4.   To obtain the factors of x^4-y^4.

First $(x^4-y^4)=(x^2+y^2)(x^2-y^2).$

Also as $x^2-y^2=(x+y)(x-y),$

we can write $x^4-y^4=(x^2+y^2)(x+y)(x-y).$

Ex. 5.   Multiplying a^2-ab+b^2 by $a+b$, the product is found to be
$$a^3+b^3.$$
$$\therefore\ a^3+b^3=(a+b)(a^2-ab+b^2).$$
Similarly $a^3-b^3=(a-b)(a^2+ab+b^2).$

The quantities $(a+b)(a^2-ab+b^2)$ are the factors of a^3+b^3, and $(a-b)(a^2+ab+b^2)$ are the factors of a^3-b^3.

Generally **a^n+b^n is divisible by a+b when n is an odd number, 1, 3, 5, etc.**   Thus, in Ex. 5, n is 3.

Also a^n-b^n is divisible by $a-b$ when n is an odd number The case of $n=3$ is shown, and by actual division, assuming n to be any odd number, the rule can be further verified.

When n is an even number, 2, 4, etc., it will be found that a^n-b^n is divisible by both $(a+b)$ and $(a-b)$.

Ex. 6.   Let $n=6$; $\therefore\ a^n-b^n$ becomes a^6-b^6.

We know that $a^6-b^6=(a^3+b^3)(a^3-b^3)$, and in Ex. 5 the factors of (a^3+b^3) and (a^3-b^3) have been obtained.

Hence the factors of a^6-b^6 are
$$(a+b)(a^2-ab+b^2)(a-b)(a^2+ab+b^2).$$
Thus a^6-b^6 is divisible by both $a+b$ and $a-b$.

When the preceding simple examples are clearly made out it
is advisable to consider the more general expression $a^n \pm b^n$, and
to find that :

$a^n + b^n$ is divisible by $a+b$ when n is odd.

$a^n - b^n$,,   ,, $a-b$,,   ,,

$a^n - b^n$,,   ,,   both $a+b$ and $a-b$ when n is even.

The cases where n equals 2, 3, 4, 6 have already been taken.
Other values of n should be used and more complete veri-
fications be obtained of the rules given.

In finding the factors of any given expression any letter or
letters common to two or more terms may be written as a
multiplier, thus, given $ac+ad$ we can write this as $a(c+d)$.

Again, $ac+bc+ad+bd = a(c+d)+b(c+d) = (c+d)(a+b)$.

By multiplying $x+2$ by $x+3$ we obtain x^2+5x+6.

$$\therefore (x+2)(x+3) = x^2+5x+6.$$

Hence, given the expression x^2+5x+6, to find the quantities
$x+2$, and $x+3$, or the factors of the given expression, we find that

The first term is the product of x and x, or x^2.

  ,,   last   ,,   ,,   ,,   2 and 3, or 6.

  ,,   middle ,,   ,,   ,,   the first term, and the sum
of 2 and 3, or $5x$.

Proceeding in this manner the factors of a given expression are
readily obtained.

Ex. 7.   Resolve into factors $x^2+8x+12$.

Here, the two numbers required must have a sum of 8 and a
product equal to 12. Of such pairs of numbers, the sum of which is
8, are 4 and 4, 7 and 1, and 6 and 2, but only the last pair have a
product of 12. Hence, the factors are $(x+2)(x+6)$.

A convenient method is to arrange the possible factors in vertical
rows, thus

$$\begin{array}{ccc} x+4 & x+7 & x+2 \\ x+4 & x+1 & x+6 \end{array}$$

These may be multiplied together as in ordinary multiplication, but it is much better to perform the process mentally,
obtaining first the product of the two first terms, then the product of the two last terms, and finally the sum of the diagonal
products.

Thus, in the second group, the product of the two first terms
is x^2, of the two last is 7 ; and the sum of the diagonal products
$7x+x = 8x$. Hence, these are the factors of x^2+8x+7.

Proceeding in like manner the product of the terms in the last is $x^2 + 8x + 12$.

Similarly $\qquad (x-6)(x-2) = x^2 - 8x + 12,$

$\qquad\qquad\quad (x+6)(x-2) = x^2 + 4x - 12.$

Also, $\qquad\qquad (x-6)(x+2) = x^2 - 4x - 12.$

All these products should be verified and in each case the process should be carried out mentally.

Or, we could write the given expression

$\qquad x^2 + 8x + 12$ as $x^2 + 2x + 6x + 12.$

Taking out the quantity common to two terms we obtain

$$x(x+2) + 6(x+2).$$

This shows that $x+2$ is common to both terms, hence we may write

$$x^2 + 8x + 12 = (x+2)(x+6).$$

Ex. 8.   In a similar manner,

$\qquad x^2 - 9x + 20 = x^2 - 5x - 4x + 20$

$\qquad\qquad = x(x-5) - 4(x-5) = (x-4)(x-5).$

Ex. 9. $x^2 + 11x + 30 = x^2 + 5x + 6x + 30 = (x+5)(x+6).$

Factors obtained by substitution.—The factors in the preceding, and in other, examples may also be found by substituting for x some quantity which will reduce the given expression to zero.   Thus, in $x^2 - 9x + 20$ the last term suggests that two of the following, 4 and 5, 10 and 2, or 20 and 1, are terms of the factors, but the middle term of our expression denoting the sum of the numbers selects -4 and -5.   To ascertain if 4 and 5 are terms of the factors, put $x = 4$; then

$$16 - 36 + 20 = 0 ;$$

thus $x - 4$ is a factor.

Similarly putting $x = 5$ we obtain $25 - 45 + 20 = 0$;

$\qquad\qquad\qquad \therefore\ x - 5$ is a factor.

Hence $\qquad\qquad x^2 - 9x + 20 = (x-4)(x-5).$

Ex. 9. $x^2 + 6x - 55.$

Put $x = -11$; this reduces the given expression to zero ;

$\qquad\qquad\qquad \therefore\ x + 11$ is a factor.

Next put $x = +5$; $x - 5$ is found to be a factor ;

$\qquad\qquad \therefore\ x^2 + 6x - 55 = (x+11)(x-5).$

EXERCISES. XIV.

Resolve into factors

1. $x^2 - 7x + 10$. 2. $x^2 - x - 90$. 3. $x^2 - 3x - 4$.
4. $x^2 + 2x - 15$. 5. $27a^3 + 8b^3$. 6. $8x^3 - 27$.
7. $x^2 - x - 30$. 8. $x^2 + 12x - 85$. 9. $x^2 - 2xy - xz + 2yz$.
10. $3x^2 - 27y^2$. 11. $x^2 + 18x - 175$. 12. $x^2 - 3xz - 2xy + 6yz$.
13. $625x^4 - y^4$. 14. $10x^2 + 79x - 8$. 15. $x^3 - 13x^2y + 42xy^2$.
16. $(a^2 + b^2 - c^2)^2 - 4a^2b^2$. 17. $(x - 2y)^3 + y^3$.
18. (i) $a^2 - b^2 + c^2 - d^2 - 2(ac - bd)$; (ii) $(p^2 + q^2 - r^2)^2 - 4p^2q^2$;

(iii) $1 - m^{\frac{1}{2}} - m + m^{\frac{3}{2}}$.

Fractions.—The rules and methods adopted in dealing with fractions in Algebra are almost identical with those in Arithmetic. In both cases fractions are of frequent occurrence and their consideration is of the utmost importance. Some little practice is necessary before even a simple fraction can be reduced to its lowest terms. Perhaps the best method in the simplification of fractions is to write out the given expressions in factors wherever possible. To do this easily the factors already referred to on pp. 73 and 74 should be learnt by heart.

When proper fractions have to be added, subtracted, or compared, it is necessary to reduce them to a common denominator, and to lessen the work it is desirable that this denominator shall be as small as possible.

Ex. 1. Add $\dfrac{1}{2} + \dfrac{1}{3x}$.

First reduce to a common denominator $6x$; mentally multiply both numerator and denominator of the first fraction by $3x$, and obtain $\dfrac{3x}{6x}$; and similarly, by multiplying $\dfrac{1}{3x}$ by 2, get $\dfrac{2}{6x}$.

$$\therefore \frac{1}{2} + \frac{1}{3x} = \frac{3x + 2}{6x}.$$

Ex. 2. Simplify $\left(\dfrac{1}{2} + \dfrac{1}{3x}\right) \div \left(9x - \dfrac{4}{x}\right)$.

$$\frac{\dfrac{1}{2} + \dfrac{1}{3x}}{9x - \dfrac{4}{x}} = \frac{\dfrac{3x + 2}{6x}}{\dfrac{9x^2 - 4}{x}} = \frac{x(3x + 2)}{6x(3x + 2)(3x - 2)} = \frac{1}{6(3x - 2)}.$$

The factors $x(3x + 2)$, which are common to both numerator and denominator, have been cancelled.

Ex. 3.   Simplify (i) $\dfrac{x^4 + x^2 y^2}{x^4 - y^4}$,   (ii) $\dfrac{x^2 + 3x + 2}{x^2 + x - 2}$.

Here,   (i) $\dfrac{x^4 + x^2 y^2}{x^4 - y^4} = \dfrac{x^2(x^2 + y^2)}{(x^2 + y^2)(x^2 - y^2)} = \dfrac{x^2}{x^2 - y^2}$.

   (ii) $\dfrac{x^2 + 3x + 2}{x^2 + x - 2} = \dfrac{(x+1)(x+2)}{(x-1)(x+2)} = \dfrac{x+1}{x-1}$.

Ex. 4.   Simplify $\dfrac{x-a}{\dfrac{1}{a} - \dfrac{1}{b}} \times \dfrac{a-b}{1 - \dfrac{a}{x}}$.

$$\dfrac{x-a}{\dfrac{1}{a} - \dfrac{1}{b}} \times \dfrac{a-b}{1 - \dfrac{a}{x}} = \dfrac{x-a}{\dfrac{b-a}{ab}} \times \dfrac{a-b}{\dfrac{x-a}{x}}$$

$$= \dfrac{(x-a)\,ab}{b-a} \times \dfrac{(a-b)\,x}{x-a} = -abx.$$

The terms common to numerator and denominator are cancelled ; the term $b - a$ being for this purpose written in the form $-(a - b)$.

Highest Common Factor.—When the denominators of two or more fractions can be written in the form of factors, the reduction of the fractions to their simplest form can be readily effected.   But the process of factorisation cannot in all cases be easily carried out, and in such cases we may proceed to find the *Highest Common Factor* (H.C.F.).   The process is analogous to that of finding the G.C.M. in arithmetic.   The H.C.F. of two or more given expressions may be defined as **the expression of highest dimensions which can be divided into each of the given expressions without a remainder.**

Ex. 5.   Simplify the fraction $\dfrac{x^4 + x^3 + 2x - 4}{x^3 + 3x^2 - 4}$.

To find the H.C.F. we proceed as follows :

$$x^3 + 3x^2 - 4\,)\,x^4 + x^3 + 2x - 4\,(\,x - 2$$
$$\underline{x^4 + 3x^3 - 4x}$$
$$\underline{\begin{array}{l} -2x^3 \qquad\quad + 6x - 4 \\ -2x^3 - 6x^2 \qquad\quad + 8 \end{array}}$$
$$6x^2 + 6x - 12$$
$$= 6(x^2 + x - 2)\ ;$$

$$x^2 + x - 2\,)\,x^3 + 3x^2 - 4\,(\,x + 2$$
$$\underline{x^3 + x^2 - 2x}$$
$$2x^2 + 2x - 4$$
$$\underline{2x^2 + 2x - 4}$$

Therefore the H.C.F. $= x^2 + x - 2$.

Hence $\dfrac{x^4+x^3+2x-4}{x^3+3x^2-4}=\dfrac{(x^2+x-2)(x^2+2)}{(x^2+x-2)(x+2)}=\dfrac{x^2+2}{x+2}.$

Least Common Multiple.—When required to add, subtract, or compare two fractions, it is often necessary to obtain the *Least Common Multiple* (L.C.M.) of the denominators, *i.e.* **the expression of least dimensions into which each of the given expressions can be divided without a remainder.**

To find the L.C.M. we may find the H.C.F. of two given expressions, *divide one expression by it and multiply the quotient by the other.* Thus the H.C.F. of the two expressions

$$x^3-3x^2-15x+25 \text{ and } x^3+7x^2+5x-25$$

is $\qquad\qquad x^2+2x-5\ ;$

dividing the first expression by this H.C.F. the quotient is $x-5$.

Hence the L.C.M. is

$$(x-5)(x^3+7x^2+5x-25) \text{ or } (x-5)(x+5)(x^2+2x-5).$$

Ex. 6. Simplify the following :

$$\frac{1}{x^3-3x^2-15x+25}-\frac{1}{x^3+7x^2+5x-25}.$$

The common denominator will be the L.C.M. of the two denominators, and the fractions become

$$\frac{x+5}{(x^2+2x-5)(x-5)(x+5)}-\frac{x-5}{(x^2+2x-5)(x-5)(x+5)}.$$

$$=\frac{10}{(x-5)(x+5)(x^2+2x-5)}.$$

Surds.—As already explained on p. 29, when surd quantities occur in the denominator of a fraction it is desirable to simplify before proceeding to find the numerical values of the fraction.

Ex. 1. Find the value of $\dfrac{20}{\sqrt{2}}$.

Unless some process of simplification is adopted it would be necessary to divide 20 by $1\cdot4142\ldots$, a troublesome operation. If, however, we multiply both numerator and denominator by $\sqrt{2}$ we obtain

$$\frac{20\sqrt{2}}{2}=10\sqrt{2}=10\times1\cdot414\ldots,$$

a result easily obtained.

A similar method is applicable when the numerator and denominator of a fraction each contain two terms. Thus,

Ex. 2. Find the value of $\dfrac{2+\sqrt{3}}{2-\sqrt{3}}$.

Here, as $\sqrt{3}=1 \cdot 732$, if we proceed to insert the value of the root

we get $\qquad \dfrac{2+\sqrt{3}}{2-\sqrt{3}}=\dfrac{2+1 \cdot 73205}{2-1 \cdot 73205}=\dfrac{3 \cdot 73205}{\cdot 26795}$,

and it would be necessary to divide $3 \cdot 73205$ by $\cdot 26795$. Instead of doing this we may *rationalise* the denominator, *i.e.* multiply both numerator and denominator by $2+\sqrt{3}$. The fraction then becomes

$$\dfrac{(2+\sqrt{3})(2+\sqrt{3})}{(2-\sqrt{3})(2+\sqrt{3})}=\dfrac{(2+\sqrt{3})^2}{4-3}=\dfrac{4+3+4\sqrt{3}}{1}=7+4\sqrt{3}.$$

In this form the necessary calculation can readily be carried out.

EXERCISES.   XV.

Simplify

1. $\dfrac{20abx}{15a^2}.$ 2. $\dfrac{a^2-x^2}{a+x}.$ 3. $\dfrac{x^3+a^3}{x^2+2ax+a^2}.$ 4. $\dfrac{4+12x+9x^2}{2+13x+15x^2}.$

5. $\dfrac{4x^3+4x^2-7x+2}{4x^3+5x^2-7x-2}.$ 6. $\dfrac{2x^2-11x+15}{x^2+3x-18} \times \dfrac{x^2+5x-6}{2x^2-3x-5}.$

7. $\dfrac{x+y}{x-y}-\dfrac{x-y}{x+y}.$ 8. $\dfrac{x^4-a^4}{(x-a)^2} \div \dfrac{x^2+ax}{x-a}.$

9. $\dfrac{x^2+4x+3}{x^2+5x+6} \times \dfrac{x^2+6x+8}{x^2+5x+4}.$ 10. $\dfrac{x^2+6x-7}{x^2+3x-4} \div \dfrac{x^2+4x-21}{2x+8}.$

11. Express as the difference of two squares $1+x^2+x^4$, and thence factorise the expression.

12. $\dfrac{168a^3b^2c}{48a^2bc^3}.$ 13. $\dfrac{a^2+(a+b)ax+bx^2}{a^4-b^2x^2}.$ 14. $\dfrac{x^4+x^2+1}{x^3-1.}$

15. $\dfrac{x^2-x-6}{x^2+4x+4} \times \dfrac{x^2-2x-8}{x^2-7x+12}.$ 16. $\dfrac{2x^2-x-15}{5x^2-13x-6}.$

17. $\dfrac{1}{2(3x-2y)}-\dfrac{1}{2(3x+2y)}+\dfrac{3x}{9x^2-4y^2}.$

18. $\dfrac{1}{x+3y}+\dfrac{6y}{x^2-9y^2}-\dfrac{1}{3y-x}.$ 19. $\dfrac{3x^3-6x^2+x-2.}{x^3-7x+6.}$

20. $\left(1-\dfrac{x-y}{x+y}\right)\left(2+\dfrac{2y}{x-y}\right).$

21. $\dfrac{1}{(x-1)}-\dfrac{1}{2(x+1)}-\dfrac{x+3}{2(x^2+1)}-\dfrac{4}{x^4-1}.$

22. $\dfrac{1}{a+b}+\dfrac{2b}{a^2-b^2}+\dfrac{1}{a-b}.$

CHAPTER VIII.

SIMPLE EQUATIONS.

Symbolical expression.—One of the greatest difficulties experienced by a beginner in Algebra is to express the conditions of a problem by means of algebraical symbols, and considerable practice is necessary before even the simplest problem can be stated. The few examples which follow are typical of a great number.

Let x denote a quantity ; then 5 times that quantity would be $5x$; the square of the quantity would be x^2 ; and a fourth part of it would be indicated by $\frac{x}{4}$.

If a sum of £50 were equally divided among x persons, then each would receive $\frac{£50}{x}$.

If the difference of two numbers is 7, and the smaller number is denoted by x, the other will be represented by $x+7$. If the larger is denoted by x, then the smaller would be represented by $x-7$.

If the distance between two towns is a miles, the time taken by a train travelling at x miles an hour would be $\frac{a}{x}$; when the numerical values of a and x are known, the time taken can be obtained. Thus, let the distance a be 200 miles, and x the velocity, or speed, be 50 miles an hour, then the time taken to complete the journey is $\frac{200}{50} = 4$ hours.

Although the letters a, x, etc., are used in algebraical operations, symbols are often employed which at once, by the letters used, express clearly the quantities indicated.

Thus, space could be denoted by s; the velocity by v; and the time taken by t; then instead of $\frac{a}{x}$ in the last example we use $\frac{s}{v}$; or, the relation between s, v, and t is given by $s=vt$. From this, when any two of the three terms are given, the remaining one may be obtained.

In the case of a body falling vertically, the relation between space described and time of falling is given by $s=\frac{1}{2}gt^2$; where s denotes the space described in feet, t the time in seconds, and g denotes 32·2 feet per second in a second, or the amount by which the velocity of a body falling freely is increased in each second of its motion. In this case, given either s or t, the remaining term may be calculated.

Equations.—An equation may in Arithmetic, or Algebra, be considered simply as a statement that two quantities are equal.

Thus, the statement that 2 added to 7 is 9, may be expressed as an equation thus $2+7=9$. In a similar manner, other statements of equality, or, briefly, other equations, could be formed; indeed, the greater part of the student's work in Arithmetic has been concerned with such equations.

All such equations, involving only simple arithmetical operations, may be called *Arithmetical Equations*, to distinguish them from such equations as $2x+7=9$, which are called *Algebraical Equations*. As in Arithmetic, the answer to any given question remains unknown until the calculation is completed. So in Algebra the *solution of an equation* consists in finding a value, or values, which at the outset are unknown.

Simple equations.—When two algebraical expressions are connected together by the sign of equality, the whole expression thus formed is called an *equation*. The use of an equation consists in this, that from the relations expressed between certain known and unknown quantities we are able under proper conditions to find the unknown quantity in terms of the known.

The process of finding the value of the unknown quantity is called **solving the equation**; the value so found is the **solution** or the **root** of the equation. This root, or solution, when substituted in the given expression makes the two sides identical.

An equation which involves the unknown quantity only to the first power, or degree, is called a **simple equation**; if it contains the square of the unknown quantity it is called a **quadratic equation**; if the cube of the unknown quantity, a **cubic equation**. Thus, the degree of an expression is the power of the highest term contained in it.

If an equality involving only an algebraic operation exists between two quantities the expression is called an **identity**, thus $(x+y)^2 = x^2 + 2xy + y^2$ is an identity.

In the equation $2x+7=9$, x represents an unknown number such that twice that number increased by 7 is equal to 9. It is of course clear that $x=1$, but we may with advantage use this simple example to explain the operation of solving an equation. Before doing so, it is necessary to note that as an equation consists of two equal members or sides, one on the left, the other on the right-hand side of the sign of equality, the results will still be equal when both sides of the equation are :

(i) equally increased or diminished, which is the same in effect as taking any quantity from one side of an equation and placing it on the other side with a contrary sign ;

(ii) equally multiplied, or equally divided ;

(iii) raised to the same power, or, the same root of each side of the equation is extracted. And also, if

(iv.) the signs of all the terms in the equated expressions are changed from + to −, both sides of the equation being altered similarly, the result will still be the same.

Thus, in the equation $2x+7=9$, subtracting 7 from each side we get
$$2x+7-7=9-7,$$
or
$$2x=2.$$
Dividing by 2, then,
$$x=1.$$

Ex. 1. Solve the equation $4x+9=37$.

Subtracting 9 from each side we get
$$4x=28$$

$$\therefore \; x=\frac{28}{4}=7.$$

To prove this, put 7 for x. Then each side is equal to 28.

Instead of subtracting we can **transpose** the 7 in the preceding example from one side of the equation to the other by changing its sign ; thus $4x=37-9=28$.

Ex. 2.   Solve $4x + 5 = 3x + 8$.

Subtract $3x$ from both sides of the equation, and we get

$$4x - 3x + 5 = 8 ;$$

next subtract 5 from each side ;

$$\therefore\; x = 8 - 5 = 3.$$

It is sufficiently clear that $+3x$ and $+5$ on the right- and left-hand sides of the equation respectively may be removed from one side to the other (or *transposed*) and appear on the opposite side with changed sign.

Hence the rule for the solution of equations is : **Transpose all the unknown quantities to the left-hand side, and all the known quantities to the right-hand side ; simplify, if necessary, and divide by the coefficient of the unknown quantity.**

Ex. 3.   Solve $$\frac{5x - 1}{2} - \frac{7x - 2}{10} = \frac{33}{5} - \frac{x}{2}.$$

Multiply both sides of the equation by 10.

$$\therefore\; 25x - 5 - 7x + 2 = 66 - 5x ;$$

or $$23x = 69 ;$$

$$\therefore\; x = 3.$$

Fractional equations.—If the attempt is made to solve all equations by fixed methods or rules, much unnecessary labour will often be entailed.   Thus, in equations containing fractions, or, as they are called, *fractional equations*, the rule usually given would be to first clear of fractions by using the L.C.M. of the denominators ; but, if this is done in all cases the multiplier may be a large number, troublesome to use.   In such cases it is better, where possible, to simplify two or more terms before proceeding to deal with the remaining part of the equation.

Ex. 4.   Solve $\dfrac{11x - 13}{25} + \dfrac{17x + 4}{21} + \dfrac{19x + 3}{7} = 28\tfrac{1}{7} + \dfrac{5x - 25\tfrac{1}{3}}{4}.$

We may with advantage simplify three of the given terms, using 21 as a multiplier, thus :

$$\therefore\; \frac{21(11x - 13)}{25} + 17x + 4 + 57x + 9 = 591 + \frac{21(5x - 25\tfrac{1}{3})}{4},$$

or $$\frac{21(11x - 13)}{25} + 74x - \frac{21(5x - \tfrac{76}{3})}{4} = 578.$$

Multiplying by 100 we obtain
$$84(11x-13)+7400x-525(5x-\tfrac{7.6}{3})=57800,$$
or
$$924x-1092+7400x-2625x+13300=57800.$$
$$\therefore\ 5699x-45592.$$
$$\therefore\ x=\tfrac{45592}{5699}=8.$$

When decimal fractions occur in an equation it is often desirable to clear of fractions by multiplying both sides of the equation by a suitable power of ten.

Ex. 5.   Solve $\cdot015x+\cdot1575-\cdot0875x=\cdot00625x$.

We can clear of fractions by multiplying every term by 100000.
$$\therefore\ 1500x+15750-8750x=625x,$$
or
$$15750=7875x.$$
$$\therefore\ x=2.$$

Ex. 6.   Solve $\qquad \dfrac{a}{bx}-a^2=b^2-\dfrac{b}{ax}.$

First remove the fractions by multiplying all through by abx
$$\therefore\ a^2-a^3bx=ab^3x-b^2,$$
transposing,$\qquad\qquad -a^3bx-ab^3x=-a^2-b^2,$

changing sign or multiplying by -1,
$$x(a^2+b^2)ab=a^2+b^2\ ;$$
$$\therefore\ x=\frac{a^2+b^2}{ab(a^2+b^2)}=\frac{1}{ab}.$$

Ex. 7.   Solve $\qquad \dfrac{\sqrt{a+x}+\sqrt{a-x}}{\sqrt{a+x}-\sqrt{a-x}}=2.$

Equations of this kind are simplified by adding the numerator and the denominator to obtain a new numerator, and then subtracting in order to find the new denominator as on p. 101.

Hence $\dfrac{\sqrt{a+x}+\sqrt{a-x}}{\sqrt{a+x}-\sqrt{a-x}}=\dfrac{\sqrt{a+x}+\sqrt{a-x}+\sqrt{a+x}-\sqrt{a-x}}{\sqrt{a+x}+\sqrt{a-x}-\sqrt{a+x}+\sqrt{a-x}}=\dfrac{2+1}{2-1}\ ;$

$$\therefore\ \frac{2\sqrt{a+x}}{2\sqrt{a-x}}=3.$$
or
$$a+x=9(a-x)\qquad\therefore\ x=\tfrac{4}{5}a.$$

EXERCISES. XVI.

Solve the equations

1. $18x+13=59-5x$.
2. $4x+16=10x-5$.

3. $3(x-2)=4(3-x)-4$.
4. $7x-3=5x+13$.

5. $3x-\dfrac{x}{3}=42-2x$.
6. $\dfrac{x}{2}+\dfrac{3x}{4}-\dfrac{2x}{3}-\dfrac{x}{8}=11$.

7. $30x + 12 + 32x - 8 = 500.$ 8. $2x + 3 = 16 - (2x - 3).$

9. $x - 7(4x - 11) = 14(x - 5) - 19(8 - x) - 61.$

10. $3(x - 5) - 5(x - 4) = 21x - 41.$ 11. $21x + 7 = 4(x - 3) + 3x + 61.$

12. $\dfrac{5x - 7}{2} - \dfrac{3x - 4}{3} = \dfrac{9 - 2x}{6}.$ 13. $6x + 4(2x - 7) - 9(7 - 2x) = 645.$

14. $5x - 7(x - 8) - 20(8 - x) = 10(2x - 19).$

15. $\dfrac{7x + 17}{12} = \dfrac{2x + 1}{9} + \dfrac{1}{4}\left\{x + 6 - \dfrac{3}{4}(3x + 19)\right\}.$

16. $\dfrac{x}{3} - \dfrac{1}{2} = \dfrac{1}{3}\left\{\dfrac{1}{6} - (x - 4\tfrac{1}{2})\right\} + 1\tfrac{7}{18}.$

17. $\dfrac{x}{a} + \dfrac{x}{b - a} = b.$ 18. $\dfrac{x}{b} = \dfrac{x - a}{a} - b + \dfrac{x(a + 1)}{ab}.$

19. $\dfrac{x - a}{a - b} + \dfrac{ax + 1}{ab + 1} = 0.$ 20. $a\left(\dfrac{x - b}{a + b}\right) - b\left(\dfrac{x - a}{a - b}\right) = \dfrac{a^2 + b^2}{2(a + b)}.$

Problems involving simple equations with one unknown quantity.—When a question or problem is to be solved, its true meaning ought in the first place to be perfectly understood, and its conditions exhibited by algebraical symbols in the clearest manner possible. When this has been done the equation can be written down and the solution obtained.

Ex. 1.   If 3 be added to half a certain number the result is equal to 7. Find the number.

Let x denote the number; then, one-half the number is $\dfrac{x}{2}$; and, 3 added to this gives the expression $\dfrac{x}{2} + 3$; but the sum is equal to 7. Hence we have $\dfrac{x}{2} + 3 = 7$ as the required equation.

Subtracting 3 from each side of the equation it becomes

$$\frac{x}{2} = 4.$$

Next multiplying the equation throughout by 2
$$\therefore x = 8.$$

Thus the required number is 8. The result in this and in all equations should be substituted in both sides. When this is done the left-hand side is seen to be equal to the right, or, the equation is said to be satisfied.

The beginner will find that simple exercises of the type shown in Ex. 1, are easily made and tend to give clear notions how to express arithmetical processes by algebraical symbols.

Ex. 2. The sum of two numbers is 100 ; 8 times the greater exceeds 11 times the smaller part by 2 ; find the numbers.

Let x denote the smaller part.

Then $\qquad\qquad\qquad 100-x=$ greater part,

and $\qquad\qquad$ 8 times the greater $=8(100-x)$.

Hence $\qquad\qquad\qquad 8(100-x)=11x+2$,

or $\qquad\qquad\qquad 800-8x=11x+2$;

$$\therefore\ 19x=798,$$
$$x=42.$$

Also $\qquad\qquad\qquad (100-x)=58.$

Hence the two numbers are 58 and 42.

Ex. 3. A post which projects 7 feet above the surface of water is found to have $\frac{1}{3}$ its length in the water and $\frac{1}{4}$ its length in the mud at the bottom ; find its total length.

Let x denote its total length in feet.

Then $\dfrac{x}{3}$ is the length in the water.

And $\dfrac{x}{4}$ is the length in the mud.

But the length in the mud, the length in the water, together with 7, is equal to the total length.

Hence $\qquad\qquad\qquad \dfrac{x}{3}+\dfrac{x}{4}+7=x,$

or $\qquad\qquad\qquad 4x+3x+84=12x$;

$$\therefore\ 5x=84,\text{ or } x=16\tfrac{4}{5}\text{ feet.}$$

Ex. 4. A rectangle is 6 feet long ; if it were 1 foot wider its area would be 30 square feet.   Find the width.

Let x denote the width in feet.

Then $x+1$ is the width when one foot wider.

The area is $6(x+1)$, but the area is 30 square feet ;

$$\therefore\ 6(x+1)=30,$$

or $\qquad\qquad\qquad 6x+6=30.$

Transposing, $\qquad\qquad\qquad 6x=24$; $\therefore\ x=4.$

A practical application.—In electrical work equations are of the utmost importance.   As a simple case we may consider what is known as *Ohm's Law*.   This law in its simplest form may be expressed by the equation

$$R=\frac{E}{C},\dotfill(1)$$

where R denotes the resistance of an electric circuit in certain units called ohms, E the electromotive force in volts, and C the current in ampères. An explanation of the law may be obtained from any book on electricity, and need not be given here. Our purpose is only to show that in (1), and in all such equations involving three terms, when two of the terms are given, the remaining one (or unknown quantity) may be found.

Ex. 5. A battery contains 30 Grove's cells united in series; a wire is used to complete the circuit. Find the strength of the current, assuming the electromotive force of a Grove's cell to be 1·8 volts, the resistance of each cell ·3 ohm, and the resistance of the wire 16 ohms.

Here   Electromotive Force $= 30 \times 1\cdot8 = 54$ volts.
$$\text{Resistance} = (30 \times \cdot3) + 16 = 25 \quad ,,$$
$$\therefore \; C = \tfrac{5 4}{2 5} = 2\cdot16 \text{ ampères.}$$

Falling bodies.—The space s described by a body falling freely from rest in a time t is given by the formula $s = \tfrac{1}{2}gt^2$. It should be noticed that as g has the value 32·2 ft. per sec. per sec., if either s or t be given the remaining term can be obtained.

Such equations, which involve three, four, or more terms, are of frequent occurrence. In all cases the substitution of numerical values for all the terms except one enables the remaining term to be obtained.

Ex. 6.   Let $s = 128\cdot8$.   Find t.
Here $128\cdot8 = \tfrac{1}{2} \times 32\cdot2 \times t^2$.
$$\therefore \; t^2 = \frac{128\cdot8 \times 2}{32\cdot2} = 8.$$

Hence $t = \sqrt{8} = 2\cdot8$ sec.

EXERCISES.   XVII.

1. Divide 75 into two parts, so that 3 times the greater shall exceed 7 times the lesser by 15.

2. Divide 25 into two parts, such that one-quarter of one part may exceed one-third of the other part by 1.

3. The sum of the fifth and sixth parts of a certain number exceeds the difference between its fourth and seventh parts by 109; find the number.

4. At what times between the hours of 2 and 4 o'clock are the hands of a watch at right angles to each other?

5. There are three balls, of which the largest weighs one-third as much again as the second, and the second one-third as much again

as the third: the three together weigh 2 lbs. 5 oz. How much do they each weigh?

6. Five years ago A was 7 times as old as B; nine years hence he will be thrice as old. Find the present ages of both.

7. Divide £111 between A, B, and C, so that A may have £10 more than B, and B £20 less than C.

8. A broker bought as many railway shares as cost him £1875; he reserved 15, and sold the remainder for £1740, gaining £4 a share on the cost price. How many shares did he buy?

9. Two pedestrians start at the same time from two towns, and each walks at a uniform rate towards the other town, when they meet; one has travelled 96 miles more than the other, and if they proceed at the same rate they will finish their journeys in 4 and 9 days respectively. Find the distance between the towns and the rates of walking per day in miles.

10. A man gives a boy 20 yards start in 100 yards, and loses the race by 10 yards. What would have been a fair start to give?

11. A father leaves £14,000 to be divided amongst his three children, that the eldest may have £1000 more than the second, and twice as much as the third. What is the share of each?

12. Divide £700 between A, B, and C, so that C may have one-fourth of what A and B have together, and that A's share may be $2\frac{1}{2}$ times that of B.

13. A cistern can be filled by two taps, A and B, in 12 hours, and by B alone in 20 hours. In what time can it be filled by A alone?

14. Two cyclists, A and B, ride a mile race. In the first heat A wins by 6 seconds. In the second heat A gives B a start of $58\frac{2}{3}$ yards and wins by 1 second. Find the rates of A and B in miles per hour.

15. A slow train takes 5 hours longer in journeying between two given termini than an express, and the two trains when started at the same time, one from each terminus, meet 6 hours afterwards. Find how long each takes in travelling the whole journey.

16. A man walks a certain distance in a certain time. If he had gone half a mile an hour faster he would have walked the distance in $\frac{4}{5}$ of the time; if he had gone half a mile an hour slower he would have been $2\frac{1}{2}$ hours longer on the road. Find the distance.

17. Two pipes, A and B, can fill a cistern in 12 and 20 minutes respectively, and a pipe C can carry off 15 gallons per minute. If all the pipes are opened together the cistern fills in two hours. How many gallons does it hold?

18. A man walks at the rate of $3\frac{1}{2}$ miles an hour to catch a train, but is 5 minutes late. If he had walked at the rate of 4 miles an hour he would have been $2\frac{1}{2}$ minutes too soon. Find how far he has to walk.

19. Two trains take 3 seconds to clear each other when passing in opposite directions, and 35 seconds when passing in the same direction. Find the ratio of their velocities.

CHAPTER IX.

SIMULTANEOUS EQUATIONS AND PROBLEMS INVOLVING THEM.

Simultaneous equations.—If an equation contains two un-known quantities denoted by x and y, then by giving definite values to one of the unknown quantities, a corresponding series of values can be obtained for the other.

Ex. 1. Solve $3x - 5y = 6$.

This means that we require to find two numbers such that five times the second subtracted from three times the first number will give 6.

By transposition, $3x = 5y + 6$; and giving values 1, 2, 3, etc., to y, we may obtain a corresponding series of values of x.

If, $\qquad y = 1$, then $3x = 11$; $\qquad \therefore \quad x = \frac{11}{3}$.

$\qquad\qquad y = 2$, then $3x = 16$; $\qquad \therefore \quad x = \frac{16}{3}$.

Proceeding in this manner, a table of values can be arranged as follows :

x	$\frac{11}{3}$	$\frac{16}{3}$	7	$2\frac{6}{3}$	....
y	1	2	3	4	....

Thus, for any assigned value of y a corresponding value of x can be obtained.

In a similar manner if values are assigned to x, corresponding values of y can be found.

If, now, we have a second equation $4x + 3y = 37$, then as before, by giving any assigned value to either x or y, a corresponding value of the other unknown is obtained, and a table of corre-sponding values of x and y can be tabulated as in the preceding case. Comparing the two sets of values so obtained it will be found that only one pair of values of x and y will satisfy both equations at once, or the two simultaneous values are $x=7$, $y=3$.

Equations such as

$$3x - 5y = 6,$$
$$4x + 3y = 37,$$

which are satisfied by the same values of the unknown quantities, are called *simultaneous equations*.

To find two unknown quantities, we must have two distinct and possible equations.

Ex. 2. $4x + 3y = 37$, and $12x + 9y = 111$.

These form two equations, but they are not distinct, as the second can be obtained from the first by multiplying by 3.

To solve simultaneous equations, we require as many distinct and independent equations as there are unknowns to be found, *i.e.* if two unknowns have to be determined, two distinct equations are required ; if three unknowns, three equations, and so on.

If only one equation connecting two unknown quantities is given, although the value of each of the unknowns cannot be determined, it is still possible to obtain the ratio of the quantities.

Ex. 3.   If $\dfrac{5x - 4y}{3x - 2y} = 4$, find the ratio of x to y.

$$\therefore\ 5x - 4y = 4(3x - 2y) = 12x - 8y.$$
$$\therefore\ 4y = 7x,$$

or

$$\frac{x}{y} = \frac{4}{7}.$$

Elimination.—When in the data of a problem the given equations are not only distinct, but are sufficient in number, it is possible from such data to obtain others, in which one or more of the unknown quantities do not occur. The process by which this is effected is called *elimination*. At the outset it is convenient, in a few simple cases, to show some of the methods which may be adopted in dealing with simultaneous equations containing two or more unknown quantities.

Solution of simultaneous equations.—In the solution of a simultaneous equation containing two unknown quantities, there are two general methods by which their values may be obtained. The first is by multiplication or division, which processes are used to make the coefficients of one of the unknowns the same in the two equations. Then, by addition, or subtraction, we can *eliminate* one unknown quantity. This leaves an equation containing only one unknown, the value of which can be found in the usual manner.

The other method is to find the value of one unknown in terms of the other unknown in one of the equations, and then to substitute the value so found in the other equation.

Ex. 4.

$$3x - 5y = 6, \dots\dots\dots\dots\dots \text{ (i)}$$
$$4x + 3y = 37. \dots\dots\dots\dots\dots\text{(ii)}$$

To apply the *first method*, multiply (i) by 3 and (ii) by 5. This will make the terms in y the same in both equations, and as these have opposite signs their sum is zero.

$$\therefore\ 9x - 15y = 18$$
$$20x + 15y = 185$$

By addition, $29x \qquad = 203$

$$\therefore\ x = \frac{203}{29} = 7.$$

Substitute this value in (i) ;

$$\therefore\quad 21 - 5y = 6 ;$$

or $5y = 21 - 6 = 15 ; \ \therefore\ y = 3.$

On substituting these values of x and y in the given equations the equations are satisfied. Thus, substituting the values in (i), we get $3 \times 7 - 15 = 6$. The values obtained should always be substituted in this manner to ensure accuracy.

By the second method :

From (i) $3x = 6 + 5y ;$

or $x = \dfrac{6 + 5y}{3} ;$

$$\therefore\ 4x = \frac{24 + 20y}{3}.$$

Substitute this value in (ii) ;

$$\therefore\ \frac{24 + 20y}{3} + 3y = 37.$$

Multiply both sides of the equation by 3 ;

or $24 + 20y + 9y = 111.$

Hence $29y = 111 - 24 = \ 87 ;$

$$\therefore\ y = \frac{87}{29} = 3.$$

Having found the value of y, then by substitution in (i) or (ii), the value of x is readily obtained.

Miscellaneous examples.—As the solution of simultaneous equations is of the utmost importance, a few miscellaneous examples are worked here.

Ex. 5.
$$6x + 3y = 33, \rbrace \quad \text{.............................. (i)}$$
$$13x - 4y = 19. \rbrace \quad \text{................................(ii)}$$

Multiplying (i) by 4, we get $24x + 12y = 132$

,,　　(ii) by 3, we get $\underline{39x - 12y = 57}$

By addition $63x = 189$

$$\therefore \quad x = \frac{189}{63} = 3,$$

and by substitution in (i), $\qquad y = 5.$

If the known quantities are represented by the letters, a, b, c, d, the solution is effected in the same manner.

Ex. 6.　Solve
$$ax + by = c, \quad \text{..................................... (i)}$$
$$bx + ay = d, \quad \text{....................................(ii)}$$

Multiplying (i) by b, we get　$abx + b^2y = bc$

,,　　(ii) by a, we get　$\underline{abx + a^2y = ad}$

By subtraction,　　　　　　$b^2y - a^2y = bc - ad$

or　　　　　　$y(b^2 - a^2) = bc - ad \, ;$

$$\therefore \quad y = \frac{bc - ad}{b^2 - a^2}.$$

To obtain x we may either substitute for y, or proceed to eliminate y from (i) and (ii).

Thus multiplying (i) by a,　　　$a^2x + aby = ac$

,,　　(ii) by b,　　　$\underline{b^2x + aby = bd}$

Subtracting the upper line from lower, $(b^2 - a^2)x = bd - ac$

$$\therefore \quad x = \frac{bd - ac}{b^2 - a^2}.$$

From the preceding examples the student will have seen that in solving two simultaneous equations, the object is to determine from the two given equations a value of one of the unknowns. Using the value so obtained we proceed to find the other. The methods which may with advantage be employed in solving equations quickly can only be seen by practice.

Simultaneous equations of more than two unknowns.— The general methods previously explained may usually be employed. The following methods are also made use of when more than two unknowns have to be found.

Ex. 7.   Solve

$$x + y + z = 53 \quad \text{..............................} \quad \text{(i)}$$
$$x + 2y + 3z = 105 \quad \text{...........................} \quad \text{(ii)}$$
$$x + 3y + 4z = 134 \quad \text{...........................} \quad \text{(iii)}$$

Subtract (i) from (ii), $\therefore y + 2z = 52 \quad \text{.........................} \quad \text{(iv)}$

,,   (ii) from (iii), $y + z = 29 \quad \text{........................} \quad \text{(v)}$

By subtracting (v) from (iv), $z = 23$

Substitute this value for z in (v),

$$\therefore y + 23 = 29 \ ;$$

or $y = 29 - 23 = 6.$

Again substituting for y and z in (i),

$$x + 6 + 23 = 53 \ ;$$
$$\therefore x = 53 - 29 = 24.$$

Hence the values are $\left. \begin{array}{l} x = 24, \\ y = 6, \\ z = 23. \end{array} \right\}$

Partial fractions.—When the denominators of two or more fractions are alike we can proceed to add, subtract, or compare the fractions ; in like manner the converse operation would be to replace a given fraction by two or more simpler fractions as in the following example :

Ex. 8.   Express $\dfrac{3x - 16}{x^2 - 7x + 12}$ as the sum of two simpler fractions.

Here the given fraction is $\dfrac{3x - 16}{(x - 3)(x - 4)}.$

We may express this as the sum of two simpler fractions $\dfrac{A}{x - 3} + \dfrac{B}{x - 4}$, and we require to find the numerical values of the numerators A and B.

Let $\dfrac{3x - 16}{(x - 3)(x - 4)} = \dfrac{A}{(x - 3)} + \dfrac{B}{(x - 4)}.$

Then clearing of fractions we have

$$3x - 16 = A(x - 4) + B(x - 3) = (A + B)x - (4A + 3B).$$

The coefficients of the terms in x are 3 on the left hand side, and $A + B$ on the right.

Equating coefficients :

$$A + B = 3 \quad \text{.............................} \quad \text{(i)}$$
$$4A + 3B = 16 \quad \text{..........................} \quad \text{(ii)}$$

Multiplying (i) by 3 and subtracting from (ii) we find

$$A = 7, \text{ and } B = -4.$$

Hence $\dfrac{3x - 16}{(x - 3)(x - 4)} = \dfrac{7}{x - 3} - \dfrac{4}{x - 4}.$

Ex. 9. Show that $\dfrac{9x+20}{x^2+5x+6}=\dfrac{2}{x+2}+\dfrac{7}{x+3}$.

Fractions of a more complicated character may be reduced to partial fractions by an extension of the previous methods. Reference must be made to more advanced works for these cases, and also for the theory of the subject.

EXERCISES. XVIII.

Solve the equations :

1. $3x+\dfrac{9y}{2}=42.$

   $\dfrac{3x}{5}+4y=27.$

2. $9x+5y=65.$

   $7x-\dfrac{5y}{2}=25.$

3. $\dfrac{x}{5}-\dfrac{y}{4}=1.$

   $3x-4y=10.$

4. $2x+3y=5.$

   $10x-6y=11.$

5. $7x+3y=10.$

   $35x-6y=1.$

6. $2x-15y=3x-24y=1.$

7. $6x-12y=1,\quad 8x+9y=18.$

8. $16x-y=4x+2y=6.$

9. $3x-7y=7,\quad 11x+5y=87.$

10. $3x-4y=25.$

    $5x+2y=7.$

11. $\dfrac{x-y}{2}+\dfrac{x+y}{3}=2\tfrac{1}{2}.$

    $\dfrac{x+y}{2}+\dfrac{x-y}{3}=4\tfrac{1}{6}.$

12. $3x+2y=118.$

    $x+5y=191.$

13. $\dfrac{1}{x}+\dfrac{1}{y}=a,\ \dfrac{1}{y}+\dfrac{1}{z}=b,\ \dfrac{1}{z}+\dfrac{1}{x}=c.$

14. $\dfrac{x}{p}+\dfrac{y}{q}=m.$

    $\dfrac{x}{q}-\dfrac{y}{p}=n.$

15. If $7(x-y)=3(x+y)$, find the ratio of x to y.

16. $\dfrac{4x+5y}{40}=x-7.$

    $\dfrac{2x-y}{3}=-3y+\tfrac{1}{2}.$

17. $\dfrac{x-a}{b}+\dfrac{y-b}{a}=0.$

    $\dfrac{x+y-b}{a}+\dfrac{x-y-a}{b}=0.$

18. $(a+p)x+(b-q)y=n.$

    $(b-q)x+(a+p)y=n.$

Problems leading to simultaneous equations.—It will be found that some practice is necessary before even the data of a simple question can be expressed in algebraic symbols,.

and it is necessary to remember that in all cases there must be as many independent equations as there are unknowns to be determined. Thus if a simultaneous equation contains two unknown quantities, then two independent equations are requisite ; if only one equation is given the *ratio* of one unknown to the other can alone be determined.

If x and y denote two numbers, then the sum of the two is $x+y$, the difference of them is $x-y$, the product xy, etc.

Ex. 1. Find two numbers the sum of which is 1♣ and their difference 3.

Let x denote one number and y the other.

Then, the sum of the numbers is $x+y$; but this, by the question, is equal to 19.

Hence $\qquad\qquad x+y=19$ .... (i)

Also $\qquad\qquad x-y=3$ .... (ii)

Adding (i) to (ii) $\qquad 2x=22$;

$$\therefore\ \ x=11.$$

Subtracting (ii) from (i) $\qquad 2y=16$;

$$\therefore\ \ y=8.$$

Hence, the two numbers are 11 and 8.   It is easy by inspection to see that when these are inserted in the equations both are satisfied.

$$\therefore\ \ 11+8=19 \text{ and } 11-8=3.$$

Ex. 2. If 3 be added to the numerator of a certain fraction, its value will be $\frac{1}{3}$, and if 1 be subtracted from the denominator, its value will be $\frac{1}{5}$.   What is the fraction?

Let x be the numerator and y the denominator of the fraction.

Add 3 to the numerator, then $\dfrac{x+3}{y}=\dfrac{1}{3}$.

Subtract 1 from the denominator, and $\dfrac{x}{y-1}=\dfrac{1}{5}$;

$$\therefore\ \ \frac{x+3}{y}=\frac{1}{3}, \text{ and } \frac{x}{y-1}=\frac{1}{5} ;$$

$$\therefore\ \ 3x+9=y, \text{ .... (i)}$$

and $\qquad\qquad\qquad 5x=y-1.$ .... (ii)

Transposing we get $\qquad y-3x=9$ ....(iii)

$\qquad\qquad\qquad\qquad\qquad y-5x=1$ ....(iv)

Subtracting (iv) from (iii), $\qquad 2x=8$; $\ \therefore\ x=4.$

Substituting this value of x in (iii),

$$y-12=9 ; \ \therefore\ y=21.$$

Hence the fraction is $\frac{4}{21}$.

Ex. 3. A number consisting of two digits is equal in value to double the product of its digits, and also equal to twelve times the excess of the unit's digit over the digit in the ten's place; find the number.

If we denote the digits by x and y, and y denote the digit in the unit's place, then the number may be represented by $10x + y$. But this is equal to double the product of the digits;

$$\therefore \quad 10x + y = 2xy. \quad \text{............................} \quad \text{(i)}$$

The excess of the unit's digit over the digit in the ten's place is $(y - x)$, and we are given that

$$12(y - x) = 2xy. \quad \text{...............................} \quad \text{(ii)}$$

Hence $\qquad\qquad 10x + y = 12y - 12x$;

$$\therefore \quad 22x = 11y \text{ or } 2x = y. \quad \text{........................} \text{(iii)}$$

Substituting this value in (i) we get

$$5y + y = y^2 ;$$
$$\therefore \quad 6y = y^2 \text{ or } y = 6 ;$$

and from (iii), $\qquad\qquad x = 3.$

Hence the number is 36.

Ex. 4. Find two numbers in the ratio of 2 to 3, but which are in the ratio of 5 to 7 when 4 is added to each.

Let x and y denote the two numbers.

Then the first condition that the two numbers are in the ratio of 2 to 3 is expressed by

$$\frac{x}{y} = \frac{2}{3}. \quad \text{..} \text{(i)}$$

Similarly, the latter condition, that when 4 is added to each of them the two numbers are in the ratio of 5 to 7, is expressed by

$$\frac{x+4}{y+4} = \frac{5}{7}. \quad \text{...............................} \text{(ii)}$$

From (i) $\qquad\qquad 3x = 2y. \quad \text{........................} \text{(iii)}$

From (ii) $\qquad\qquad 7x + 28 = 5y + 20 \text{ or } 7x + 8 = 5y. \quad \text{..........} \text{(iv)}$

Multiplying (iii) by 5 and we obtain $\qquad 15x = 10y$

,, (iv) by 2 ,, $\qquad 14x + 16 = 10y$

Subtracting, $\qquad\qquad\qquad\qquad\qquad \overline{\quad x - 16 = \ 0}$
$$\therefore \quad x = 16.$$

From (iii), $\qquad\qquad y = \tfrac{3}{2}x = 24.$

Hence the two numbers are 16 and 24.

The unknown quantities to be found from a simultaneous equation are not necessarily expressed as x and y. It is frequently much more convenient to use other letters. Thus *pressure, volume,* and *temperature* may be denoted by p, v, and t respectively.

Also, *effort* and *resistance* may be indicated by the letters E and R.

It will be obvious that letters consistently used in this manner at once suggest, by mere inspection, the quantities to which they refer.

Some applications.—It is often necessary to express the relation between two variable quantities by means of a formula, or equation. The methods by which such variable quantities are *plotted* and the law obtained have already been explained but practice in solving a simultaneous equation is necessary before any such law can be determined.

Ex. 5.   The law of a machine is given by
$$R = aE + b, \quad\text{...............................(i)}$$
and it is found that when R is 40, E is 10, and when R is 220, E is 50; find a and b.

Substituting the given values in (i) we get
$$220 = 50a + b \quad\text{...............................(ii)}$$
$$40 = 10a + b \quad\text{...............................(iii)}$$

Subtracting,
$$180 = 40a$$
$$\therefore\ a = \frac{180}{40} = 4\cdot5.$$

Substituting this value in (iii),
$$b = 40 - 10 \times 4\cdot5 = -5.$$
Hence the required law is $R = 4\cdot5E - 5$.

EXERCISES.   XIX.

1. Find the fraction to the numerator of which, if 16 be added, the fraction becomes equal to 4, and if 11 be added to the denominator the fraction becomes $\frac{1}{4}$.

2. The difference of two numbers is 14, their quotient is 8.   Find them.

3. What fraction is that which, if the denominator is increased by 4, becomes $\frac{1}{3}$; but, if the numerator is increased by 27, becomes 2?

4. Find that number of two digits which is 8 times the sum of its digits, and the half of which exceeds by 9 the same number with its digits reversed.

5. Find a fraction which will become $\frac{1}{2}$ if 1 is added to its denominator, and $\frac{1}{3}$ if 3 is taken from its numerator.

6. Two numbers differ by 3, and the difference of their squares is 69. Find them.

7. If 1 be added to the numerator and 1 subtracted from the denominator of a certain fraction, the value of the fraction becomes $\frac{2}{5}$; if 2 be added to the numerator and 2 subtracted from the denominator, the value becomes $\frac{3}{4}$. What is the fraction?

8. Find two numbers such that the first increased by 15 is twice the other when diminished by 3; while a half of the remainder when the former is subtracted from the latter, is an eighth of that sum.

9. A, B, and C travel from the same place at the rates of 4, 5, and 6 miles an hour respectively; and B starts 2 hours after A. How long after B must C start in order that they may overtake A at the same instant?

10. If six horses and seven cows cost in all £276, while five horses and three cows cost £179, what is the cost of a horse and what is the cost of a cow?

11. There is a fraction such that when its numerator is increased by 8 the value of the fraction becomes 2, and if the denominator is doubled, its value becomes $\frac{3}{4}$; find the fraction.

12. Twenty-one years ago A was six times as old as B; three years hence the ratio of their ages will be $6 : 5$; how old is each at present?

13. There are two coins such that 15 of the first and 14 of the second have the same value as 35 of the first and 6 of the second. What is the ratio of the value of the first coin to that of the second?

14. Each of two vessels, A and B, contains a mixture of wine and water, A in the ratio of 7 to 3, and B in the ratio of 3 to 1; find how many gallons from B must be put with 5 gallons from A in order to give a mixture of wine and water in the ratio of 11 to 4.

15. Eliminate t from the equations
$$v = u + ft.$$
$$s = ut + \tfrac{1}{2}ft^2.$$

16. A racecourse is 3000 ft. long; A gives B a start of 50 ft., and loses the race by a certain number of seconds; if the course had been 6000 ft. long, and they had both kept up the same speed as in the actual race, A would have won by the same number of seconds. Compare the speed of A with that of B.

17. The receipts of a railway company are apportioned as follows: 49 per cent. for working expenses, 10 per cent. for the reserved fund, a guaranteed dividend of 5 per cent. on one-fifth of the capital, and the remainder, £40,000 for division amongst the holders of the rest of the stock, being a dividend at the rate of 4 per cent. per annum. Find the capital and the receipts.

CHAPTER X.

RATIO, PROPORTION, AND VARIATION.

Ratio and proportion.—It has already been seen that **ratio may be defined as the relation with respect to magnitude which one quantity bears to another of the same kind.**

By means of algebraical symbols the ratio between two quantities can be expressed in a more general manner than is possible by the methods of Arithmetic.

Thus, the ratio between two quantities a and b may be expressed by $a:b$ or $\dfrac{a}{b}$; and the ratio is unaltered by multiplying or dividing both terms by the same quantity.

Proportion.—When two ratios $a:b$ and $c:d$ are equal, then the four quantities a, b, c, d are said to be in *proportion* or are *proportionals*.

Hence $\qquad\qquad a:b=c:d$ or $\dfrac{a}{b}=\dfrac{c}{d}.$ ....(i)

The two terms b and c are called the *means*, and a and d the *extremes*.

When four quantities are in proportion the product of the means is equal to the product of the extremes.

Thus if $a:b=c:d$ then $b\times c=a\times d$.

This important rule can be proved as follows : As the value of a ratio is unaltered by multiplying both terms by the same quantity we may, in Eq. (i), multiply the first ratio by d and the second by b.

Then, we have $\dfrac{ad}{bd}=\dfrac{bc}{bd}$; hence $bc=ad$.

In the proportion $\frac{a}{b}=\frac{c}{d}$; by adding unity to each side we get

$$\frac{a}{b}+1=\frac{c}{d}+1,$$

or

$$\frac{a+b}{b}=\frac{c+d}{d} \quad \text{...........................(ii)}$$

· In a similar manner subtracting 1 from each side we obtain

$$\frac{a-b}{b}=\frac{c-d}{d} \quad \text{...........................(iii)}$$

Dividing (ii) by (iii) then $\frac{a+b}{a-b}=\frac{c+d}{c-d}$,

a result often required in both Algebra and Trigonometry.

The most general form of the above may be written

$$\frac{ma+nb}{pa+qb}=\frac{mc+nd}{pc+qd},$$

whatever m, n, p, and q may be ; this can also be obtained as follows :

In Eq. (i) we have $\frac{a}{b}=\frac{c}{d}$.

Multiplying both sides by p we obtain

$$\frac{pa}{b}=\frac{pc}{d}.$$

Again dividing both sides by q,

$$\therefore \frac{pa}{qb}=\frac{pc}{qd}$$

Adding 1 to each side,

$$\frac{pa+qb}{qb}=\frac{pc+qd}{qd},$$

or

$$\frac{pa+qb}{pc+qd}=\frac{qb}{qd}=\frac{b}{d}.$$

In a similar manner we can obtain $\frac{ma+nb}{mc+nd}=\frac{b}{d}$

Hence

$$\frac{pa+qb}{pc+qd}=\frac{ma+nb}{mc+nd} ;$$

$$\therefore \frac{ma+nb}{pa+qb}=\frac{mc+nd}{pc+qd}.$$

This important proposition should be tested by substituting simple numbers for the letters.

Thus, let $a=3$, $b=4$, $c=1\cdot5$, $d=2$,

Then $\frac{a}{b}=\frac{c}{d}$ becomes $\frac{3}{4}=\frac{1\cdot5}{2}$.

Now let $m=5$, $n=6$, $p=9$, $q=10$.

Then

$$\frac{ma+nb}{pa+qb}=\frac{mc+nd}{pc+qd}$$

becomes

$$\frac{5\times3+6\times4}{9\times3+10\times4}=\frac{5\times1\cdot5+6\times2}{9\times1\cdot5+10\times2}\ ;$$

$$\therefore\ \frac{39}{67}=\frac{19\cdot5}{33\cdot5}.$$

And the ratio of the first two numbers is equal to the ratio of the second. Other simple numbers should be inserted in each case, when it will be found that the two ratios remain equal to each other, or, in other words, the four quantities are proportionals.

Mean proportional.—When the second term of a proportion is equal to the third, each is said to be a mean proportional to the other two. Thus 6 is said to be a mean proportional to 4 and 9.

Geometrical mean (written G.M.).—The mean proportional between two quantities is also called the geometrical mean, and is equal to the square root of the product of the quantities.

Thus the G.M. of 4 and 9 is $\sqrt{4\times9}=6$.

Similarly if $a:b=b:c$ then $b=\sqrt{ac}$.

Arithmetical mean (written A.M.) is half the sum of two quantities. Thus, the A.M. of 4 and 9 is $\frac{4+9}{2}=6\cdot5$.

The arithmetical mean of a and c is $\frac{a+c}{2}$.

Third proportional.—When three quantities are in proportion and are such that the ratio of the first to the second is the same as the second to the third, then the latter is called a third proportional to the other two.

Variation.—When two quantities are related to each other in such a manner that any change in one produces a corresponding change in the other, then one of the quantities is said to vary directly as the other.

The symbol $\propto$ is used to denote variation. Thus, the statement that x is proportional to y, or, that x varies as y, may be written $x\propto y$.

For many purposes, especially to obtain numerical values, it is necessary to replace the sign of variation by that of equality, hence we may write that y *multiplied by some constant* (k) *is equal to x*, $\therefore$ $x = ky$.

The value of k can be obtained when x and y are known. Having obtained the value of k, then, given either x or y, the value of the other can be found.

Nearly all the formulae required by the engineer are concerned with the sign of variation. As there are so many applications to choose from it is a difficult matter to make a selection. The following are a few typical cases :

Ex. 1. The space described by a falling body varies as the square of the time. If a falling body describes a distance of 64·4 feet in 2 seconds, find the distance moved through in 5 seconds.

Here, denoting the space by s and the time by t, then $s \propto t^2$,

$$\therefore \ s = kt^2.$$
$$64 \cdot 4 = k \times 2^2 \ ; \ \text{or,} \ k = 16 \cdot 1.$$

Hence, in 5 seconds $s = 16 \cdot 1 \times 5^2 = 402 \cdot 5$ feet.

Stress and Strain.—*Stress is directly proportional to strain.*

Hence stress $\propto$ strain, or $\dfrac{\text{stress}}{\text{strain}} = $ constant. This is known as

Hooke's Law ; the word *stress* denoting the *force per unit area* or the ratio of load to area, and *strain the ratio of alteration of length to original length*. The constant is called the *modulus of elasticity* for the substance and is usually denoted by the letter E.

Ex. 2. The area of cross-section of a bar of metal is 2 sq. in., and when a load of 10,000 lbs. is applied the alteration in the length of the bar is ·0288″, find E. The length of the bar is 12 feet.

Here $\quad$ Stress $= \dfrac{\text{load}}{\text{area}} = \dfrac{10,000}{2} = 5,000$ lbs. per sq. in.

$$\text{Strain} = \frac{\text{alteration in length}}{\text{original length}} = \frac{\cdot 0288}{12 \times 12} = \cdot 0002,$$

$$\therefore \ E = \frac{5000}{\cdot 0002} = 25,000,000 \ \text{or} \ 2 \cdot 5 \times 10^7 \ \text{lbs. per sq. in.}$$

Ex. 3. The heat H, in calories, generated by a current of C ampères in a circuit, varies as the square of the current, the resistance of the circuit R and the time t in seconds during which the current passes,

$$\therefore \ H \propto C^2 Rt, \quad \text{or} \ H = kC^2 Rt.$$

When H is 777600, C is 10, R is 18 ohms, and $t=30$ minutes, find H when $C=20$, $R=60$, $t=60$.

Here $777600 = k \times 10^2 \times 18 \times 30 \times 60$,

$$\therefore\ k = \frac{777600}{100 \times 18 \times 30 \times 60} = {\cdot}24.$$

Hence $H = {\cdot}24 \times 20^2 \times 60 \times 60$,

$$\therefore\ H = 345600.$$

Inverse proportion and variation.—One quantity is said to vary inversely as another when the product of the two quantities is always constant.   Or,

$$xy = k\ ;$$
$$\therefore\ x = \frac{k}{y}.$$

Hence, *as one quantity increases the other decreases in the same ratio* ; or, *one quantity varies as the reciprocal of the other.*

The reciprocal of a quantity is unity divided by the quantity; thus, the reciprocal of y is $\dfrac{1}{y}$.

For a *given quantity of gas the force exerted varies inversely as the volume when the temperature remains constant.*   This relation is known as **Boyle's Law** for a gas.

Denoting the pressure and corresponding volume of a gas by p and v, the law gives $p \propto \dfrac{1}{v}$,

or $pv = \text{constant} = k$.............................(i)

Ex. 4.   When the pressure of a gas is 60 lbs. per sq. in. the volume is 2 cub. ft.   If the gas expands according to Boyle's Law, find the pressure when the volume is 3 cub. ft.

From (i) we have $60 \times 2 = k$; $\therefore\ k = 120$.

If the volume change to 3 cub. ft., then the pressure is given by

$$p = \frac{k}{v} = \frac{120}{3} = 40 \text{ lbs. per sq. in.}$$

The load W that a bar, or beam, of length l, breadth b, and depth d will carry, varies directly as the breadth, as the square of the depth and inversely as the length, or

$$W \propto \frac{bd^2}{l}.$$

Ex. 5.   A bar of fir 10 in. long, 1 in. broad, and 1 in. deep will carry a load of 540 lbs.; find the depth of a bar of fir similarly loaded to carry a load of $\frac{3}{4}$ ton when the breadth is 2 in. and the length 5 feet.

$$W = k\frac{bd^2}{l}.$$

To find k we have

$$540 = \frac{k \times 1 \times 1}{10} ; \quad \therefore \ k = 5400.$$

Hence $\qquad \frac{3}{4} \times 2240 = \frac{5400 \times 2 \times d^2}{60},$

or $\qquad d^2 = \frac{3 \times 2240 \times 60}{4 \times 5400 \times 2} ; \quad \therefore \ d = 3\cdot05 \ \text{in.}$

Simple and compound proportion.—By introducing algebraical symbols questions which involve arithmetical difficulties are readily solved, as in the following example :

Ex. 6. If 40 men working 9 hours a day can build a wall 50 ft. long in 16 days, find how many men will be required to build a similar wall 25 ft. long in 20 days working 8 hours a day.

Here the number of men required will vary directly as the length of wall to be built, and inversely as the number of days and the number of hours per day.

Denoting by m, l, d, and h, the number of men, length of wall, number of days, and number of hours per day.

Then the statement is, $m \propto \dfrac{l}{dh},$

or $\qquad\qquad\qquad m = k \cdot \dfrac{l}{dh}$ ....(i)

To find the value of k it is only necessary to substitute the given values for m, l, d, and h.

$$\therefore \ 40 = k\frac{50}{16 \times 9}, \ \text{or} \ k = \frac{40 \times 16 \times 9}{5}.$$

To find m the number of men required substitute the known values for the right-hand side. Then

$$m = \frac{4 \times 16 \times 9}{5} \times \frac{25}{20 \times 8}$$
$$= 18.$$

EXERCISES. XX.

1. The rents of an estate should be divided between A and B in the proportion $5:3$; £470, however, is paid to A, and £280 to B. Which has been overpaid, and by how much ?

2. If the ratio of $2x + y$ to $6x - y$ equals the ratio of 2 to 3, what is the ratio of x to y ?

3. If an express train, travelling at the rate of 55 miles an hour, can accomplish a journey in $3\frac{1}{2}$ hours, how long will it take a slow train to travel two-thirds of the distance, its rate being to that of the express train as 4 to 9 ?

4. A's rate of working is to B's as 4 to 3, and B's is to C's as 2 to 1. How long will it take C to do what A would do in 6 days?

5. A gas is expanding according to the law $pv=$const., if when $p=100$ lbs. per square inch v is 2 cubic feet, find the pressure when v is 8 cubic feet.

6. It is known that x varies directly as y and inversely as z; it is also known that x is 500 when y is 300 and z is 14 ; find the value of z when x is 574 and y is 369.

7. If x varies as the square of y, and if x equals 144 when y equals 3, find the value of y when $x=324$.

8. A person contracts to do a piece of work in 30 days, and employs 15 men upon it. At the end of 12 days one-fourth only of the work is finished. How many additional hands must be engaged in order to perform the contract ?

9. If 30 men working 9 hours a day can build a certain length of wall in 16 days, find how many youths must be employed to build a similar wall of half that length in 20 days, working 8 hours a day, the work of 4 youths being equal to that of 3 men.

10. If 360 men working $10\frac{1}{2}$ hours a day can construct a road 1089 yards long in 35 days; how long would the job take 420 men working 9 hours a day ?

11. A garrison of 1500 men has provisions for 12 weeks at the rate of 20 ounces for each man per day ; how many men would the same provisions maintain for 20 weeks, each man being allowed 18 ounces per day ?

12. If 20 men can build a wall 70 ft. long, 8 ft. high, and 4 ft. thick in five days, working 7 hours a day, how many hours per day must 30 men work to build a wall 120 ft. long, 12 ft. high, and 3 ft. thick in the same time ?

13. The volume of a sphere varies as the cube of the diameter. If a solid sphere of glass 1·2 inches in diameter is blown into a shell bounded by two concentric spheres, the diameter of the outer sphere being 3·6 inches, show that the thickness of the shell is 0·0225 inches (nearly).

14. The expenses of a certain public school are partly fixed and partly vary as the number of boys. In a certain year the number of boys was 650 and the expenses were £13,600 ; in another year the number of boys was 820 and the expenses were £16,000. Find the expenses for a year in which there were 750 boys in the school.

CHAPTER XI.

INDICES.   APPROXIMATIONS.

Indices.—The letter or number placed near the top and to the right of a quantity which expresses the power of a quantity is called the index.   Thus in a^5, a^7, a^9, the numbers 5, 7, and 9 are called the indices of a, and are read as "a to the power five," "a to the power seven," etc.   Similarly a^b denotes a to the power b.

There are three so-called index rules or laws.

First index rule.—To multiply together different powers of the same quantity add the index of one to the index of the other.   To divide different powers of the same quantity subtract the index of the divisor from the index of the dividend.

$$\text{Thus } a^3 \times a^2 = (a \times a \times a)(a \times a) = a^{3+2} = a^5.$$

Ex. 1. $a^3 \times a^5 = a^{3+5} = a^8.$

Ex. 2. $a^2 \times a^3 \times a^4 = a^{2+3+4} = a^9.$

This may be expressed in a more general manner as follows :

$$a^m = (a \times a \times a \ldots \text{to } m \text{ factors})$$

and
$$a^n = (a \times a \times a \ldots \text{to } n \text{ factors}),$$

$$\therefore a^m \times a^n = (a \times a \times a \ldots \text{to } m \text{ factors})(a \times a \times a \ldots \text{to } n \text{ factors})$$
$$= (a \times a \times a \ldots \text{to } m+n \text{ factors})$$
$$= a^{m+n}.$$

This most important rule has been shown to be true when $m=3$ and $n=2$.   Other values of m and n should be assumed, and a further verification obtained.

Also
$$\frac{a^5}{a^3} = \frac{a \times a \times a \times a \times a}{a \times a \times a} = a^{5-3} = a^2.$$

Similarly
$$\frac{a^m}{a^n} = \frac{a \times a \times a \text{ to } m \text{ factors}}{a \times a \times a \text{ to } n \text{ factors}} = a^{m-n}.$$

Ex. 3.   Explain why the product is a^7 when a^3 is multiplied by a^4, and why the quotient is a when a^4 is divided by a^3.

$$a^4 \times a^3 = (a \times a \times a \times a) \times (a \times a \times a) = a^{4+3} = a^7.$$

Also
$$\frac{a^4}{a^3} = \frac{a \times a \times a \times a}{a \times a \times a} = a.$$

It is often found convenient to use both fractional and negative indices in addition to those just described.

The meaning attached to fractional and negative indices is such that the previous rule holds for them also.   When one fractional power of a quantity is multiplied by another fractional power the fractional indices are added, and when one fractional power is divided by another the fractional indices are subtracted.

$$a^{\frac{1}{2}} \times a^{\frac{1}{2}} = a^{\frac{1}{2}+\frac{1}{2}} = a^1 = a,$$

$$a^{\frac{1}{3}} \times a^{\frac{1}{3}} = a^{\frac{2}{3}}, \quad a^{\frac{1}{3}} \times a^{\frac{1}{3}} \times a^{\frac{1}{3}} = a^{\frac{1}{3}+\frac{1}{3}+\frac{1}{3}} = a^1 = a.$$

Hence, the meaning to attach to $a^{\frac{1}{2}}$ is the square root of a; to $a^{\frac{2}{3}}$ is the cube root of a squared, and to $a^{\frac{1}{3}}$ the cube root of a.

Thus
$$\sqrt{a} \text{ can be written as } a^{\frac{1}{2}},$$
$$\sqrt[3]{a} \text{ can be written as } a^{\frac{1}{3}}.$$

Also
$$\frac{1}{\sqrt{a}} = a^{-\frac{1}{2}},$$

and
$$\frac{1}{\sqrt[3]{a}} = a^{-\frac{1}{3}}.$$

Again
$$\frac{a^{\frac{1}{3}}}{a^{\frac{1}{2}}} = a^{\frac{1}{3}} \times a^{-\frac{1}{2}} = a^{\frac{1}{3}-\frac{1}{2}} = a^{-\frac{1}{6}}.$$

Also
$$\frac{a^{\frac{1}{3}}}{a^{\frac{1}{3}}} = a^{\frac{1}{3}-\frac{1}{3}} = a^0.$$

Similarly
$$\frac{a^3}{a^3} = \frac{a \times a \times a}{a \times a \times a} = a^{3-3} = a^0.$$

Generally, since $a^m \times a^n = a^{m+n}$ is true for all values of m and n.   If n be 0, then

$$a^m \times a^0 = a^{m+0} = a^m \; ;$$

$$\therefore \; a^0 = \frac{a^m}{a^m} = 1.$$

The second index rule.—To obtain a power of a power multiply the two indices.

Ex. 1.   Thus to obtain the cube of a^2 we have
$$(a^2)^3 = (a \times a)(a \times a)(a \times a) = a^{2 \times 3} = a^6,$$
where the index is the product of the indices 2 and 3.

Ex. 2.   Find the value of $(2 \cdot 15^2)^3$.
$$(2 \cdot 15^2)^3 = 2 \cdot 15^{2 \times 3}$$
$$= 2 \cdot 15^6 = 98 \cdot 72,$$
or, expressing this rule as a formula, we have
$$(a^m)^n = a^{mn},$$
∴ *a quantity a^m may be raised to a power n by using as an index the product mn.*

The third index rule.—To raise a product to any power raise each factor to that power.

Ex. 1. $(abcd)^m = a^m \times b^m \times c^m \times d^m.$

Ex. 2.   Let $a=1$, $b=2$, $c=3$, $d=4$, and $m=2$.

Then $(abcd)^m = (1 \times 2 \times 3 \times 4)^2 = 1^2 \times 2^2 \times 3^2 \times 4^2$
$$= 24^2 = 576.$$

In fractional indices the index may be written either in a fractional form or the root symbol may be used. The general form is $a^{\frac{m}{n}}$. This may be written in the form $\sqrt[n]{a^m}$, which is read as *the nth root of a to the power m.*

Ex. 6. $2^{\frac{5}{3}} = \sqrt[3]{2^5} = \sqrt[3]{32} = 3 \cdot 174.$

Ex. 7.   Find the values of $8^{\frac{2}{3}}$, $64^{-\frac{1}{2}}$, $4^{-\frac{3}{2}}$.

Here $8^{\frac{2}{3}} = \sqrt[3]{8^2} = \sqrt[3]{64} = 4.$

$$64^{-\frac{1}{2}} = \frac{1}{\sqrt{64}} = \frac{1}{8}.$$

$$4^{-\frac{3}{2}} = \frac{1}{4^{\frac{3}{2}}} = \frac{1}{\sqrt{64}} = \frac{1}{8}.$$

Ex. 8.   Find the value of $64^{\frac{1}{2}} + 4^{1 \cdot 5} + 2^{2 \cdot 5} + 27^{\frac{1}{3}}$.

Here $64^{\frac{1}{2}} = 8$, $4^{1 \cdot 5} = 4^{\frac{3}{2}} = 64^{\frac{1}{2}} = 8$,

$$2^{2 \cdot 5} = 2^{\frac{5}{2}} = 32^{\frac{1}{2}} = 5 \cdot 656,$$

$$27^{\frac{1}{3}} = 3.$$

Hence $64^{\frac{1}{2}} + 4^{1 \cdot 5} + 2^{2 \cdot 5} + 27^{\frac{1}{3}} = 24 \cdot 656.$

Ex. 9.   Find to two places of decimals the value of $x^2 - 5x^{\frac{1}{2}} + x^{-2}$ when $x = 5$.

Here
$$x^2 - 5x^{\frac{1}{2}} + x^{-2} = 25 - 5\sqrt{5} + \frac{1}{5^2}$$

$$= 25 - \frac{10}{2} \times 2 \cdot 236 + \cdot 04 = 13 \cdot 86.$$

Powers of 10.—Reference has already, on p. 25, been made to a convenient method of writing numbers consisting of several figures.

Thus the number $\cdot 6340000$ is $6 \cdot 34 \times 1000000$, or, $6 \cdot 34 \times 10^6$.
Similarly $6340 = 6 \cdot 34 \times 10^3$,

$$\cdot 634 \quad = \frac{6 \cdot 34}{10} = 6 \cdot 34 \times 10^{-1},$$

$$\cdot 000634 = \frac{6 \cdot 34}{10000} = 6 \cdot 34 \times 10^{-4}, \text{ etc.}$$

Suffix.—A small number or letter placed at the right of a letter but near the bottom is called a *suffix* and it is important to notice the difference between an index and a suffix. Thus P^2 means $P \times P$, but P_2 is merely a convenient notation to avoid the use of a number of letters, each of which may refer to different magnitudes of similar quantities. In this manner the letters P_0, P_1, P_2 ..., etc., may each refer to forces, etc., of different magnitudes, and in different directions.

Binomial theorem.—We have already found that
$$(a+b)^2 = a^2 + 2ab + b^2,$$
and by multiplying again by $a+b$ we obtain
$$(a+b)^3 = a^3 + 3a^2b + 3ab^2 + b^3.$$

It is seen at once that some definite arrangement of the coefficients and indices of such expressions may be made so that another power, say $(a+b)^4$, can be written down : the method used, and called the *Binomial Theorem*, is very important. The rule should be applied to the operation of expanding several simple expressions, such as $(a+b)^3$, $(a+b)^4$, etc., and afterwards committed to memory.

$$(a+b)^n = a^n + \frac{na^{n-1}b}{1} + \frac{n(n-1)}{1 \cdot 2}a^{n-2}b^2 + \dots.$$

Take $n = 2$, then
$$a^n = a^2, \text{ and } \frac{na^{n-1}b}{1} = 2ab.$$

Hence $(a+b)^2 = a^2 + \frac{2ab}{1} + \frac{b^2 \cdot (2 \cdot 1)}{1 \cdot 2} = a^2 + 2ab + b^2.$

Take $n=3$; here $a^n = a^3$; $\dfrac{na^{n-1}b}{1} = 3a^2b$, etc. ;

$$\therefore (a+b)^3 = a^3 + \frac{3a^2b}{1} + \frac{3 \cdot 2ab^2}{1 \cdot 2} + \frac{3 \cdot 2 \cdot 1}{1 \cdot 2 \cdot 3}a^0b^3$$
$$= a^3 + 3a^2b + 3ab^2 + b^3.$$

As a handy check, the reader should notice, that in each term the sum of the powers of a and b is equal to n. Thus, when $n=3$, in the second term a is raised to the power 2, and b to the power 1. Therefore sum of powers $=3$. Also, each coefficient has for its denominator a series of factors $1 \cdot 2 \cdot 3 \ldots r$, where r has the same numerical value as the power of b in that term. Thus, in the term containing b^3, a must be raised to the power $n-3$, and the coefficient must be

$$\frac{n(n-1)(n-2)}{1 \cdot 2 \cdot 3}.$$

Writing down terms in the numerator to be afterwards cancelled by corresponding numbers in the denominator, may appear to the beginner to be an unnecessary process, but to avoid mistakes it is better to write out in full, as above, and afterwards to cancel any common factors in the numerator and denominator.

Approximation.—The expansion of

$$(1+a)^n \text{ is } 1 + \frac{na}{1} + \frac{n(n-1)a^2}{1 \cdot 2} + \ldots$$

when a is a very small quantity, the two first terms are for all practical purposes sufficient ; thus, when a is small

$$(1+a)^n = 1 + na \text{ (approximately)}.$$

Similarly when a and b are small quantities

$$(1+a)^n (1+b)^m = 1 + na + mb \text{ (approximately)}.$$

Thus if $a = {\cdot}01$ and $n = 2$, then

$$(1 + {\cdot}01)^2 = 1 + 2 \times {\cdot}01 = 1{\cdot}02.$$

Ex. 1. $(1 + {\cdot}05)^2 = 1 + 2 \times {\cdot}05 = 1{\cdot}1$;
more accurately $(1{\cdot}05)^2 = 1{\cdot}1025{\cdot}$

Ex. 2. $(1 + {\cdot}05)^3 = 1 + 3 \times {\cdot}05 = 1{\cdot}15.$

Ex. 3. $\sqrt[3]{(1{\cdot}05)} = (1 + {\cdot}05)^{\frac{1}{3}} = 1 + \frac{1}{3} \times {\cdot}05 = 1{\cdot}0167{\cdot}$

Ex. 4. $\dfrac{1}{\sqrt[3]{1{\cdot}05}} = (1 + {\cdot}05)^{-\frac{1}{3}} = 1 - \frac{1}{3} \times {\cdot}05 = 1 - {\cdot}0167 = {\cdot}9833.$

Ex. 5. Find the superficial and cubical expansion of iron, taking a, the coefficient of linear expansion, as $\cdot000012$, or $1\cdot2\times10^{-5}$.

If the side of a square be of unit length, then when the temperature is increased by $1°$ C., the length of each side becomes $1+a$, and area of square is $(1+a)^2=1+2a+a^2$.

$$\therefore\ (1+a)^2=1+2\times\cdot000012+(\cdot000012)^2.$$

As a is a very small quantity its square is negligible. Hence the coefficient of superficial expansion is $2a=\cdot000024$, or $2\cdot4\times10^{-5}$.

Again, $(1+a)^3$ may be written as $1+3a$, neglecting the terms in a^2 and a^3.

$$\therefore\ \text{coefficient of cubical expansion}=3a=\cdot000036=3\cdot6\times10^{-5}.$$

EXERCISES. XXI.

1. Multiply $a^{\frac{2}{3}}+b^{\frac{4}{3}}+c^2-cb^{\frac{2}{3}}-ca^{\frac{1}{3}}-a^{\frac{1}{3}}b^{\frac{2}{3}}$ by $a^{\frac{1}{3}}+b^{\frac{2}{3}}+c$.

2. Find the value of (i) $\sqrt{64}+\sqrt{4^3}+2^{\frac{5}{2}}+\sqrt[3]{27}-9^{\frac{1}{2}}$.

(ii) $\sqrt{4^3}+\sqrt{2^5}+\sqrt[3]{8^2}$.

Simplify

3. $\sqrt[3]{a^2bc}\times\sqrt[4]{abc^2}\div a^{\frac{11}{12}}b^{\frac{7}{12}}c^{\frac{2}{3}}$.

4. $\left(\dfrac{a^{\frac{3}{2}}b^{\frac{7}{4}}}{a^{\frac{2}{3}}b^{\frac{1}{2}}}\right)^{-\frac{1}{6}}\times\left\{\sqrt[3]{a^{-2}}\ \sqrt[6]{b^{-1}}\right\}^2$. **5.** $\left(a^{1+\frac{q}{p}}\right)^{\frac{p}{p+q}}+\sqrt[p]{\dfrac{a^{2p}}{(a^{-1})^{-p}}}$.

6. $\dfrac{(a^{p-q})^{p+q}\times(a^q)^{q+x}}{(a^p)^{p-q}}$. **7.** $\dfrac{a^{m-n}a^{m-3n}a^{m-5n}}{a^{n-m}a^{n-3m}a^{n-5m}}$. **8.** $(x^m-x^n)^4$.

9. If a glass rod 1 inch long at $0°$ C. is $1\cdot000008$ inches long at $1°$ C., find the increase in the volume of 1 cubic inch of the glass when heated from $0°$ C. to $1°$ C.

10. How much error per cent. is there in the assumption
$$(1+a)(1+b)=1+a+b \text{ when } a=\cdot003,\ b=\cdot005?$$

11. Using the rule $(1+a)^n=1+na$, find the values of $\sqrt[3]{1\cdot003}$ and $(\cdot996)^2$, and find the error per cent. in the latter case.

12. How much error per cent. is there in the assumption that
$$(1+a)(1+\beta)=1+a+\beta, \text{ when } a=-\cdot002,\ \beta=-\cdot004?$$

13. Having given $10^{\frac{1}{2}}=3\cdot1623$, and $10^{\frac{1}{8}}=1\cdot3336$, find the values of $10^{\frac{5}{8}}$ and $10^{\frac{3}{8}}$ to five significant figures.

14. If $x=1\cdot002$ and $y=0\cdot997$, write down the values of x^3 and $y^{\frac{1}{3}}$ correct to three places of decimals.

15. Given that

$$10^{\frac{1}{2}}=3 \cdot 1623, \qquad 10^{\frac{1}{16}}=1 \cdot 1548,$$
$$10^{\frac{1}{4}}=1 \cdot 7783, \qquad 10^{\frac{1}{32}}=1 \cdot 0746,$$
$$10^{\frac{1}{8}}=1 \cdot 336,$$

find to five significant figures the values of $10^{\frac{3}{32}}$, $10^{\frac{7}{32}}$, $10^{\frac{3}{8}}$.
Explain how you would illustrate that $10^0=1$.

16. Divide $(x^3 y^{mn})^{\frac{1}{m}}$ by $(x^2 y^{mn})^{\frac{1}{n}}$.

17. Raise $\{a^3 b (a^3 bc)^{\frac{1}{5}}\}^{\frac{1}{6}}$ to the 7th power.

18. Find the mth root of $2^{\frac{1}{m}} a^m b^{2m} c^2$.

19. Simplify $\left\{ \dfrac{\sqrt[3]{a}}{\sqrt[4]{b^{-1}}} \cdot \left(\dfrac{b^{\frac{1}{4}}}{a^{\frac{1}{3}}} \right)^2 \div \dfrac{a^{-\frac{1}{3}}}{b^{-\frac{1}{2}}} \right\}^6$.

20. $\dfrac{\{(a^m)^{\frac{1}{r}}(a^q)^{\frac{1}{n}}\}^{nr}}{\{\sqrt[q]{b^n} \cdot (\sqrt[m]{b})^r\}^{mq}} \div \left\{ \left(\dfrac{a}{b} \right)^q \right\}^r$.

21. (i) $\sqrt{(a^{-\frac{5}{3}} b^3 c^{-\frac{2}{3}})} \div (\sqrt[3]{a^{\frac{1}{2}} b^4 c^{-1}})$. (ii) $\dfrac{ax^{-1}+a^{-1}x+2}{a^{\frac{1}{3}}x^{-\frac{1}{3}}+a^{-\frac{1}{3}}x^{\frac{1}{3}}-1}$:

22. Find the value of $x^2 - \frac{3}{4}x^{\frac{3}{2}} + x^{-1}$ when $x=3$.

23. $\sqrt{x^{-\frac{5}{3}}y^3 z^{-\frac{2}{3}}} \div \sqrt[3]{x^{\frac{1}{2}}y^4 z^{-1}}$.

24. Find the value of

$$2^{-2} 3^{\frac{1}{2}} x^{\frac{5}{6}} + 2^{-3} 3^{-\frac{1}{2}} x^{\frac{2}{3}} - 10(27x)^{-\frac{1}{6}}$$

when $x=64$.

CHAPTER XII.

BRITISH AND METRIC UNITS OF LENGTH, AREA, AND VOLUME. DENSITY AND SPECIFIC GRAVITY.

Measurement.—The measurement of a quantity is known when we have obtained a number which indicates its magnitude.

It is necessary, therefore, to select some definite quantity of the same kind, as a unit, and then to proceed to find how many times the unit is contained in the quantity to be measured. The number of times that the unit is contained in the given quantity is the numerical value of the quantity.

Units of length.—In order that length may be measured there must be both *a unit* and *a standard*. The unit is a certain definite distance with which all other distances can be compared ; and a standard is a bar on which the unit is clearly, accurately, and permanently marked. The two units most generally adopted are the **yard** and the **metre**.

The British System.—In this system the *unit of length* is the *yard*. It may be defined as the distance between two lines on a particular bronze bar when the bar is at a certain temperature (62° F.). The bar is deposited at the Standards Office of the Board of Trade.

British Measures of Length.

[The unit is divided by 3 and 36, etc. ; also multiplied by 2, 5½, 220, and 1760.]

12 inches = 1 foot.	40 poles, or 220 yards = 1 furlong.
3 feet = 1 **yard** (unit).	8 furlongs ⎫
2.yards = 1 fathom.	1760 yards ⎬ = 1 mile.
5½ yards = 1 rod, or pole.	or 5280 feet ⎭
6080 feet = 1 knot, or nautical mile.	

. **The French or Metric System.**—The Metric System is extensively used for all scientific, and in many cases for commercial purposes, and for many purposes is better and simpler than the British method.

The metre is divided into 10 equal parts called *decimetres* ; the decimetre is divided into 10 equal parts each called a *centimetre* : hence a centimetre is one hundredth of a metre, and this submultiple of the unit is the most commonly used of the metric measures of length. The centimetre is divided into 10 equal parts each known as a *millimetre*.

The metre is equal in length to 39·37 inches, and is thus slightly longer than our yard. Its length is roughly 3 feet 3⅓ inches, which number can be easily remembered as it consists throughout of threes.

The foot is equal in length to 30·48 centimetres.

It will be seen on reference to Fig. 49, which represents one end of a steel scale, that a length of 10 cm. is approximately

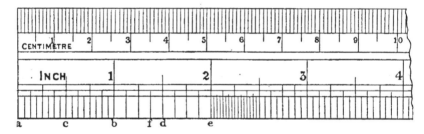

FIG. 49.—Comparison of inches and centimetres. The inches and centimetres are not drawn full size, but their comparative dimensions may be seen.

equal to 4 inches. A more accurate relation to remember is that a length of 25·4 centimetres, or 254 millimetres, is equal to the length of 10 inches. Thus, the distance from *a* to *b* may be expressed as 1 inch, 2·54 centimetres, or 25·4 millimetres.

The following approximate relations are worth remembering :

35 yards = 32 metres.

10 metres = 11 yards, or 20 metres equals the length of a cricket pitch = 1 chain.

5 miles = 8 kilometres.

British to Metric Measures of Length.

1 inch = 2·54 centimetres.
1 foot = 30·48 centimetres.
1 yard = 0·914 metre.
1 mile = 1609·33 metres.

Metric to British.

1 millimetre = ·039 inch.
1 centimetre = ·394 inch.

$$1 \text{ metre} = \begin{cases} 39·371 \text{ inches.} \\ 3·28 \text{ feet.} \\ 1·094 \text{ yards.} \end{cases}$$

1 kilometre = 0·621 mile.

Abbreviations.—The following abbreviations are generally used, and should be carefully remembered; this may be easily effected by taking the precaution to use the abbreviations on all possible occasions.

Length.
  in. is used to denote inch or inches.
  ft. „ „ feet.
  kilom. „ „ kilometres.
  dcm. „ „ decimetre or decimetres.
  cm. „ „ centimetre or centimetres.
  mm. „ „ millimetre or millimetres.
  gm. „ „ gram or grams.
  kilog. „ „ kilogram.

Unit of Area.—Measurement of area, or square measure, is derived from, and calculated by means of, measures of length. **Thus, the unit of area is the area of a square the side of which is the unit length.**

Area of a square yard, or unit area.—If the unit length be a yard proceed as follows: Make AB equal to 3 feet, as in Fig. 50, and upon AB construct a square. Divide AB and BC each into 3 equal parts, and draw lines parallel to AB and BC, as in the figure. The unit area is thus seen to consist of 9 smaller squares, every side of which represents a foot; thus, the unit area, the square yard, contains 9 square feet.

The smaller measures of length, the foot and the inch, are much more generally used than the yard. If the unit of length AE (Fig. 51) be 1 foot, the unit of area AEF is 1 square foot. In a similar manner, when the unit of length is 1 inch, the unit of area is 1 square inch. If the unit of length be 1 centimetre, the unit of area is 1 square centimetre (Fig. 51).

If the side of the square on AE (Fig. 50) represent, on some convenient scale, 1 foot, then by dividing AE and AF each into

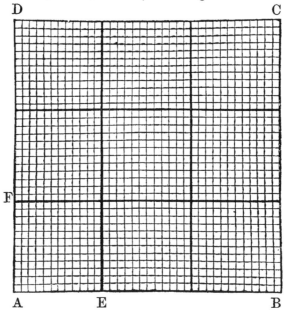

FIG. 50.—1 square yard equals 9 square feet, or 9×144 square inches.

12 equal parts, the distance between consecutive divisions would denote an inch. If through these points lines be drawn parallel to AE and AF respectively, it will be found that there are 12 rows of squares parallel to AE, and 12 squares in each of these 12 rows. **Hence, the area of a square foot represents 144 square inches.**

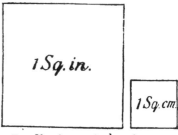

FIG. 51.—Square inch and square centimetre.

British Measures of Area or Surface.

[*Unit area* = 1 *square yard*. Larger and smaller units obtained by multiplying by 4840 and dividing by 9 and 1296.]

144 square inches	= 1 square foot.
1296 square inches or 9 sq. ft.	= 1 square yard.
4840 square yards	= 1 acre.
640 acres	= 1 square mile.

When comparatively large areas, such as the areas of fields, have to be estimated, the measurements of length, or linear measurements, are made by using a chain 22 yards long. Such a chain is subdivided into 100 links. The square measurements, or areas, are estimated by the *square chain*, or 484 (22×22) square yards in area. Or the area of a square, the length of one side of which is 22 yards, is $100 \times 100 = 10000$ sq. links; for each chain consists of 100 links. Hence we have the relation:

$$\text{1 chain} \qquad = \quad \text{22 yards} \qquad = \quad \text{100 links.}$$
$$\text{1 square chain} = \text{484 square yards} = \text{10000 sq. links.}$$
$$\text{10 square chains} = \text{4840 square yards} = \qquad \text{1 acre.}$$

144	square inches (sq. in.)	=1 square foot (sq. foot).
9	square feet	=1 square yard (sq. yd.).
$30\frac{1}{4}$	square yards	=1 square perch, rod, or pole (sq. po.).
40	square poles	=1 rood (r.).
4	roods	=1 acre (ac.)=4840 square yards.
640	acres	=1 square mile (sq. m.).

Metric measures of area.—As the metric unit of length is the metre, **the unit of area** (Fig. 52) is a square *ABDE*, having the length of its edge equal to 1 metre, and its area consequently equal to **1 square metre.**

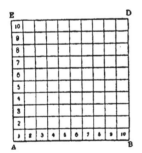

FIG. 52.—Representing a square metre divided into 10 decimetres. Scale $\frac{1}{40}$.

If *AB* and *BD* are each divided into 10 equal parts and lines drawn parallel to *AB* and *BD*, as shown, the unit area is divided into 100 equal squares, each of which is a square decimetre.

In scientific work the centimetre is the unit of length usually selected, and **the unit of area is one square centimetre** (Fig. 51).

Metric Measures of Area.

100	square millimetres=	1 square centimetre.
10000	„ „	=100 sq. cm.=1 sq. decim.
100	„ decimetres=	1 square metre.

Conversion Table.

British to Metric.		Metric to British.	
1 sq. in.	= 6·451 sq. cm.	1 sq. cm. =	0·155 sq. m.
1 sq. ft.	= 929 sq. cm.	1 sq. m. =	10·764 sq. ft.
1 sq. yard	= 8361·13 sq. cm.	1 sq. m. =	1·196 sq. yard.
1 acre	= 4046·7 sq. metres.	1 sq. km. =	0·3861 sq. mile.
1 sq. mile	= 2·59 sq. km.		
	= 2·59 × 10¹⁰ sq.cm.		

EXERCISES. XXII.

1. Find the number of square metres in (i) 10 square feet, (ii) 10 square yards.

2. Find the number of square metres in a quarter of an acre.

3. Find the number of square metres in 1000 square yards.

4. Express 2 sq. ft. 25 sq. in. as the decimal of a square metre.

5. Reduce 1000 square inches to square metres.

6. Find the number of square miles in 25,898,945 square metres.

7. Find which is greater, 10 sq. metres or 12 sq. yards, and express the difference between these areas as a decimal of a square metre.

Units of capacity and volume.—In the British system an arbitrary unit, the **gallon**, is the *standard unit of capacity and volume*, and is defined as **the volume occupied by 10 lbs. of pure water.**

A larger unit is the volume of a cube on a square base of which the length of each side is 1 foot and the height also 1 foot. The volume of such a cube is one cubic foot. A good average value for the weight of a cubic foot of water is 62·3 lbs. For convenience in calculations, a cubic foot is sometimes taken to be 6¼ gallons, and its weight 1000 oz., or 62·5 lbs. Hence the weight of a pint is 1¼ lbs.

The connection between length, area, and volume, may be shown by a diagram as in Fig. 53. Let

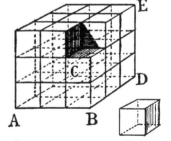

FIG. 53.—Showing a cubic yard and a cubic foot.

ABCD represent a square having its edge 1 yard, the area of the square is 9 square feet. If the vertical sides, one of which is shown at *DE*, be divided into three equal parts, and the remaining lines be drawn parallel to *DE* and the base respectively. Then, as will be seen from the figure, there are nine perpendicular rows of small cubes, the sides being 1 foot in length, area of base 1 square foot, and volume 1 cubic foot. Also there are three of these cubes in each row, making in all $3 \times 9 = 27$. Thus, 1 cubic yard = 27 cubic feet, *i.e.* $3 \times 3 \times 3 = 27$, and the weight of a cubic yard of pure water would therefore be $27 \times 62 \cdot 3$ lbs. $= 1682 \cdot 1$ lbs.

In this example, and also in considering the weight of a gallon, the student should notice that the specification " pure " water is necessary, for if the water contains matter either in solution or mixed with it, its weight would be altered. Thus, the weight of a cubic foot of salt water is usually taken to be 64 lbs., and the weight of a gallon of muddy water may be 11 or 12 lbs. instead of 10 lbs.

Units of Volume and Weight.

1728 cubic inches = 1 cubic foot.

27 cubic feet   = 1 cubic yard.

1 gallon   = $\cdot 1605$ cub. ft. = $277 \cdot 3$ cub. in.

One fourth part of a gallon is a quart and an eighth part is a pint.

Metric measures of volume.—We proceed in a similar way when we wish to measure volumes by the metric system.

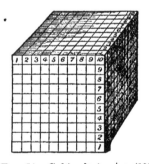

FIG. 54.—Cubic decimetre (1000 cubic centimetres) or 1 litre holds 1 kilogram or 1000 grams of water at 4° C.

A block built up with cubes representing cubic centimetres is shown in Fig. 54.

Each side of the cube measures 10 centimetres, and its volume is therefore a cubic decimetre. There are 10 centimetres in a decimetre, so the edge of the decimetre cube is 10 centimetres in length; the area of one of its faces is $10 \times 10 = 100$ square centimetres; and its volume is $10 \times 10 \times 10 = 100 \times 10 = 1000$ cubic centimetres.

This unit of volume is called a Litre. At ordinary temperature it is very nearly a cubic decimetre, or 1000 cubic centimetres (Fig. 54), and is equal to 1·76 English pints.

We have found that the unit of area is, for convenience, taken to be one square centimetre, the corresponding unit of volume is the cubic centimetre (c.c.).

For all practical purposes a litre of pure water weighs 1 kilogram or 1000 grams. Thus we have the relation

1 gram = weight of 1 cubic centimetre of water.
1 litre = 1 cubic decimetre = 1000 grams.

It is advisable to remember that there are 453·59 grams in a pound ; that 1 gram = 15·432 grains and that a kilogram = 2⅕ lbs.

The unit of weight is one pound. A smaller unit is obtained by dividing by 7000 and larger units by multiplying by 14, 112, and 2240, as follows ·

7000 grains = 1 lb.
14 lbs. = 1 stone.
112 lbs. = 1 hundred-weight, 1 cwt.
20 cwts. = 1 ton = 2240 lbs.

Conversion Table.

1 cub. in. = 16·387 cub. cm.		1 c.cm. = ·061 cub. in.	
1 „ ft. = 28316 „		1 litre = 61·027 „	
1 „ yard = 764535 „		= 1·76 pint or	
1 pint = 567·63 „		= ·22 gallon.	
1 gallon = 4541 „			
1 grain = ·0648 gm.		1 gram = 15·43 grains.	
1 ounce avoirdupois = 28·35 gm.		1 kilo = 2·2 lbs.	
7000 grains ⎫ 1 pound (lb.) ⎭ = 453·59 gm.			
1 ton = $1 \cdot 01605 \times 10^6$ gm.			

10 milligrams = 1 centigram.
10 centigrams = 1 decigram.
1000 grams = 1 kilogram.

Density.—**The density of a substance is the weight of** unit **volume.** Assuming the density to be uniform, the **density** of a substance, when the unit of weight is one pound and the unit of volume one cubic foot, is **the number of pounds in** a **cubic foot of substance.**

In the cases where metric units are adopted, the **density is the number of grams in a cubic centimetre of the substance.**

Density of water.—The weight of a cubic foot of water is 62·3 lbs., of a cubic centimetre 1 gram, and of a litre 1 kilogram.

Relative density.—**The relative density of a substance is the ratio of its weight to the weight of an equal volume of some substance assumed as a standard.** It is necessary that the standard substance should be easily obtainable at any place in a pure state. Pure water fulfils these conditions.

Specific gravity.—**The relative density of a substance is usually called its specific gravity.** The specific gravities of various substances are tabulated in Table I.

If s = specific gravity of a body, then the weight of 1 cub. ft.

$$= s \times 62\cdot3 \text{ lbs.}$$

If the substance is a liquid, then 1 gallon = $s \times 10$ lbs.

Again, as a litre of water weighs 1 kilogram, weight in kilograms = $s \times$ volume in litres.

A vessel containing 1 cub. ft. of water would when the water is replaced by mercury weigh $13\cdot596 \times 62\cdot3$ lbs.

If for cast iron s is $7\cdot2$, then the weight of a cub. ft.

$$= 7\cdot2 \times 62\cdot3 \text{ lbs.} = 448\cdot56 \text{ lbs.}$$

The weight of a cub. centimetre will be $7\cdot2$ grams.

The weight of V cubic feet of water will be $V \times 62\cdot3$ lbs. or Vw, where w is the weight of unit volume of water.

Hence if V denote the volume of a body in cub. ft. its weight will be Vws. In this manner it is customary to define **specific gravity as the ratio of the weight of a given volume of a substance to the weight of the same volume of water.**

If the volume of the body is obtained in cubic inches then w will denote the weight of one cubic inch (the weight of one cubic inch of water = $62\cdot3 \div 1728 = \cdot036$ lbs.).

Principle of Archimedes.—The method of obtaining the specific gravity of a solid (not soluble in water) depends on what is known as the " Principle of Archimedes " : **When a body is immersed in a liquid it loses weight equal to the weight of the liquid which it displaces ;** that is, if the weight of a body is obtained, first in air, and next when immersed in water, the difference in the weights is the weight of an equal volume of water :

$$\therefore \text{ specific gravity of body} = \frac{\text{weight in air}}{\text{weight in air} - \text{weight in water}}.$$

Ex. 1. A piece of metal weighs 62·63 grains in air and 56 grains in water. Find its specific gravity.

$$\text{s.g.} = \frac{62 \cdot 63}{62 \cdot 63 - 56} = 9 \cdot 4.$$

Ex. 2. A piece of metal of specific gravity 9·8 weighs in water 56 grains. What is its true weight?

Let w denote its true weight.

Then
$$9 \cdot 8 = \frac{w}{w - 56} ;$$

$$\therefore \ 9 \cdot 8w - 9 \cdot 8 \times 56 = w ;$$

$$\therefore \ w = \frac{9 \cdot 8 \times 56}{8 \cdot 8} = 62 \cdot 36 \text{ grains.}$$

Ex. 3. If 28 cubic inches of water weigh a pound, what will be the specific gravity of a substance, 20 cubic inches of which weigh 3 lbs. ?

As 20 cubic inches weigh 3 lbs., 1 cub. in. $= \frac{3}{20}$;

$$\therefore \ 28 \text{ cub. in.} = 28 \times \frac{3}{20} \text{ lbs.} = 4\frac{1}{5} \text{ lbs.}$$

But the same volume of water weighs 1 lb. ;

$$\therefore \ \text{specific gravity} = \frac{4\frac{1}{5}}{1} = 4\frac{1}{5}.$$

TABLE I. RELATIVE WEIGHTS.

Name.	Weight of Unit Volume in pounds.		Relative Density, or Specific Gravity.
	Cub. ft.	Cub. in.	
Water, - - - -	62·3	·036	1
Cast Iron, - - -	450	·26	7·2
Wrought Iron, - -	480	·28	7·698
Steel, - - - -	490	·29	7·85
Brass, - - - -	515	·298	8·2
Copper, - - - -	552	·3192	8·9
Lead, - - - -	712	·4121	11·418
Tin, - - - -	462	·267	7·4
Antimony, - - -	418	·242	6·72

EXERCISES. XXIII.

1. What is the specific gravity of a substance, 20 cubic inches of which weigh 3 lbs.?

2. A body A has a volume 1·35 cub. ft. and a specific gravity of 4·4, a second body B has a volume of 10·8 cub. in. and a specific gravity of 19·8 ; what ratio does the quantity of matter in A bear to that in B?

3. If the specific gravity of iron be 7·6, what will be the apparent weight of 1 cwt. of iron when weighed in water, and what weight of wood, of specific gravity 0·6, must be attached to the iron so as just to float it?

4. A piece of iron weighing 275 grams floats in mercury of density 13·59 with $\frac{5}{9}$ of its volume immersed. Determine the volume and density of the iron.

5. A ship weighing 1000 tons goes from fresh water to salt water. If the area of the section of the ship at the water line be 15,000 sq. feet, and the sides vertical where they cut the water, find how much the ship will rise, taking the specific gravity of sea water as 1·026.

6. A cubic cm. of mercury weighs 13·6 grams ; obtain the equivalent of a pressure of 760 mm. of mercury in inches of mercury, in feet of water, in pounds per square inch and per square foot, and in kilograms per square cm.

7. A cubical vessel, each side of which is a decimetre, is filled to one-fourth of its height with mercury, the remaining three-fourths with water, find the total weight of the water and the mercury.

8. Find the number of kilograms in ·7068 of a ton.

9. The area of a pond is half an acre when frozen over ; find the weight of all the ice if the mean thickness be assumed to be 2 inches. Specific gravity of ice ·92.

10. A body weighs 80 lbs. in air, its apparent weight in water is 56 lbs. and 46 lbs. in another liquid. Find the specific gravity of the liquid.

11. Three pints of a liquid whose specific gravity is 0·6 are mixed with four pints of a liquid specific gravity 0·81, and there is no contraction ; find the specific gravity of the mixture.

CHAPTER XIII.

MULTIPLICATION AND DIVISION BY LOGARITHMS.

Logarithms.—By the use of logarithms computations involving multiplication, division, involution, and evolution are made much more rapidly than by ordinary arithmetical processes. Many calculations which are very difficult, or altogether impossible, by arithmetical methods are, moreover, readily made by the help of logarithms.

Logarithms of numbers consist of an integral part called the **index** or **characteristic**, and a decimal part called the **mantissa**. If the reader will refer to Table III., he will find that opposite each of the numbers from 10 to 99 four figures are placed. These four figures are positive numbers, and each set of four is called a *mantissa*: the *characteristic*, which may be either positive or negative, has to be supplied in a way to be presently described when writing down the logarithm of any given number.

Logarithmic tables of all numbers from 1 to 100000 have been calculated with seven figures in the mantissa, but for ordinary purposes, and where only approximate calculations are required, such a table as that shown in Table III., at the end of this book, and known as *four-figure logarithms*, is very convenient.

By means of the numbers 10 to 99, with (*a*) those at the top of the table, and (*b*) those in the difference column on the right, the logarithm of any number consisting of four significant figures can be written down.

In logarithms all numbers are expressed by the powers of some number called the *base*.

The logarithm of a number to a given base is the index showing the power to which that base must be raised to give the number.

If N denote any number and a the given base, then by raising a to some power x we can get N. This is expressed by the equation

$$N = a^x.$$

Any number can be used as the base, but, as we shall find, the system of logarithms in which the base is 10 is that commonly used.

Thus, if the base be 2, then as $8 = 2^3$, 3 is the logarithm of 8 to the base 2. This can also be expressed by writing $\log_2 8 = 3$.

In a similar manner, if the base be 5, then 3 is the logarithm of 125 to the base 5 ;

$$\therefore \ \log_5 125 = 3.$$

Also $\qquad\qquad 64 = 2^6 = 4^3 = 8^2.$

Similarly,   6 is the log of 64 to the base 2 ;

           3 is the log of 64 to the base 4 ;

           2 is the log of 64 to the base 8, etc. ;

$$\therefore \ \log_2 64 = 6 \ ; \ \log_4 64 = 3 \ ; \ \log_8 64 = 2, \text{ etc.,}$$

using in each case the abbreviation *log* for logarithm.

Logarithms to the base 10.—It is most convenient to use 10 as the base for a system of logarithms. It is then only necessary to print in a table of such logarithms the decimal part or mantissa ; the characteristic can, we shall see, be determined by inspection. The tables are in this way less bulky than would otherwise be the case. When calculated to a base 10, logarithms are known as Common Logarithms.

$$\text{Since } N = 10^x \ ;$$

$$\therefore \ \log_{10} N = x.$$

Or by definition, substituting positive numbers for N,

as $\qquad\qquad 1 = 10^0 \ ; \quad \therefore \ \log \ \ 1 = 0.$

  Also, $\qquad\quad 10 = 10^1 \ ; \quad \therefore \ \log \ \ 10 = 1.$

  Again, $\qquad 100 = 10^2 \ ; \quad \therefore \ \log 100 = 2, \text{ etc.}$

In the chapter on Indices (p. 110) we found that $\cdot 1$, or $\frac{1}{10}$, can be written as 10^{-1} ; also $\cdot 01$, or $\frac{1}{100}$, can be written as 10^{-2}.

Hence $\qquad\qquad \log \cdot 1 = \log \frac{1}{10} = -1,$

and $\qquad\qquad\quad \log \cdot 01 = -2, \text{ etc.}$

The *mantissa is always positive*, and instead of writing the negative sign in front of the number, it is customary in logarithms to place it over the top; thus $\log \cdot 1$ is not written -1 but as $\bar{1}$, and $\log \cdot 01 = \bar{2}$.

In the preceding logarithms we have only inserted the characteristic; each mantissa consists of a series of ciphers.

Thus,
$$\log \quad 1 = 0 \cdot 0000,$$
$$\log \quad 10 = 1 \cdot 0000,$$
$$\log 100 = 2 \cdot 0000, \text{ and so on.}$$

As the logarithm of 1 is zero, and log 10 is 1, it is evident that the logarithms of all numbers between 1 and 10 will consist only of a certain number of decimals.

Thus, $\log 2 = \cdot 3010$ indicates, that if we raise 10 to the power $\cdot 3010$ we shall obtain 2, or $10^{3010} = 2$.

In a similar manner, the logarithm of $200 = 2 \times 100$ might be written as $10^2 \times 10^{\cdot 3010}$,
$$\therefore \ 200 = 10^{2 \cdot 3010}.$$

Hence we write $\qquad \log 200 = 2 \cdot 3010.$

Also $\qquad \cdot 0002 = \frac{2}{10000} = 2 \times 10^{-4}.$

Hence $\qquad \cdot 0002 = 10^{\bar{4} \cdot 3010}.$

$$\therefore \ \log \cdot 0002 = \bar{4} \cdot 3010.$$

The characteristic of a logarithm.—Referring to Table III., opposite the number 47 we find the *mantissa* $\cdot 6721$, and as 47 lies between 10 and 100 the characteristic is 1. Hence the log of 47 is $1 \cdot 6721$.

Again, the number 470 lies between 100 and 1000, and therefore the characteristic is in this case 2 ;

$$\therefore \ \log 470 = 2 \cdot 6721.$$

In a similar manner, the. logarithms of 4700 and 47000 are $3 \cdot 6721$ and $4 \cdot 6721$ respectively ; in each case the mantissa is the same, but the characteristic is different.

The rule by which the characteristic is found may be stated as follows : **The characteristic of any number greater than unity is positive, and is less by one than the number of figures to the left of the decimal point. The characteristic of a number less than unity is negative, and is greater by one than the number of zeros which follow the decimal point.**

Ex. 1. To write down log ·047

Here the two significant figures are 47, and the mantissa is the same as before. As one zero follows the decimal point, the characteristic is $\bar{2}$;

$$\therefore \ \log ·047 = \bar{2}·6721.$$

Again, to obtain the log of ·00047·

There are three zeros following the decimal point, the characteristic is $\bar{4}$, and

$$\therefore \ \log ·00047 = \bar{4}·6721·$$

Similarly, in the case of log ·47· Here the rule will give $\bar{1}$ for the characteristic ;

$$\therefore \ \log ·47 = \bar{1}·6721.$$

Another method of determining the characteristic is to treat any given number as follows :

$$470 = 4·7 \times 100 = 4·7 \times 10^2.$$

Hence as before the characteristic is 2.

Similarly, $\quad 4700 = 4·7 \times 10^3, \qquad ·47 = 4·7 \times 10^{-1},$
$$·047 = 4·7 \times 10^{-2}, \quad ·0047 = 4·7 \times 10^{-3}.$$

If all numbers are written in this convenient form the characteristic is the index of the multiplier 10. If this method be applied to all numbers it will save the trouble of remembering rules.

To obtain the logarithm of a number consisting of four figures.

Ex. 1. Find the log of 3768.

First look in Table III. for the number 37, then the next figure 6 is found at the top of table, so that the mantissa of

$$\log 376 = ·5752·$$

At the extreme right of the table will be seen a column of differences, as they are called ; thus, under the figure 8 on a horizontal line with 37 is found the number 9. This must be added to the mantissa previously obtained.

Hence we have mantissa of log 376 = 5752
$$\text{Add difference,} \qquad 9$$
$$\therefore \ \text{mantissa for log } 3768 = \overline{5761}$$

Hence $\qquad\qquad \log 3768 = 3·5761,$
also $\qquad\qquad \log ·003768 = \bar{3}·5761,$
and $\qquad\qquad \log ·3768 = \bar{1}·5761,$ etc.

To find the number corresponding to a given logarithm or the antilogarithm of a number.

Ex. 2. Given the log 2·4725, to find the number.

From Table IV. of antilogarithms.

Opposite the mantissa ·472 we have 2965. In the difference column under the number 5, and on the horizontal line 47, we have the figure 3.

Hence the corresponding mantissa = 2968, and as the characteristic is 2, the number required is 296·8.

If the given logarithm had been $\bar{2}$·4725 the required number would be ·02968·

Multiplication by logarithms.—Add the logarithms of the multiplier and multiplicand together : the sum is the logarithm of their product. The number corresponding to this logarithm is the product required.

Let a and b be the numbers.

Let $\qquad \log a = x$ and $\log b = y$;

$\qquad\qquad \therefore \ a = 10^x, \ b = 10^y.$

$\qquad\qquad a \times b = 10^{x+y},$

or $\qquad\qquad \log_{10} ab = x + y = \log a + \log b.$

Ex. 1. Multiply 2·784 by 6·85.

$$\text{From Table III. } \log 278 = \quad 4440$$
$$\text{Diff. col. for } 4 = \quad\underline{\quad 6}$$
$$\therefore \ \log 2\text{·}784 = \quad \text{·}4446$$
$$\text{Also } \log 6\text{·}85 \ = \quad \underline{\text{·}8357}$$
$$\therefore \ \text{logarithm of product} = \overline{1\text{·}2803}$$

$$\text{From Table IV. antilog } 280 = 1905$$
$$\text{Diff. col. for } 3 = \quad \underline{\quad 1}$$
$$\text{antilog } 2803 = \overline{1906}$$

Hence $\qquad\qquad 2\text{·}784 \times 6\text{·}85 = 19\text{·}06.$

Ex. 2. Multiply ·002885 by ·0915.

$$\log \text{·}002885 = \bar{3}\text{·}4602$$
$$\text{,, } \text{·}0915 \ = \underline{\bar{2}\text{·}9614}$$
$$\overline{\bar{4}\text{·}4216}$$

Hence $\qquad\qquad \text{·}002885 \times \text{·}0915 = \text{·}000264\text{·}$

Ex. 3.   Find the numerical value of $a \times b$ when $a = 32\cdot4$, $b = \cdot000467$.

$$\log 32\cdot4 \qquad = 1\cdot5105$$
$$,, \quad \cdot000467 = \overline{4}\cdot6693$$
$$\overline{\dot{2}\cdot1798}$$
$$\therefore \;\; a \times b = \cdot01513.$$

Using the data of Ex. 1, to prove the rule, then by the definition of a logarithm $\cdot4446$ is the index of the power of 10, which is equal to $2\cdot784$

$$\therefore \;\; 10^{\cdot4446} = 2\cdot784. \quad \text{Similarly } 10^{\cdot8357} = 6\cdot85,$$
$$\therefore \;\; 2\cdot784 \times 6\cdot85 = 10^{\cdot4446} \times 10^{\cdot8357} = 10^{1\cdot2803}.$$

EXERCISES. XXIV.

Multiply

1. $\cdot000257$ by $3\cdot01$.   2. $\cdot000215$ by $\cdot0732$.   3. $\cdot0032$ by $23\cdot45$.

4. $3\cdot413$ by $10\cdot16$.   5. $\cdot05234$ by $3\cdot87$.

6. $4\cdot132$ by $\cdot625$ and $\cdot1324$ by $\cdot00562$.   7. $4\cdot017$ by $\cdot00342\cdot$

8. $\cdot003 \times 17 \times \cdot004 \times 20000$.   9. $76\cdot05$ by $1\cdot036$.

10. Find the numerical value of $a \times b$ when

(i) $a = 14\cdot95$, $b = \cdot00734\cdot$
(ii) $a = 420\cdot3$, $b = 2\cdot317$.
(iii) $a = 5\cdot617$, $b = \cdot01738$.
(iv) $a = \cdot01342$, $b = \cdot0055$.

11. Calculate (i) $23\cdot51 \times 6\cdot71$.
(ii) $168\cdot3 \times 2\cdot476$.

12. Why do we add the logarithms of numbers to obtain the logarithm of their product?

Division by logarithms.—Subtract the logarithm of the divisor from the logarithm of the dividend and the result is the logarithm of the quotient of the two numbers. The number corresponding to this logarithm is the quotient required.

Let a and b be the two numbers.

Let $\qquad\qquad \log a = x$ and $\log b = y$;
$$\therefore \; a = 10^x, \; b = 10^y.$$

Hence $\qquad\qquad \dfrac{a}{b} = \dfrac{10^x}{10^y} = 10^{x-y},$

or $\qquad\qquad \log_{10}\dfrac{a}{b} = x - y = \log a - \log b.$

Using this rule for division, it is an easy matter to write down the logarithm of a number less than unity, and to verify the rule given on p. 127.

Thus, log ·047 = $\bar{2}$·6721.

This may be verified by noting that $·047 = \dfrac{4·7}{100}$;

$$\therefore \ \log ·047 = \log\dfrac{4·7}{100} = \log 4·7 - \log 100 = ·6721 - 2 = \bar{2}·6721.$$

In a similar manner $·47 = \dfrac{4·7}{10}$;

$$\therefore \ \log ·47 = \log\dfrac{4·7}{10} = \log 4·7 - \log 10 = ·6721 - 1 = \bar{1}·6721.$$

Ex. 1. Divide 3·048 by ·00525.

From Table III.   log 304 =   4829
                          Diff. col. for 8,   11

$\therefore$ log 3·048 =   ·4840...........................(i)

Also   log ·00525 = $\bar{3}$·7202........................(ii)

Subtracting (ii) from (i),   2·7638

From Table IV.   antilog 763 = 5794
                          Diff. col. for 8,   11

$\therefore$ antilog of 7638 = 5805

Hence   log 2·7638 = 580·5 ;
                          $\therefore$ 3·048 ÷ ·00525 = 580·5.

Ex. 2. Divide ·00525 by 3·048.

Here as in Ex. 1 subtracting the log of 3·048 from log ·00525 we obtain log $\bar{3}$·2362. From Table IV. antilog corresponding to this is 1723.

$$\therefore \ ·00525 \div 3·048 = ·001723.$$

In some cases it should be noticed that when four-figure logarithms are used the fourth significant figure, although not always quite exact, is usually not far wrong : three significant figures are necessarily accurate. Thus, in Ex. 1—3·048 ÷ ·00525 = 580·571..., and thus, as on p. 4, the fourth figure should be 6, not 5. Again, ·00525 ÷ 3·048 = ·0017234..., and the four significant figures are correct.

Evaluation by logarithms involving multiplication and division.—It is easily possible to evaluate any arithmetical calculation true to three significant figures.

Ex. 1.   Evaluate $\dfrac{a \times b}{c}$ when $a = 1.986$, $b = .1188$, $c = .5046$.

Substituting the given numbers we obtain

$$\frac{1.986 \times .1188}{.5046},$$

$$\therefore \ \log 1.986 + \log .1188 - \log .5046 = .2980 + \bar{1}.0749 - \bar{1}.7029$$

$$= \bar{1}.6700.$$

antilog $6700 = 4677$; $\therefore \ \dfrac{1.986 \times .1188}{.5046} = .4677,$

or $\dfrac{a \times b}{c} = .4677.$

EXERCISES.   XXV.

Divide

1. 30 by 6·25.   2. ·325 by 1300 and 3250 by ·013.   3. ·00062 by 64.

4. Why is it that we subtract the logarithms of two numbers to obtain the logarithm of their quotient?

Divide

5. 429 by ·0026·

6. (i) (·02 − ·002 + ·305) by ·016 × ·016.   (ii) ·05675 by ·0705.

7. ·05344 by 83·5.   8. ·00729 by ·2735.   9. ·0009481 by ·0157.

10. Calculate $a \times b \div c$ when

(i) $a = 619.3$, $b = .117$, $c = 1.43$.

(ii) $a = 6.234$, $b = .05473$, $c = 756.3$.

11. Calculate $a \div b$ when

(i) $a = .0004692$, $b = .000365$.   (ii) $a = 94.78$, $b = 2.847$.

(iii) $a = 907.9$, $b = 17.03$.

12. Calculate (i) $23.51 \div 6.78$.   (ii) $23.51 \div .0678$.

13. Compute (i) $16.83 \div 24.76$.   (ii) $.1613 \div .002476$.

14. If $\phi = c \times \dfrac{\pi}{4} d^2 \sqrt{2gh}$, find the numerical value of c, given $\phi = 1.811$, $d = .642$, $g = 32.2$, $h = 1.249$.

15. Calculate ab and $a \div b$ when $a = .5642$, $b = .2471$.

16. (i) Compute $4.326 \times .003457$.   (ii) $.01584 \div 2.104$.

17. (i) Compute $30.56 \div 4.105$.   (ii) $.03056 \times 0.4105$.

CHAPTER XIV.

INVOLUTION AND EVOLUTION BY LOGARITHMS.

Involution by logarithms.—To obtain the power of a number, multiply the logarithm of the number by the index representing the power required; the product is the logarithm of the number required.

Let $\qquad\qquad \log a = x.$

Then $\qquad\qquad a = 10^x.$

And $\qquad\qquad a^n = (10^x)^n$;

$$\therefore \ \log_{10} a^n = nx = n \log a.$$

Ex. 1. Find $\log a^3$, also $\log a^{\frac{1}{2}}$.

In the first case the index is 3, and, hence $\log a^3$ is three times the logarithm of a.

Similarly, $\log a^{\frac{1}{2}}$ is one-half the logarithm of a.

These examples are illustrations of the general rule, viz. :

$$\log a^n = n \log a,$$

where n is any number, positive or negative, integral or fractional.

Ex. 2. Find by logarithms the value of $(·05)^3$. The process is as follows :

Write down the log of the number as is shown on the right; multiply the log by the index 3, and in this way obtain for the mantissa ·0970, and for the characteristic 4. This result is arrived at by saying when obtaining the

$$\begin{array}{r} \bar{2}·6990 \\ 3 \\ \hline \bar{4}·0970 \end{array}$$

characteristic $3 \times 6 = 18$ plus 2 carried from last figure gives 20, and we write down 0. Next $3 \times -2 = -6$, but -6 added to $+2$ carried from the previous figure gives -4, which is written $\bar{4}$.

$$\therefore \ \log (·05)^3 = \bar{4}·0970.$$

It is seen from Table IV. that antilog ·097 = 1250.

The characteristic 4 indicates that three cyphers precede the first significant figure. Hence the required number is ·000125.

$$\therefore \ ·05^3 = ·000125.$$

Contracted multiplication may be used with advantage when the given index consists of three or more figures.

Ex. 3.   Calculate the value of $(9)^{3.76}$.

$$\log 9 = \cdot9542 ;$$
$$\therefore \quad \cdot9542$$
$$\underline{673}$$
$$2\cdot8626$$
$$6679\!\!4$$
$$\underline{5725\!\!2}$$
$$3\cdot5877$$

$$\text{antilog } 587 = 3864$$
$$\text{Diff. col. for } 7 = \quad 6$$
$$\therefore \quad \text{antilog } 5877 = 3870$$

Hence $\qquad\qquad (9)^{3\cdot76} = 3870.$

When the index of a number not only consists of several figures, but the number itself is less than unity, so that the characteristic of the logarithm of the number is negative, it is advisable to convert the whole logarithm into a negative number before proceeding to multiply by the index.

Ex. 4.   Calculate $(\cdot578)^{-3.76}$.

$$\log \cdot578 = \bar{1}\cdot7619, \text{ or, } -1 + \cdot7619 = -\cdot2381.$$

The product of $-\cdot2381$ and $-3\cdot76$ is $\cdot8952$.

$$\text{antilog } 8952 = 7856 ;$$
$$\therefore \quad (\cdot578)^{-3\cdot76} = 7\cdot856 \cdot$$

When the mantissa of a logarithm is positive, and the index a negative number, the resulting product is negative. If such a result occurs the mantissa must be made positive before reference is made to Table IV.

Ex. 5.   Calculate the value of $(8\cdot4)^{-1\cdot97}$.

$$\log 8\cdot4 = \cdot9243.$$
$$-1\cdot97 \times \cdot9243 = -1\cdot8208.$$

As the mantissa $\cdot8208$ is negative, it must be made positive, by subtracting it from unity and prefixing $\bar{1}$ for the characteristic, *i.e.*,

$$-\cdot8208 = \bar{1}\cdot1792.$$

Hence $\qquad\qquad -1\cdot8208 = \bar{2}\cdot1792 \cdot$

$$\text{antilog } 1792 = 1511 ;$$
$$\therefore \quad (8\cdot4)^{-1\cdot97} = \cdot01511.$$

This may be verified, if necessary, by writing $(8\cdot4)^{-1\cdot97}$ in its equivalent form, $\dfrac{1}{(8\cdot4)^{1\cdot97}}.$

In any given expression when the signs + and − occur it is necessary to calculate the terms separately and afterwards to proceed to add or subtract the separate terms. The method adopted may be shown by the following example :

Ex. 6.   Evaluate $2a^3 + (b^2)^{3\cdot76} + 2c^{-3\cdot76} - d^{-1\cdot97}$.

When $\qquad a = \cdot07,\ b = 3,\ c = \cdot578,\ d = 8\cdot4$.

Let x denote the value required, then

$$x = 2a^3 + (b^2)^{3\cdot76} + 2c^{-3\cdot76} - d^{-1\cdot97}.$$

Here the four terms must be separately calculated.

$\qquad\qquad \cdot07^3$ is found to be $\cdot000343 \quad \therefore\ 2a^3 = \cdot000686$.

From *Ex.* 3. $(3^2)^{3\cdot76}$,,   ,,   3870.

From *Ex.* 4. $\cdot578^{-3\cdot76}$,,   ,, $7\cdot856$ $\therefore\ 2c^{-3\cdot76} = 15\cdot712$.

From *Ex.* 5. $8\cdot4^{-1\cdot97}$,,   ,, $\cdot01511$.

Hence $\quad x = \cdot000686 + 3870 + 15\cdot712 - \cdot01511 = 3885\cdot727696$.

Evolution by logarithms.—**The logarithm of the number, the root of which is required, is divided by the number which indicates the root.**

No difficulty will be experienced when the characteristic and mantissa are both positive. But, although the characteristic of the logarithm may be negative, the mantissa remains positive. Hence the characteristic, when negative, usually requires a little alteration in form before dividing by the number, in order to make such logarithm exactly divisible by the number.

The methods adopted can best be shown by examples.

Ex. 1.   Find the cube root of 475.

From Table III., mantissa of log 475 = 6767 ;

$$\therefore\ \log 475 = 2\cdot6767.$$

To obtain the cube root it is necessary to divide the logarithm by 3 ; we thus obtain

$$\frac{2\cdot6767}{3} = \cdot8922.$$

From Table IV. we get

$$\text{antilog } 892 = 7798$$
$$\text{Diff. col. for 2,} \qquad 4$$
$$\therefore\ \text{antilog } 8922 = 7802$$

Hence $\qquad\qquad\qquad \sqrt[3]{475} = 7\cdot802.$

When the given number is less than unity, the characteristic of its logarithm is negative, and a slight adjustment must be made before the division is performed.

Ex. 2. Find the value of $\sqrt[3]{\cdot475}$.

$$\log \cdot475 = \bar{1}\cdot6767.$$

To obtain the cube root it is necessary to divide $\bar{1}\cdot6767$ by 3; before doing so the negative characteristic is, by adding -2, made into -3, so as to be exactly divisible by 3. Also $+2$ is added to the mantissa, thus $\bar{1}\cdot6767$ becomes $\bar{3}+2\cdot6767$.

Hence $\frac{1}{3}(\bar{3}+2\cdot6767) = \bar{1}\cdot8922.$

As in the preceding example, the corresponding antilog is 7802:

$$\therefore \sqrt[3]{\cdot475} = \cdot7802.$$

The adjustment indicated in the preceding example should be performed mentally, although at the outset the beginner may find it advisable to write down the numbers.

In dividing a logarithm by a given number it is necessary, when the divisor is greater than the first term in the mantissa, to prefix a cipher.

Ex. 3. Find the fifth root of **3.**

$$\log 3 = \cdot4771,$$
and $$\tfrac{1}{5}(\cdot4771) = \cdot0954.$$
$$\text{antilog } 0954 = 1246;$$
$$\therefore 3^{\frac{1}{5}} = 1\cdot246.$$

In this example, since the divisor 5 is greater than the first term 4 in the mantissa, a cipher is prefixed. Then, by ordinary division, we have 5 into 47 gives 9; the remaining two figures 5 and 4 are obtained in a similar manner.

Ex. 4. Find the fourth root of 0·007 or $(\cdot007)^{\frac{1}{4}}$.

$$\log \cdot007 = \bar{3}\cdot8451.$$
$$\tfrac{1}{4}(\bar{3}\cdot8451) = \tfrac{1}{4}(\bar{4}+1\cdot8451)$$
$$= \bar{1}\cdot46127;$$
$$\therefore \log(\cdot007)^{\frac{1}{4}} = \bar{1}\cdot4613.$$

Corresponding to the mantissa 461 we find the antilogarithm = 2891

Diff. col. for 3 = 2

2893

$\therefore$ the antilogarithm corresponding to the logarithm $\bar{1}\cdot4613$ is 2893.

Hence $$(0\cdot007)^{\frac{1}{4}} = \cdot2893.$$

Ex. 5.   Find the seventh root and the seventh power of 0·9306.

$$\log \cdot 9306 = \bar{1}\cdot 9688,$$
$$\log \text{ of seventh root} = \tfrac{1}{7}(\bar{7} + 6\cdot 9688) = \bar{1}\cdot 9955.$$

antilog 995 = 9886

Diff. col. for 5,   11

∴ antilog of ·9955 = 9897

The characteristic $\bar{1}$ indicates that the number is less than unity.
Hence   seventh root = ·9897.

Let x denote the seventh power of ·9306.

Then $x = (0\cdot 9306)^7$;

∴ $\log x = 7 \log \cdot 9306$
$$= 7 \times \bar{1}\cdot 9688 = \bar{1}\cdot 7816\cdot$$

antilog 781 = 6039

Diff. col. for 6,   8
                        ─────
                        6047

Hence $x = \cdot 6047.$

Ex. 6.   Evaluate $a^{\frac{2}{3}}b^{\frac{5}{6}}(a+b)^{-\frac{7}{3}} \times (a-b)^{\frac{1}{5}}$
when $a = 3\cdot 142,\ b = 2\cdot 718.$

In this example the signs + and − must be eliminated before
logarithms can be used.   This elimination is effected by first finding
the values of $a+b$ and $a-b$.   Thus $a+b = 3\cdot 142 + 2\cdot 718 = 5\cdot 86$ and
$a - b = 3\cdot 142 - 2\cdot 718 = \cdot 424\cdot$

Hence denoting the given expression by x we have
$$x = (3\cdot 142)^{\frac{2}{3}} \times (2\cdot 718)^{\frac{5}{6}} \times (5\cdot 86)^{-\frac{7}{3}} \times (\cdot 424)^{\frac{1}{5}},$$
$$\log x = \tfrac{2}{3}\log 3\cdot 142 + \tfrac{5}{6}\log 2\cdot 718 + \tfrac{1}{5}\log \cdot 424 - \tfrac{7}{3}\log 5\cdot 86$$
$$= \tfrac{2}{3} \times \cdot 4972 + \tfrac{5}{6} \times \cdot 4343 + \tfrac{1}{5} \times \bar{1}\cdot 6274 - \tfrac{7}{3} \times \cdot 7679$$
$$= \cdot 3315 + \cdot 3619 + \bar{1}\cdot 9255 - 1\cdot 7917 = \bar{2}\cdot 8272.$$

antilog 8272 = 6717 ;

∴ $x = \cdot 06717.$

Miscellaneous examples.—As logarithms enable calculations involving the arithmetical processes of multiplication,
division, involution, and evolution to be readily performed,
a few miscellaneous examples involving formulae frequently
required are here given.

Ex. 1.   The collapsing pressure of a furnace flue (in lbs. per sq. in.) may be found from the formula :

$$P = \frac{174000 \times t^2}{d \times \sqrt{l}}.$$

Given $d = 33$; $t = \frac{3}{8}$; and $l = 15$; find P.
Substituting the given values we have

$$P = \frac{174000 \times (\frac{3}{8})^2}{33 \times \sqrt{15}}.$$

$\therefore$ $\log P = \log 174000 + \log 3^2 - (\log 33 + \frac{1}{2} \log 15 + \log 8^2)$;
$\therefore$ $\log P = \log 174000 + 2 \log 3 - (\log 33 + \frac{1}{2} \log 15 + 2 \log 8)$
$\qquad = 5 \cdot 2405 + \cdot 9542 - (1 \cdot 5185 + \cdot 5880 + 1 \cdot 8062)$
$\qquad = 6 \cdot 1947 - 3 \cdot 9127 = 2 \cdot 2820.$
$$\text{antilog } 282 = 1914 ;$$
$$\therefore \ P = 191 \cdot 4 \cdot$$

Ex. 2.   If $\qquad 2 \cdot 718^x = 148 \cdot 4$, find x.
$$x \log 2 \cdot 718 = \log 148 \cdot 4 ;$$
$$\therefore \ x \times \cdot 4343 = 2 \cdot 1715.$$
Hence $\qquad\qquad x = \dfrac{2 \cdot 1715}{\cdot 4345} = 5.$

The numerical values of equations in which the ratio of one variable to another is required, can also be obtained by logarithms.

Ex. 3.   Given $x^5 = \dfrac{7 \cdot 4 (x^6 - y^6)}{x}$ find the value of $\dfrac{y}{x}$.

Multiplying both sides by x, then
$$x^6 = 7 \cdot 4 x^6 - 7 \cdot 4 y^6,$$
or $\qquad\qquad 7 \cdot 4 y^6 = 6 \cdot 4 x^6 ;$

$$\therefore \ \frac{y}{x} = \sqrt[6]{\frac{6 \cdot 4}{7 \cdot 4}} ;$$

$$\log \frac{y}{x} = \frac{1}{6} (\log 6 \cdot 4 - \log 7 \cdot 4)$$

$$= \frac{1}{6} (\cdot 8062 - \cdot 8692)$$

$$= \bar{1} \cdot 9895 ;$$

$$\therefore \ \frac{y}{x} = \cdot 9761.$$

Ex. 4.   Given $7^x = 3^{x+1} \div 2^{x-2}$, find x.
Here $\qquad\qquad\qquad 7^x \times 2^{x-2} = 3^{x+1} ;$
$$\therefore \ x \log 7 + (x - 2) \log 2 = (x + 1) \log 3.$$
From which we find $\qquad\qquad x = 1 \cdot 614.$

Ex. 5. If Q denotes the quantity of water passing over a V-shaped notch per second and h the height of the water above the bottom of the notch (Fig. 55) then $Q \propto h^{\frac{5}{2}}$. If Q is 7·26 when h is 1·5 find h when Q is 5·68.

$$Q = k h^{\frac{5}{2}} ; \quad \therefore \quad 7 \cdot 26 = k \times (1 \cdot 5)^{\frac{5}{2}} ;$$
$$\log k = \log 7 \cdot 26 - \tfrac{5}{2} \log 1 \cdot 5 = \cdot 4207 ;$$
$$\therefore \quad k = 2 \cdot 634.$$

Hence when Q is 5·68 we have
$$5 \cdot 68 = 2 \cdot 634 \times h^{\frac{5}{2}} ;$$
$$\therefore \quad h = 1 \cdot 359.$$

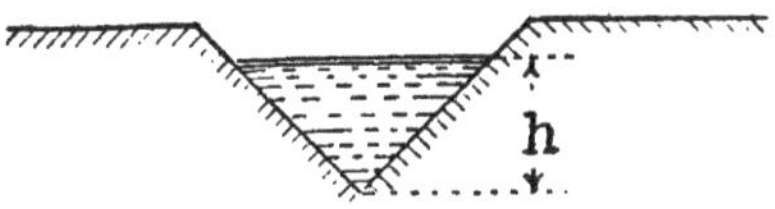

FIG. 55.

Ex. 6. If V is the speed of a steam vessel in knots, D the displacement in tons, and HP the horse-power, then $HP \propto V^3 D^{\frac{2}{3}}$ or $HP = k V^3 D^{\frac{2}{3}}$. Given $V = 17$, $D = 19700$, and $HP = 13300$, find k.

Here $\quad 13300 = k \times 17^3 \times 19700^{\frac{2}{3}} ;$
$$\therefore \quad \log k = \log 13300 - 3 \log 17 - \tfrac{2}{3} \log 19700 = \bar{3} \cdot 5697 ;$$
$$\therefore \quad k = \cdot 003713.$$

Hence the equation becomes $HP = \cdot 003713 \, V^3 D^{\frac{2}{3}}$.

Napierian Logarithms.—The system of logarithms employed by Napier the discoverer of logarithms, and called the **Napierian** or **Hyperbolic system,** is used in all theoretical investigations and very largely in practical calculations. The base of this system is the number which is the sum of the series

$$1 + 1 + \tfrac{1}{2} + \frac{1}{2 \times 3} + \frac{1}{2 \times 3 \times 4} + \dots ;$$

this sum to five figures is 2·7183. Usually the letter e is used to denote a hyperbolic logarithm, as for example $\log 2$ to base 10 would be written $\log_{10} 2$ or more simply as $\log 2$, but the hyperbolic logarithm of 2 is written as $\log_e 2$.

Transformation of logarithms.—A system of logarithms calculated to a base a may be transformed into another system in which the base is b.

Let N be a number. Its logarithms in the first system we may denote by x and in the second system by y.

Then $\qquad N = a^x = b^y$ or $b = a^{\frac{x}{y}} ;$

$$\therefore \quad \frac{x}{y} = \log_a b \text{ and } \frac{y}{x} = \frac{1}{\log_a b}.$$

Hence, if the logarithm of any number in the system in which the base is a be multiplied by $\dfrac{1}{\log_a b}$, we obtain the logarithm of the number in the system in which the base is b.

The common logarithms, or, as they are usually called simply logarithms, have been calculated from the Napierian logarithms. Let l and L be the logarithms of the same number in the common and Napierian systems respectively, then

$$l = \frac{1}{\log_e 10} L,$$
$$\log_e 10 = 2{\cdot}30258509 = 2{\cdot}3026 \text{ approx.},$$

and
$$\frac{1}{2{\cdot}30258509} = {\cdot}43429448 = {\cdot}43429 \text{ approx.}$$

Hence, the common logarithm of a number may be obtained by multiplying the Napierian logarithm of the same number by ·43429....

To convert common into Napierian logarithms multiply by 2·3026 instead of the more accurate number 2·30258509.

The preceding rules will be best understood by a careful study of a few examples.

Ex. 1.   Log 10 to base e is 2·3026.
$$\therefore \quad \log_e 10 = 2{\cdot}3026,$$
or
$$e^{2{\cdot}3026} = 10.$$

From this relation any number which is a power of 10 may be expressed as a power of e.   Thus in Table III. $\log 19{\cdot}5 = 1{\cdot}29$.
$$\therefore \quad 19{\cdot}5 = 10^{1{\cdot}29} = e^{2{\cdot}3026 \times 1{\cdot}29} = e^{2{\cdot}9703},$$
or
$$\log_{10} 19{\cdot}5 = 1{\cdot}29,\ \log_e 19{\cdot}5 = 2{\cdot}9703.$$

Ex. 2.   Find $\log_e 3$ and $\log_e 8{\cdot}43$.
$$\therefore \quad \log_e 3 = {\cdot}4771 \times 2{\cdot}3026 = 1{\cdot}0986,$$
$$\log 8{\cdot}43 = {\cdot}9258 \ ;$$
$$\therefore \quad \log_e 8{\cdot}43 = {\cdot}9258 \times 2{\cdot}3026 = 2{\cdot}1317.$$

Ex. 3.   Find $\log 13$ to base 20.
Here $\log 13 = 1{\cdot}1139$, also $\log 20 = 1{\cdot}3010$.
$$\therefore \quad \log_{20} 13 = \frac{1{\cdot}1139}{1{\cdot}3010} = {\cdot}8562.$$

EXERCISES.   XXVI.

Find the value of

1. $\cdot 03571 \times \cdot 2568 \cdot$

2. $\dfrac{8352 \times 3{\cdot}69}{(30{\cdot}57)^3}.$

3. $\dfrac{1{\cdot}265 \times {\cdot}01628}{2{\cdot}283 \times 64{\cdot}28}.$

Divide

4.   (i) $\bar{5}{\cdot}3010$ by 9.   (ii) $\bar{4}{\cdot}4771$ by 11.

Find the logarithms of

5. $(\cdot025)^{\frac{1}{5}}$.

6. $\left(\dfrac{28}{5}\right)^{\frac{3}{4}}$.

7. $\sqrt[3]{\dfrac{3^2 \times 5^4}{\sqrt{2}}}$.

8. $\sqrt[3]{\dfrac{2^3 5^5}{\sqrt{3}}}$

9. $(1500)^{\cdot001}$.

10. Find the tenth root of $\cdot0234$.

11. Find $\log \sqrt[5]{\cdot00054 \times 3\cdot6}$ and $\log \dfrac{1}{324 \sqrt[5]{125}}$.

12. Find the value of $\sqrt{2543 \times \cdot1726}$.

13. Why do we divide the logarithm of a number by 3 to obtain the logarithm of its cube root. Find the cube root of $44\cdot6$.

14. Find the value of E from the equation
$$E = \frac{80 \times 33\cdot62 \times 16}{1\cdot5 \times (\cdot375)^2 \times \cdot7854 \times \cdot0166}.$$

15. Find the fourth proportional $\sqrt[3]{8\cdot37}$, $\cdot84$, and $\sqrt[5]{\cdot05432}$.

16. Find the value of N from the equation
$$N = \frac{71\cdot12 \times (1\cdot25)^2}{\cdot1406 \times (\cdot25)^4 \times \cdot022}.$$

17. Find the hyperbolic logarithms of $1\cdot5$, 2, $2\cdot5$, 3, $3\cdot5$, 4, $4\cdot5$, 5.

Find the value of x from

18. $\left(\dfrac{7}{3}\right)^x = 100$.

19. $(1\cdot05)^x = (8\cdot25)^{\frac{1}{x}}$.

20. Find the eleventh root of $(39\cdot2)^2$.

21. Find the seventh root of $\cdot00324$.

22. Find a mean proportional to $\sqrt{4\cdot756}$ and $(\cdot0078)^{\frac{2}{3}}$.

23. Prove that $\log_a b + \log_b a = 1$.

24. If $a = 10\cdot4$, $b = 2\cdot38$, $x = \cdot2064$, and $y = \cdot0986$, determine the values of a^b and $a^{\frac{1}{3}} b^{-\frac{2}{5}} (x^2 - y^2)^{\frac{1}{2}}$.

25. Find the value of $p_1(kr^{-1} - r^{-k}) - p_3$, when $p_1 = 80$, $p_3 = 3$, $k = \cdot8$, $r = 1\cdot667$.

26. If s varies as $t^{\frac{2}{3}}$ and s is $20\cdot4$ when t is $1\cdot36$, find t when $s = 32\cdot09$.

27. If Q varies as $H^{\frac{5}{2}}$ and Q is $7\cdot26$ when H is $1\cdot5$, find Q when H is $1\cdot359$.

28. Having given the equation $y = ax + bz^{\frac{2}{3}} x^2$, find the values of a and b, if $y = 49\cdot5$ when $x = 1$, and $z = 8$ and $y = 356$ when $x = 1\cdot5$ and $z = 20$. What is the value of y when $x = 2$ and $z = 20$?

29. From the two given formulae
$$D = 1\cdot86 C^2 + 1\cdot08, \quad N = \frac{20\cdot29}{D},$$
find the values of N when C has the values $1\cdot23$, $2\cdot89$, $4\cdot63$, $6\cdot48$,

and 8·06.   Arrange the work so as to carry out the computation with the least trouble.

30. Calculate the values of

   (i) $0\ 252^{2\cdot19}$.   (ii) $\sqrt[5]{\cdot00054\times3\cdot6}$.

   (iii) If $m=ar^{-1\cdot16}$, find r when $m=2\cdot263$ and $a=\cdot4086$·

31. In a horse-shoe magnet the following relation is found to hold very nearly, $P=c\times d^2\times10^{-7}$, find P.   Where P is the pull in lbs. per sq. in., c is a constant $=5\cdot77$, d is the density in the air gap $=6000$ per square centimetre.

Find the value of x from the equations,

32. $x=\left(\dfrac{1\cdot03\times10^{-5}\times9300\times1\cdot05\times1\cdot1}{240}\right)^{\frac12}$.

33. $x^5=\dfrac{11\cdot6\times\cdot4785}{\cdot0278}$.

MISCELLANEOUS EXAMPLES.   XXVII.

Calculate

1. $\sqrt[3]{23\cdot51}$·   **2.** $6\cdot78^{2\cdot34}$.   **3.** $\cdot678^{-1\cdot301}$.

4. Work out the values when $s=\cdot95$ and $r=1\cdot75$ of

   (i) $(sr^{-1}-r^{-s})\div(s-1)$.   (ii) $(1+\log_e r)\div r$.

5. If $pu^{1\cdot0646}=479$ find p when u is $12\cdot12$, and find u when p is $60\cdot4$.

Compute

6. $1\cdot683^{3\cdot65}$.   **7.** $\cdot01683^{-0\cdot26}$.

8. If H is proportional to $D^{\frac23}v^{3}$, and if H is 871 when D is 1330 and v is 12, find H when D is 1200 and v is 15.

9. Calculate the ratio of d_0 to d_1 from the equation :

$$d_0{}^3=1\cdot3\,\frac{d_0{}^4-d_1{}^4}{d_0}.$$

10. Find the ratio of a to b when $a^3=\dfrac{17(a^4-b^4)}{a}$.

11. Find the ratio of y to x from the equation $x^3=\dfrac{0\cdot6(x^4+y^4)}{x}$.

12. Find the ratio of x to y from $y^4=\dfrac{0\cdot8(x^5+y^5)}{y}$.

13. If $Q\propto H^{\frac85}$ and when H is $8\cdot5$, Q is $557\cdot1$; find Q when H is 4 25.

14. (i) If H is proportional to $D^{\frac23}v^{3}$, and if D is 1810 and v is 10 when H is 620, find H if D is 2100 and v is 13.

   (ii) If $y=ax^{\frac12}+bxz^2$.

       And $y=62\cdot3$ when $x=4$ and $z=2$.

       Also $y=187\cdot2$ when $x=1$ and $z=1\cdot46$; find a and b, and find the value of y when x is 9 and z is $0\cdot5$.

CHAPTER XV.

SLIDE RULE.

Slide rule —It will be clear already to the reader who has followed the section dealing with logarithms, that by their use the multiplication of two or more numbers is affected by adding the logarithms of the factors, and their division by the subtraction of the logarithms of the factors. Or, shortly, by the use of logarithms multiplication is replaced by addition, and division by subtraction.

Hence, if instead of the equal divisions of a scale (Fig. 56), unequal divisions corresponding to logarithms are employed, then, when performed graphically, in the manner to be immediately described, multiplication will correspond to addition and division to subtraction.

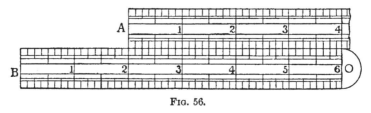

Fig. 56.

It is an easy matter to add together two linear dimensions by means of an ordinary scale or rule. Thus, to add 2 and 3 units together. Assume the scale B (Fig. 56) to slide along the edge of the scale A, then the addition of the numbers 2 and 3 is made when the 2 on B is coincident with 0 on A ; the sum of the two numbers is found to be 5 opposite the number 3 on the scale A.

If the scales on A and B are not divided in the proportion of the numbers, but of the logarithms of the numbers, then, using this graphic method, we can by sliding one scale along

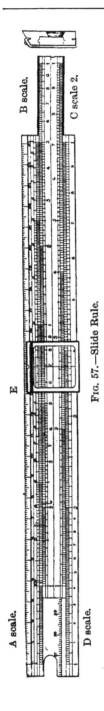

FIG. 57.—Slide Rule.

the other perform the operation of addition; but, as the scales are logarithmic, the result would correspond to the product of the numbers added.

Similarly, the number corresponding to the difference would be a quotient.

Construction of slide rule.—The object of the slide rule is to perform *arithmetical calculations* in a simple manner. There is a great saving of time and labour effected by its use, as it solves at sight all questions depending on ratio.

It consists of a fixed part or rule containing a groove in which a smaller rule slides.

Reference to Fig. 57 shows that the upper part of the rule contains two scales exactly alike, while the lower part of the rule contains only one scale, its length being double that of the upper one. As the upper part contains two scales, it will be convenient to refer to the division 1 in the centre of the rule, shown at E as the *left-hand* 1, the other to the right of it as the *right-hand* 1.

There are two scales on the smaller rule or *slide*, as we may call it, at B; and at C one double the length. These scales on the slide correspond to those on the rule.

It will be convenient to refer to the four scales by the letters A, B, C, D.

There is in addition to the parts mentioned a movable frame or thin metal runner, held in position on the face of the rule by sliding in two grooves. This is shown both at E and in the end view. Although it slides freely along the instrument, any shake which might otherwise occur is prevented by a small steel spring placed at the upper part of the carrier.

The principle of action is the same in all slide rules, although the arrangement of the lines depends upon the purpose to which the rule is to be applied. The modified form of the calculating rule, which we propose to explain, is one of the most accurate instruments of the kind that can be obtained. The instrument, with the exception of the runner E, is usually made of boxwood or mahogany. The wood may be faced with white celluloid, the black division lines showing more clearly on the white background.

Graduation of slide rules.—In Fig. 58 it will be seen that the distance apart of the divisions are by no means equal. The divisions and subdivisions are not equidistant as in an ordinary scale, but are proportional to the logarithms of the numbers and are set off from the left or commencing unit.

In studying Indices it was found, p. 107, that

if 10^3 be multiplied by 10^4 the result is 10^{3+4} or 10^7.

From the definition of a logarithm,

2 is the logarithm of 100, since $10^2 = 100$.

Or, as 10 raised to the power 2 gives 100, the logarithm of 100 is 2.

In a similar manner if 10 be raised to a power ·4771 we obtain the result 3 ;

$$\therefore \ \log 3 = ·4771·$$

P.M.B. K

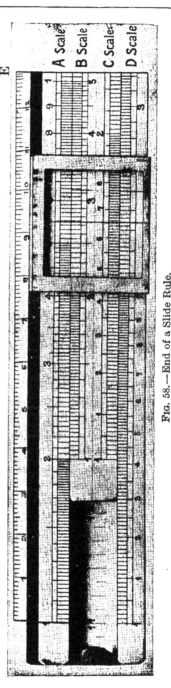

Fig. 58.—End of a Slide Rule.

Also since $10^{3010} = 2$; $\therefore$ $\log 2 = \cdot 3010$.

Hence $\cdot 7781$ is the log of $6 = 10^{3010 + 4771}$.

With the rule this addition is effected by drawing the slide to the right until the left-hand index of scale B is coincident with 2 on scale A. Then over 3 on the scale B is found 6 on the scale A.

Again
$$\frac{10^5}{10^2} = 10^{5-2} = 10^3.$$

So also
$$\frac{10^{7781}}{10^{3010}} = 10^{7781 - 3010} = 10^{4771}.$$

Or more simply, to divide 6 by 2.

$$\log 6 - \log 2 = \cdot 7781 - \cdot 3010$$
$$= \cdot 4771 ;$$

and $\cdot 4771$ is the log of 3.

This is obviously only the converse operation to that already described. Thus, set 3 on B to 6 on A ; then, coincident with the index 1 on B is the answer 2 on A.

A model slide rule.—Simple exercises similar to the above will be found very useful as a first step, and such practice will enable the student to deal with numbers with certainty and ease. It is an excellent practice to make a slide rule, using two strips of cardboard or thick paper.

Assuming any length, such as from 1 to E, scale A, to be 10 inches long and to be divided into 10 parts, then the distance from 1 of any intermediate number (from 1 to 10) is made proportional to its logarithm. Or, as the length of the scale is to be 10 inches, the distance in inches of any number from 1 is equal to the logarithm of the number multiplied by 10.

To find the position of the 2nd division, since $\log 2 = \cdot 301$, $\cdot 301$ parts, or $3 \cdot 01$ inches from 1, would indicate its position.

In like manner the 3rd division would be $\cdot 477$ parts, or $4 \cdot 77$ inches ; the 4th, $\cdot 602$ parts, or $6 \cdot 02$ inches ; the 5th, $6 \cdot 99$ inches, etc.

Denoting the distance of any division from point 1 by x, if l denote the length of the scale from 1 to E, and L the logarithm of the number indicating the division required, then

$$x = l \cdot L.$$

When the upper scale A is set out, the scales B and C on the slide and the scale D may be similarly marked from it.

The excellence of any slide rule depends upon the skill with which these division lines ha/e been constructed. In good rules they are as accurate as it is possible to make them. In dealing with a carefully made slide rule we deal with the *effect* of a considerable amount of labour and thought which have been expended in its construction.

Although a knowledge of logarithms is not essential before a slide rule is used, any more than it is necessary that a man should be able to make a watch before he is allowed to use. one, or that he should understand the nature of an electric current before using an electric bell, it is much better to clearly understand the principles underlying the construction of any instrument.

Multiplication with a slide rule. – In (Fig. 58) putting the units' figure of the slide opposite the 2 on the fixed scale A, we get registered the products of all the numbers on the slide and 2 above. Thus

$$2 \times 1 = 2, \ 2 \times 1\cdot75 = 3\cdot5, \ 2 \times 2 = 4, \text{ etc.}$$

Or, we may use the two lower scales C and D. In each case we make the index 1 on the slide coincide with either of the factors read on scales A or D, and the product will be found coincident with the other factor read on the slide. The use of the two upper scales enables a much larger series of values to be read with one motion of the slide, but as the scales on C and D are double the length of those on A and B it is obvious that the former are more suitable to obtain accurate results.

It should be carefully noticed that the values attached to the various divisions on the scales depend entirely upon the value assumed for the left-hand index figure. Thus, the left-hand index, or units' figure, may denote 0·1, 1, 10, 100, or any multiple of 10 ; and, when in any calculation the initial value is assumed, it must be maintained throughout. Thus, in Fig. 58, the products may be read off as

$$2 \times 1 = 2, \text{ etc.} \ ; \ 2 \times 10 = 20, \text{ etc.} \ ; \ \text{or } 2 \times 100 = 200, \text{ etc.}$$

If the product cannot be found when the left-hand index is used the right-hand index must be employed.

Division with a slide rule.—Set the divisor on B under the dividend on A, and read the quotient on A over the index of B ;

or, set the divisor on *C* over the dividend on *D*, and read the quotient on *D* under the units' figure of *C*.

Ratio with a slide rule.—This, as already indicated, is only a convenient method of expressing division. One of the simplest applications of ratio is to convert a vulgar fraction to a decimal fraction.

Thus, the decimal equivalent of $\frac{3}{8}$ is found by placing 8 on the scale *B* opposite to 3 on the scale *A* ; then, coincident with the index on *B*, is the result ·375 on *A*, The two lower scales *C* and *D* may, of course, be used instead of *A* and *B*.

The circumference of a circle is obtained by multiplying its diameter by 3·1416. Hence if the index on the scale *B* be put into coincidence with this number 3·1416 (marked π on scale *A*), then against any division representing the diameter of a circle on *B*, on *A* the division indicating the circumference of the circle is found. Conversely, the diameter may be obtained when the circumference is given.

As proportion is simply the equality of two ratios, the rules for performing proportion by the help of the slide rule follow at once from those already given for ratio.

Proportion with a slide rule.—Making use of the two upper scales *A* and *B*, operate so as to find the quotient, and *without reading off the answer*, look along the rule for the product of the quotient by the third factor in the proportion.

Ex. 1.　3 : 4 = 9 : *x*.

Read off the answer 12 by the process described, or put the proportion in the form of ratios; thus $\frac{3}{4} = \frac{9}{x}$.

Place 4 on *B* under 3 on *A*, and under 9 on *A* read off the answer *x* = 12 on *B*.

Involution with the slide rule.—On inspection, the numbers on the upper scale *A* are seen to be the squares of the numbers on the lower scale *D*.

To obtain the square of a fractional number some difficulty would be experienced in noting the coincidence of divisions on *A* and *D*, separated as they are by the slide; in this case we can make use of the runner, thus :

Set the runner to coincide with the given number on the scale *D*, and, by its means, read off the square of the number on scale

A. In this manner, having obtained the square, the fourth, or any even power, can be obtained.

Square root by the slide rule.--In extracting square roots the process for involution is reversed.

The number, the root of which is required, is found on the scale *A*, and its root is, as before, found directly below. The runner enables the coincidence of the two divisions denoting the number and its root to be readily obtained.

As shown on p. 225, the area of a circle is $3\cdot1416 \times r^2$, or $\cdot7854d^2$, where *r* is the radius and *d* the diameter of the given circle. Conversely, if the area of a circle is given, the diameter can be obtained from :

$$\text{diameter} = \sqrt{\frac{\text{area}}{\cdot7854}}.$$

Ex. 1. Find the area of a circle 3″ diameter.

The mark on the runner is set to the 3 on the lower scale ; the upper mark over the top scale registers 9. Then moving the slide to the right until its 1 coincides with the mark, we have coincident with ·7854 (which is marked on the scale) the required area 7·06 square inches.

Ex. 2. Find the area of a circle 2·5″ diameter.

The square of 2·5 is seen to be 6·25, and multiplied by ·7854 the area is 4·9 square inches.

To obtain the cube of a number with a slide rule. —*First method*. Bring the right-hand 1 of scale *C* to the given number on *D*. Then over the same number on the scale *B* read off the required cube on *A*.

Second method. The slide may be inverted, keeping the same face upwards. The scale *B* will now move along scale *D*. Put in coincidence on the scales *B* and *D* the two marks indicating the number the cube of which is required, then opposite the right-hand 1 on the slide the cube required will be found on scale *A*.

Cube root with the slide rule.—*First method*. Set the given cube on scale *A* and coincident with it on scale *B* any rough approximation to the root, move the slide to the right or left until on *B* coincident with the cube on *A*, the same number is simultaneously found on *D* opposite the right or left hand index of *C*.

Second method. The inverted slide is used. This is placed with the right-hand 1 of the slide coincident with the number the cube root of which is required. Now find what number on the scale D coincides with the same number on the inverted scale of B; this number is the cube root required.

Ex. 1. Find the cube root of 64.

Move the slide from right to left, and it will be found that 4 on scale B coincides with 64 on scale A, simultaneously with 4 on D and 1 on C. Hence 4 is the cube root required.

Invert the slide, keeping the same face uppermost. Set 1 on C inverted, to 64 on scale A, the division on B which coincides with D is 4. Hence 4 is the cube root.

When the product or quotient of several numbers is wanted the runner may be used with advantage to record the results of the partial operations.

Ex. 2. Find the value of $\dfrac{3\cdot4 \times 2\cdot8}{1\cdot7 \times \cdot3}$.

We may use either scales A and B or C and D.

Using C and D move the slide until the graduation 1 on scale C coincides with 3·4 on scale D. Opposite the division 2·8, on C, is found 9·52 on D. The line on the runner can be made to coincide with this. Next move the slide until the graduation 1·7 is coincident with the line. If necessary opposite 1 on C the result of $\dfrac{3\cdot4 \times 2\cdot8}{1\cdot7}$ could be read off on D; but instead of doing so the line on the runner is made to coincide with the 1 on scale C, and the slide is moved until ·3 is coincident with the line. The answer 18·6 is now read off opposite the index on C. In this manner in any complicated calculation the runner may be used to record any operation.

CHAPTER XVI.

RATIOS, SINE, COSINE, AND TANGENT.

Measurement of an angle in radians.—A very convenient method for measuring angles, which is especially useful when dealing with angular velocity, is obtained by estimating the arc subtended by a given angle, and *dividing the length of the arc by the radius of the circle.*

Thus, in Fig. 59, the measure of the angle BOA in radians is the ratio of the length of arc AB to the radius OA, or

$$\text{Measure of angle in radians} = \frac{\text{length of an arc } AB}{\text{length of its radius } OA} \quad \text{.....(i)}$$

Evidently this measure of an angle will be unity when the arc measured along the curve is equal to the radius, **or, the unit angle is that angle at the centre of a circle subtended by an arc equal in length to the radius.** This unit is called a radian. The circle contains 4 right angles or 360°. This may be expressed in radians by the ratio of

$$\frac{\text{circumference of circle}}{\text{radius of circle}},$$

but the circumference of the circle as shown on p. 222 is $2\pi r$. Where r is the radius OA and π denotes the number

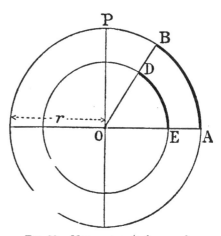

FIG. 59.—Measurement of an angle in radians.

of times that the diameter of a circle is contained in the circumference, the value of π is 3·1416, or more accurately 3·14159.

Hence the measure of 4 right angles $=\dfrac{2\pi r}{r}=2\pi$. Thus 4 right angles may be expressed in two ways, viz., as 360°, or as 2π radians, and one radian $=\dfrac{360}{2\pi}=\dfrac{180}{\pi}=57°\!\cdot\!29578$ (taking $\pi=3\!\cdot\!1416$). For many purposes the approximate value 57°·3 is used instead of the more accurate value.

It will be seen from Fig. 59 that if any circle be drawn with centre O and cutting the lines OA and OB in any two points such as D and E, then the angle $EOD=$ angle AOB. It follows, therefore, that the unit angle is independent of the size of the circle and is an invariable unit, being, as already indicated, equal to $\dfrac{180}{\pi}$, or 57°·2958.

It is advisable not only to be able to define the two units, the *degree* and the *radian*, but also to realise their relative magnitudes. Thus an angle 5° denotes an angle of 5 *degrees*, but an angle simply denoted by 5 contains $5\times\dfrac{180}{\pi}$ degrees.

∴ From (i) it follows that if an arc of a circle is five times as long as the radius, the angle subtended at the centre is five radians, if the arc is one-third the radius, the angle is one-third of a radian, etc. To find the length of arc subtending a given angle it is only necessary to write (i) **arc = angle × radius.**

Also when π refers to a number it denotes 3·1416, but applied to an angle, then the angle contains π radians and is 180 degrees.

Ex. 1.   An angle is $\tfrac{2}{3}$ radians, what is its value in degrees?

In this example, as the angle is $\tfrac{2}{3}$ in circular measure, its value in degrees will be $\tfrac{2}{3}\times$ unit angle, or $\tfrac{2}{3}\times57°\!\cdot\!2958=38°\!\cdot\!1972$ or 38°·2 using 57°·3.

Ex. 2.   (*a*) What is the numerical value of a right angle in radians?   (*b*) Find the radian measure of an angle of 112° 43′. (*c*) Find the length of an arc which subtends an angle of 112° 43′ at the centre of a circle whose radius is 153 feet.

(*a*) The measure of 4 right angles is 2π radians, therefore the measure of 1 right angle is

$$\frac{2\pi}{4}\text{ radians}=\frac{\pi}{2}\text{ radians}=1\!\cdot\!5708\text{ radians.}$$

(b) In each degree there are 60 minutes, hence $112° \ 43' = 6763$ minutes.   Also $180° \times 60 = 10800$ minutes ;

$$\therefore \ \frac{6763 \times 3 \cdot 1416}{10800} = 1 \cdot 967 \text{ radians.}$$

(c) In radian measure, angle $= \dfrac{\text{length of arc}}{\text{radius}}$;

$$\therefore \ \text{length of arc} = \text{angle} = \text{radius} = 1 \cdot 967 \times 153$$
$$= 301 \text{ feet nearly.}$$

EXERCISES XXVIII.

1. If an arc of 12 feet subtend at the centre of a circle an angle of 50°, what is the radius of the circle ?

2. Explain the different methods of measuring angles. Find which is greater, an angle of 132° or an angle whose radian measure is 2·3 radians.

3. Find the number of degrees in the angle whose radian measure is ·1.

4. A train is travelling on a curve of half a mile radius at the rate of 20 miles an hour ; through what angle has it turned in 10 secs. ? Express the angle in radians and in degrees.

5 Define the radian measure of an angle. Find the measure of the angle subtended at the centre of a circle of radius $6\frac{4}{11}$ inches by an arc of 1 ft.

6. The radius of a circle is 10 ft. Find the angle subtended at the centre by an arc 3 ft. in length.

7. If one of the acute angles of a right-angled triangle be 1·2 radians, what is the other acute angle ?

8. Two angles of a triangle are respectively $\frac{1}{2}$ and $\frac{2}{7}$ of a radian ; determine the number of radians and degrees in the third angle.

9. A certain arc subtends an angle of 1·5 radians at the centre of a circle whose radius is 2·5 feet ; what will be the radius of the circle at the centre of which an arc of equal length subtends an angle of 3·75 radians ?

Functions of angles.—We have already explained the use of the protractor, and a table of chords in setting out and measuring angles. It is now necessary to refer to another useful method of forming an estimate of the magnitude of angles, that namely by means of the so-called functions of angles, the sine, cosine, tangent, etc. The values of these functions have been tabulated for every degree and every minute up to 90°. It is possible, in addition, by the help of columns of differences to calculate these functions to decimal parts of a degree, or in seconds and fractional parts of a second. It will be an advantage to understand what information is derivable from such tables, and how to use them with facility.

If at any point B along a straight line AD (Fig. 60) a line BC be drawn perpendicular to AD, then ABC is a right-angled triangle. One angle at B, a right angle, is known; the other two, and the three sides, can be determined by the data of any given question. We may denote the angle BAC by the letter A (the letter at the angular point), and the remaining two angles may be referred to as the angles B and C. If one of the sides be given, and one of the two remaining angles, the other sides and the remaining angle can be found either by construc-

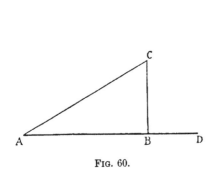

FIG. 60.

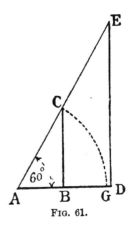

FIG. 61.

tion or by calculation. Also any two of the three sides will give sufficient data to enable the triangle to be drawn and the two angles and the remaining side to be found.

It is not necessary to give the actual lengths of two of the sides; it will answer the same purpose if the ratio of AB to BC be known, for if any line such as DE be drawn parallel to BC (Fig. 61) it will cut off lengths AD and DE from the two sides AB and AC produced, such that the ratio of AD to DE is equal to the ratio of AB to BC.

Also the equality is unaltered if BC and DE be replaced by AC and AE respectively.

This statement can be written $\dfrac{AB}{AC} = \dfrac{AD}{AE}$.

To obtain the value of the ratio $\frac{AB}{AC}$ when the angle A is $60°$.— Draw any line AD as base (Fig. 61), make AC equal to 10 units and at $60°$ to AD.

From C draw CB perpendicular to AD and meeting AD at B; ascertain as accurately as possible the lengths of AB and BC.

AB will be found to be 5 units, and BC to be 8·66 units.

Hence the ratio $\frac{AB}{AC} = \frac{5}{10} = \frac{1}{2}$, and $\frac{BC}{AC} = \frac{8·66}{10}$.

This ratio of $\frac{BC}{AC}$ is called the **sine** of the angle BAC or (as only one angle is formed at A) sine A.

It will be seen from the above that the **sine** of an angle (which is abbreviated into *sin*) is formed by the **ratio of two sides of a right-angled triangle; the side opposite the angle being the numerator, and the hypotenuse or longest side of the triangle (adjacent to the angle) the denominator.**

Let the three sides of the triangle be represented by the letters a, b, and c, where a denotes the side opposite the angle A, b the side opposite the angle B, and c the side opposite the angle C.

Then, sine $60° = \dfrac{\text{side } BC}{\text{side } AC} = \dfrac{a}{b} = \dfrac{8·66}{10} = ·866$.

Referring to Table V., opposite the angle $60°$ this value will be found, and the length of the side BC, or a, can be obtained by calculation. Thus, in the right-angled triangle ABC (Fig. 61) we have

$$a^2 = b^2 - c^2 = 10^2 - 5^2 = 75,$$
$$\therefore \ a = \sqrt{75} = 8·66.$$

The ratio of $\frac{AB}{AC}$ or $\frac{c}{b}$ is called the **cosine** of the angle BAC (cosine is abbreviated into cos),

$$\therefore \ \cos A = \cos 60° = \frac{c}{b} = \frac{5}{10} = ·5.$$

Angle of 30°.—The sum of the three angles of a triangle is $180°$. As one of the angles in Fig. 61 is $90°$ and the other $60°$, the remaining angle is $30°$. Also,

$$\sin 30° = \frac{c}{b} = \cos 60° = ·5$$

and $\qquad \cos 30° = \dfrac{a}{b} = \sin 60° = \cdot 866.$

Again referring to Table V., these calculated values are found opposite sin 30° and cos 30° respectively.

The tangent.—Of the three sides of the triangle AB, BC, and CA (Fig. 62) we have already taken the ratio of $\dfrac{BC}{AC}$ and $\dfrac{AB}{AC}$, the former is called the sine and the latter the cosine of θ. One other ratio, and a most important one, is the ratio of $\dfrac{BC}{AB}$. This ratio is called the **tangent** of BAC, or, denoting BAC by θ,

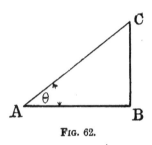

FIG. 62.

and using the abbreviation tan for tangent, we have $\tan \theta = \dfrac{BC}{AB}$.

Ex. 1.   Construct angles of 30°, 45°, and 60°.   In each case make the hypotenuse $AC = 10$ units on any convenient scale.   Measure the lengths AB and BC to the same scale, and tabulate as follows :

	Lengths of :		Numerical values of :			sin.	cos.	tan.
	AB	BC	$\dfrac{BC}{AC}$	$\dfrac{AB}{AC}$	$\dfrac{BC}{AB}$			
Angle of 30°								
Angle of 45°								
Angle of 60°								

It has already been seen (p. 47) that two angles are said to be **complementary** when their sum is a right angle.

Referring to Fig. 61, the sin of $30° = \cos 60°$, the ratio in each case being $\dfrac{BC}{AC}$.   Hence **the sine of an angle is the cosine of the complement of that angle ; and the cosine of an angle is the sine of the complement of that angle.**

Prove these statements by reference to the tabulated values obtained by measurement.

The supplement of an angle is the angle by which it falls short of two right angles (180°); thus, the supplement of an angle of 60° is 120°; the supplement of an angle of 30° is 150°; or two angles are said to be **supplementary** when the sum of the two angles is 180°.

It is important to be able to readily write down the values of the sine, cosine, etc., of angles of 60° and 30°.

To do this, the best method is to draw, or *mentally picture*, an equilateral triangle ABC (Fig. 63) each side of which is 2 units in length.

From the vertex C let fall a perpendicular CD on the base AB. As shown on p. 47, the point D bisects AB, and $AD=DB$. Also the angle ACD is equal to DCB; each of these equal angles is one-half the angle ACB, and is therefore 30°; or, as the angle at D is 90°, and the angle at C is equal to 60°, the remaining angle ACD must be 30°.

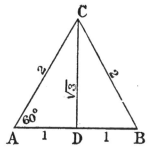

FIG. 63.—Equilateral triangle.

From the right-angled triangle ADC, we have

$$DC^2 = AC^2 - AD^2 = 4 - 1 = 3;$$
$$\therefore DC = \sqrt{3}.$$

Thus
$$\sin 60° = \frac{DC}{AC} = \frac{\sqrt{3}}{2};$$

$$\cos 60° = \frac{AD}{AC} = \frac{1}{2};$$

$$\tan 60° = \frac{DC}{AD} = \frac{\sqrt{3}}{1} = \sqrt{3} = \frac{\sin 60°}{\cos 60°}.$$

Hence
$$\sin 30° = \frac{AD}{AC} = \frac{1}{2} = \cos 60°;$$

$$\cos 30° = \frac{DC}{AC} = \frac{\sqrt{3}}{2} = \sin 60°;$$

$$\tan 30° = \frac{AD}{DC} = \frac{1}{\sqrt{3}} = \cot 60°.$$

It should be noticed that

$$\frac{\sin 60°}{\cos 60°} = \frac{DC}{AC} \div \frac{AD}{AC} = \frac{DC}{AD} = \tan 60°.$$

In a similar manner, for any angle A,

$$\frac{\sin A}{\cos A} = \tan A.$$

Instead of attempting to remember the important numerical values for the ratios, it will be found much better to use the triangle, as described in Fig. 63, its angles 90°, 60°, and 30°, and its sides in the proportion of 2, 1, and $\sqrt{3}$.

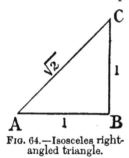

FIG. 64.—Isosceles right-angled triangle.

To ascertain the numerical values of the sine, cosine, and tangent of 45°, a similar method may be used. Thus, if AB and BC (Fig. 64) form two sides of a right-angled triangle in which $BA = BC$ and each is one unit in length, the angle $BAC = BCA$, and as the sum of the two angles is 90° each angle is 45°.

$$\text{Length of } AC = \sqrt{AB^2 + BC^2} = \sqrt{2}.$$

Hence the three sides of the triangle ABC are in the ratio of 1, 1, and $\sqrt{2}$;

$$\sin 45° = \frac{BC}{AC} = \frac{1}{\sqrt{2}} \; ; \quad \cos 45° = \frac{AB}{AC} = \frac{1}{\sqrt{2}} \; ;$$

or

$$\sin 45° = \cos 45° \; ;$$

$$\tan 45° = \frac{BC}{AB} = 1.$$

Angles greater than 90°.—On p. 33 we have found that an angle is expressed by the amount of turning of a line such as AB (Fig. 65). If the movable radius, or line, occupies the positions AC, AC', AD, and AE, then it is seen that as $B'C' = BC$ and the remaining sides of one triangle are equal to the corresponding sides of the other that the triangle BAC is equal to $B'AC'$. Hence

angle $B'AC = (180° - 30°) = 150°$, or $\sin 150° = \sin 30°$;

or, generally, $\sin(180° - A) = \sin A$.

If the line AB be assumed to rotate in a negative direction until it reaches a point E, a negative angle equal to $-30°$ is described; thus, the angle BAE may be written either as $330°$ or $-30°$.

In addition to the convention that all angles are measured in an anti-clockwise manner, the following rules are adopted :

All lines measured in an upward direction from BB′ are positive, and all lines measured from A′A towards B are positive ; those in the opposite directions, i.e. downwards, or from A′A to B′ are negative. The movable radius, or line AC, or AC', is always positive.

Hence, if BAC denote any angle A, then, since BC and $B'C'$ are both in an upward direction, we have, as before

$$\sin(180° - A) = \sin A.$$

But AB and AB' are lines drawn in opposite directions;

$$\therefore \cos(180° - A) = -\cos A.$$

In the case of the angle formed by producing $C'B'$ to D, both $B'D$ and AB' are negative, hence the sine and cosine of the angle are negative ; when the angle is formed by producing CB to E, the sine of the angle is negative, the cosine is positive.

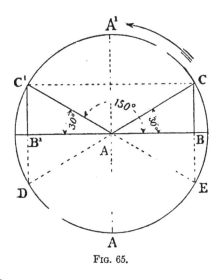

Fig. 65.

On reference to Table V., it will be found that the functions of an angle—sine, cosine, etc.—are only tabulated for values from $0°$ to $90°$; but from these the value of any angle can be obtained by means of the above conventions. Thus, the numerical value of the sine of $30°$ is $\frac{1}{2}$ or $\cdot 5$, and this is also the value of $\sin 150°$.

The cosine of $30°$ is $\dfrac{\sqrt{3}}{2}$ or $\cdot 866$, and $\cos 150°$ is $- \cdot 866\cdot$

As the tangent is $\dfrac{\text{sine}}{\text{cosine}}$, its sign, positive or negative, will depend upon the signs of the sine and cosine of the angle ; when these are alike the tangent is positive, and negative when they are unlike. Further, when the numerator is zero, the value of the tangent is 0 ; the value is indefinitely great when the denominator is 0 ; this is written as ∞.

The following important relations should be proved by drawing a right-angled triangle to scale, as already described :

$$\sin 60° = \frac{\sqrt{3}}{2} = \sin 120° ;$$

$$\cos 60° = \frac{1}{2}, \quad \cos 120° = -\frac{1}{2}.$$

$$\tan 60° = \sqrt{3}, \quad \tan 120° = -\sqrt{3}.$$

These results and those previously arrived at are collected in the following table, the table should be extended to include angles up to 360° ; each result should be expressed as a decimal fraction, and afterwards verified by reference to Table V.

	0°	30°	45°	60°	90°	120°	135°	150°	180°
sin	0	$\frac{1}{2}$	$\frac{1}{\sqrt{2}}$	$\frac{\sqrt{3}}{2}$	1	$\frac{\sqrt{3}}{2}$	$\frac{1}{\sqrt{2}}$	$\frac{1}{2}$	0
cos	1	$\frac{\sqrt{3}}{2}$	$\frac{1}{\sqrt{2}}$	$\frac{1}{2}$	0	$-\frac{1}{2}$	$-\frac{1}{\sqrt{2}}$	$-\frac{\sqrt{3}}{2}$	-1
tan	0	$\frac{1}{\sqrt{3}}$	1	$\sqrt{3}$	∞	$-\sqrt{3}$	-1	$-\frac{1}{\sqrt{3}}$	0

From the figures already drawn and also from these tabulated values, it will be seen that as an angle increases from 0° to 90°, the sine of the angle increases from 0 to 1, but the cosine decreases from 1 to 0. Conversely from 90° to 180°, the sine of an angle decreases from 1 to 0, the cosine increases from 0 to 1.

Other ratios of an angle.—In any right-angled triangle, ABC (Fig. 66) the angle BAC is denoted in the usual manner (by A).

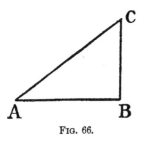

FIG. 66.

Then,

$$\text{cosecant A} = \frac{1}{\sin A} = \frac{1}{\frac{BC}{AC}} = \frac{AC}{BC};$$
$$\text{secant A} = \frac{1}{\cos A} = \frac{1}{\frac{AB}{AC}} = \frac{AC}{AB};$$

$$\text{cotangent A} = \frac{1}{\tan A} = \frac{1}{\frac{BC}{AB}} = \frac{AB}{BC}.$$

The above are usually designated as **cosec A**, **sec A**, and **cot A** respectively.

Some important relations.

(i) $\sin^2 A + \cos^2 A = 1.$

For
$$\sin A = \frac{BC}{AC}; \quad \cos A = \frac{AB}{AC}.$$

Then $(\sin A)^2 + (\cos A)^2 = \dfrac{BC^2}{AC^2} + \dfrac{AB^2}{AC^2} = \dfrac{BC^2 + AB^2}{AC^2} = 1$;

or $(\sin A)^2 + (\cos A)^2 = 1.$

Usually, $(\sin A)^2$ is written $\sin^2 A$.

And in a similar manner $(\cos A)^2 = \cos^2 A$.

(ii) $\sec^2 A = 1 + \tan^2 A.$

$$\sec A = \frac{AC}{AB};$$

$$\sec^2 A = \frac{AC^2}{AB^2} = \frac{AB^2 + BC^2}{AB^2} = 1 + \frac{BC^2}{AB^2} = 1 + \tan^2 A.$$

Also $\operatorname{cosec}^2 A = 1 + \cot^2 A.$

For $\operatorname{cosec}^2 A = \dfrac{AC^2}{BC^2} = \dfrac{BC^2 + AB^2}{BC^2} = 1 + \dfrac{AB^2}{BC^2} = 1 + \cot^2 A.$

Construction of an angle from one of its functions.—
Given the sine of an angle to construct the angle.

Given $\sin \theta = \frac{3}{7}$.

Draw a line AB (Fig. 67), and at point B erect a perpendicular BC to any convenient scale 3 units in length, with C as centre and radius 7 units, describe an arc cutting AB in A. Join A to C; then BAC is the angle required. Measure the angle, and verify by referring to Table V.

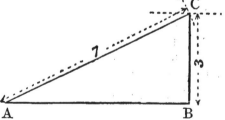

Fig. 67.—Given the sine of an angle to construct the triangle.

P.M.B. L

Given the cosine of an angle to construct the angle.

Given cos $\theta = \frac{4}{7}$. .

Set out $AB = 4$ units (Fig. 68), and erect a perpendicular BC. With A as centre, radius 7 units, describe an arc cutting BC in C; BAC is the angle required. Measure the angle by the protractor or the table of chords, p. 36, and verify by Table V.

Given the tangent of an angle to construct the angle.

Given tan $\theta = \frac{7}{5}$.

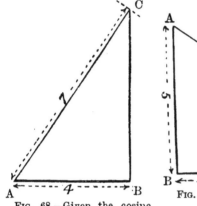

FIG. 68.—Given the cosine of an angle to construct the triangle.

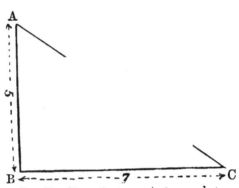

FIG. 69.—Given the tangent of an angle to construct the triangle.

Draw a line AB, 5 units in length (Fig. 69), and at B draw BC perpendicular to AB and equal to 7 units ; join A to C ; then BAC is the angle required. Verify as in the preceding cases.

Inverse ratios.—A very convenient method of writing sin $\theta = \frac{3}{7}$ is to write it as $\theta = \sin^{-1}\frac{3}{7}$, which is read as the angle the sine of which is $\frac{3}{7}$. In a similar manner we may write cos $\theta = \frac{4}{7}$ and tan $\theta = \frac{7}{5}$, as $\theta = \cos^{-1}\frac{4}{7}$ and $\theta = \tan^{-1}\frac{7}{5}$ respectively.

Small angles.—Referring to Table V., and also to Fig. 84, it will be seen that, when the size of an angle is small, the values of the sine, tangent, and the radian measure of an angle differ but little from each other.

EXERCISES. XXIX.

1. Draw an angle of 35°, and make the hypotenuse 10 units. Find by measurement the values of the sine, cosine, and tangent of the angle ; compare the values obtained with those in Table V.

Using the values, ascertain if the following statements are true:

$$\sin^2 35° + \cos^2 35° = 1 \; ; \quad \frac{\sin 35°}{\cos 35°} = \tan 35°.$$

$$\sec^2 35° = 1 + \tan^2 35° \; ; \quad \operatorname{cosec}^2 35° = 1 + \cot^2 35°.$$

2. If $\sin \theta = \frac{3}{5}$, find $\cos \theta$ and $\tan \theta$.

3. The cosine of an angle is $\frac{20}{101}$, find the sine and tangent of the angle.

4. If $\tan a = \dfrac{2}{\sqrt{5}}$, find $\sin a$ and $\cos a$.

5. If $\tan A = \frac{1}{2}$, find the value of the following:
    (i) $\cos^2 A - \sin^2 A$.
    (ii) $\operatorname{cosec}^2 A - \sec^2 A$.
    (iii) $\cot^2 A + \sin^2 A$.
    (iv) Show that $\operatorname{cosec}^2 A - \cos^2 A = \cot^2 A + \sin^2 A$.

6. If $A = 90°$, $B = 60°$, $C = 30°$, $D = 45°$, show that
$$\sin B \cos C + \sin C \cos B = \sin A,$$
and that $\qquad\qquad \cos^2 D - \sin^2 D = \cos A.$

7. The tangent of an angle is $0·675$; draw the angle, without using tables, and explain your construction.
Along the lines forming the angle set off lengths $OA = 4·23''$ and $OB = 3·76''$. Find the length AB, either by measurement or by calculation.

8. The lengths of two sides of a triangle are $3·8$ and $4·6$ inches, and the angle between them is $35°$; determine by drawing or in any way you please (1) the length of the third side, and (2) the area of the triangle.

9. In a triangle ABC, A is $35°$, C is $55°$, and AC is $3·47$ ft. Find AB and BC.

10. The sides a, b, c of a triangle are $1·2$, $1·6$, and 2 feet respectively. Find the number of degrees in the angle A, and determine the area of the triangle ABC.

Use of tables.—Having explained how the sine, cosine, tangent, etc., of such angles as $30°, 45°$, and $60°$, can be obtained, it remains now only to indicate how the trigonometrical ratios of any angle can be found. The method may be understood by a reference to Table V., in which the values of the sine, cosine, etc., of various angles are tabulated. On the extreme left of the table angles $0°$ to $45°$ are found, and from this column and the columns marked at the top by the words *sine, tangent*, etc., the value of any of these ratios for a given angle may be seen. At the extreme right the angles are continued from $45°$ to $90°$. The ratios for these angles are indicated at the bottom of each column.

Thus, given an angle of 25°, we find 25 in the column marked **angle**, and corresponding to this in the column marked **sine** we have the value ·4226, hence, sin 25° = ·4226·

By referring to the columns marked **tangent** and **cosine** we obtain tan 25° = ·4663, cos 25° = ·9063·

To find the value of sin 65°, look out 65 in the right-hand column, and in the column marked sine (at the bottom) we find corresponding to an angle 65° the value ·9063· Hence sin 65° = cos 25°.

This result agrees with that on p. 156, that the sine of an angle is equal to the cosine of the complement of the angle.

Tables are obtainable in which the values of ratios consisting of degrees, minutes, and seconds, or degrees and decimal parts of a degree, are to be found, but Table V. may also be used for such a purpose.

Ex. 1.   Find the value of sin 25° 12' and cos 25° 12'.
We find $\qquad\qquad$ sin 26° = ·4384
$\qquad\qquad\qquad\qquad$ sin 25° = ·4226
$\qquad\qquad$ Difference for 1° or 60' = ·0158
Hence difference for 12 = ·0158 × 12 ÷ 60 = ·0032.
$\qquad\qquad$ ∴ sin 25° 12' = ·4226 + ·0032 = ·4258·

As an angle increases the value of the cosine of the angle decreases (p. 160).
Thus $\qquad\qquad\qquad$ cos 25° = ·9063
$\qquad\qquad\qquad\qquad$ cos 26° = ·8988
$\qquad\qquad$ Difference for 60' = ·0075
Hence diff. for 12' = ·0075 × 12 ÷ 60 = ·0015.
$\qquad\qquad$ ∴ cos 25° 12' = ·9063 − ·0015 = ·9048.

Ex. 2.   Take out from Table V. the tangent of 3° 15' and calculate the cube root of the tangent.
$\qquad\qquad$ tan 4° = ·0699.   tan 3° = ·0524.
Hence difference for 1° or 60' = ·0175.
$\qquad\qquad$ ∴ diff. for 15' = ·0175 × 15 ÷ 60 = ·0044
$\qquad\qquad\qquad$ tan 3° 15' = ·0524 + ·0044 = ·0568
$\qquad\qquad\qquad$ log ·0568 = $\bar{2}$·7543
$\qquad\qquad\qquad$ $\bar{2}$·7543 ÷ 3 = $\bar{1}$·5847.
$\qquad\qquad$ antilog $\bar{1}$·5847 = ·3843
$\qquad\qquad$ ∴ $\sqrt[3]{\tan 3° 15'}$ = ·3843.

Ex. 3.   Find the value of
$$\sin A \cos B - \cos A \sin B \ \dots\dots\dots\dots\dots\dots\dots\dots (i)$$
when A is 65° and B is 20°.

Substituting values from Table V., in (i) we have
$$\cdot9063 \times \cdot9397 - \cdot4226 \times \cdot3420 = \cdot8516 - \cdot1445 = \cdot7071 \cdot$$

In a similar manner $\sin 65° \cos 20° + \cos 65° \sin 20°$ is found to be $\cdot9960 \cdot$

The reader familiar with elementary trigonometry will see that
$$\sin (65° - 20°) = \cdot7071 = \sin 45°,$$
$$\text{and } \sin (65° + 20°) = \cdot9962 = \sin 85°.$$

In like manner the sum, or difference, of two cosines can be obtained.

When an angle is given in radians, it is necessary either to multiply the given angle by 57°·3 or, from Table V., to ascertain the magnitude of the angle in degrees before proceeding to use the ratios referred to.

Ex. 4.
$$y = 2\cdot3 \sin \left(\cdot2618x + \frac{\pi}{6} \right),$$
find y when (i) $x = 1$,   (ii) find x when $y = 1\cdot9716$.

Here when $x = 1$ we have $\cdot2618 + \frac{\pi}{6} = \cdot7854 \cdot$ As $\cdot7854$ denotes the number of radians in the angle, to find the number of degrees we multiply by 57°·3.

∴ $\cdot7854 \times 57°\cdot3 = 45°$, or, from Table V., in the column marked radians corresponding to $\cdot7854$, we have 45°, and $\sin 45° = \cdot7071 \cdot$

∴ $2\cdot3 \times \sin 45° = 2\cdot3 \times \cdot7071 = 1\cdot6263.$

(ii)
$$1\cdot9716 = 2\cdot3 \sin \left(\cdot2618x + \frac{\pi}{6} \right)$$
$$\text{or } \sin (\cdot2618x + \cdot5236) = \frac{1\cdot9716}{2\cdot3} = \cdot8572 \cdot$$

Referring to Table V. this is found to correspond to $1\cdot0297$ radians,
$$\therefore \ \cdot2618x = 1\cdot0297 - \cdot5236$$
or
$$x = \frac{\cdot5061}{\cdot2618} = 1\cdot934.$$

Functions of angles by slide rule.—Given the numerical value of the sine or tangent of an angle, the number of degrees in the angle can be found ; or, conversely, given the number of degrees in the angle, the numerical value can be ascertained :

Ex. 1.   To find the numerical value of sin 30°.

Reverse the rule as shown in Fig. 70, placing sin 30° opposite the upper mark.

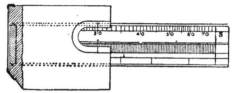

FIG. 70.—Slide rule reversed.

Again reverse the rule, and opposite the right-hand 1 on scale *A* the numerical value ·5 on slide is obtained on scale *B*.

Similarly, placing sin 60° opposite the mark, the value ·866 is obtained.

Also, sin 50° = ·766 may be read off.

For practice other values should be selected, and their numerical values written down and afterwards verified.

Ex. 2.   Obtain and write down the numerical values of the sines of angles of 15°, 20°, 25°, 30°, and for intervals of 5° up to sin 70°.

Conversely, when the numerical value of the sine is given, the corresponding angle can be found.

Tangents.—At the opposite end of the rule a gap similar to the one just described is to be found ; this may be used to find the numerical value of tangents.

Ex. 3.   Find the value of tan 30° and tan 20°.

Move slide until 30° is opposite the mark, then on upper scale coincident with 1 on the slide the value ·577 is obtained.

In like manner tan 20° = ·364·

Ex. 4.   Obtain and tabulate the numerical values of the tangents of angles from 10° to 40°.

Ex. 5.   Write down the values of the sine and cosine of 5°, 10°, etc., up to and including 45°.   Verify by reference to Table V.

Logarithms of numbers.—Logarithms can be obtained by using the transverse mark on the lower edge of the gap.

To obtain log 2, set the index on scale *C* to coincide with 2 on lower scale *D*, reverse the rule, and opposite the lower mark, the log of 2 = ·301 is read off ; also setting it opposite 3, log 3 = ·477, etc.

Ex. 1.   Obtain and write down the logarithms of all numbers from 1 to 10.   Verify the results by reference to Table III.

EXERCISES. XXX.

1. Take out from Table V. log tan 16° 6′ and calculate the square root of the tangent.

2. Find log tan 81° 12′.

3. Calculate the numerical value of $(\tan 50° \tan 22° 30′)^{\frac{1}{3}}$.

4. Find log tan 35° 15′ and the numerical value of $\sqrt[3]{(\frac{1}{3}\sin 44°)}$.

5. Evaluate $\sqrt[5]{\tan 40° \div 65}$.

6. Find the numerical value of the seventh root of
$$\tan 53° 30′ \div 32.$$

7. Find log tan 58° 5′, also the value of the cube root of
$$\tan 52° 30′ \div 15.$$

8. Find the logarithms of $(\sin 26° 13′)^{-\frac{1}{4}}$.

9. $(\sin 18° 37′)^{-\frac{1}{2}}$.

10. Evaluate $\sqrt[3]{\sin 50° \tan^2 38° 20′}$.

11. If $y = 2\cdot3 \sin\left(\cdot2618 x + \dfrac{\pi}{6}\right)$, find y when
$$x = 0, 2, 3, 5, 6, 7.$$
Also find x when $y = 2\cdot049$.

12. Find the value of $e^{bt} \sin(at)$ when
$$b = -0\cdot7, \; t = 1\cdot2, \text{ and } a = 3\cdot927.$$

13. Find the numerical value of $c^{\frac{2}{3}}(a^2 - b^2)\tan\theta$, where
$c = 25\cdot2, \; a = 90, \; b = 49\cdot6, \; \theta = \sin^{-1}(\cdot528)$. ($\theta$ is less than 90°.)

14. Find to the nearest integer the value of the expression
$$700\sqrt{11}\left\{\frac{49}{96}\sin^{-1}(0\cdot1426) + \frac{1}{8\sqrt{3}}\right\}$$
when the given angle is positive and less than $\dfrac{\pi}{2}$.

15. If $p = 11\cdot78, \; q = 5\cdot67, \; \theta = \cdot4712$ radians, calculate the value of
$$p^{\frac{3}{5}}\sin\theta\,(p^2 + q^2)^{-\frac{1}{2}};$$
also the value of $\sqrt{p^q}$.

16. In the following formula, $a = 25, \; b = 8\cdot432, \; c = 0\cdot345, \; \theta = 0\cdot4226$ radians; find the value of
$$a^{1\cdot157}\,b^{-\frac{1}{3}} \div \theta\,(c^3 + a\log_e b \,.\, \tan\theta).$$

17. $\sin 162° \tan^2 140° \div \sqrt[3]{\sec 105°}$.

18. Find the value of $a^{-\frac{1}{3}}(a^2 - b^2)^{\frac{1}{2}} \div \sin\theta\,\log_e\dfrac{a}{b}$.
If a is $9\cdot632, \; b = 2\cdot087, \; \theta$ is $0\cdot384$ radians.

19. Find the value of $a^{\frac{2}{3}}b \div \sqrt{c^2 + \sin^3 A}$
when $\qquad a = 4\cdot268, \quad b = \cdot0249,$
$$c = 3\cdot142, \quad A = 26°.$$

Applications to problems on heights and distances.—It is not always convenient, or possible, to measure directly the height of a given object, nor to find the distance of two objects apart.

Instruments are used for measuring purposes by which the angle between any two straight lines which meet at the eye of the observer can be measured. For this purpose what are called *Sextants* and *Theodolites* are used. By means of these the cross-wires of a telescope can be made to coincide with considerable accuracy with the image of an observed object, and by means of a vernier attached, the readings of the observed angles can be made to a fraction of a minute.

The angle contained between a horizontal line and the line which meets a given object is called the angle of elevation when the object is above the point of observation ; and the angle of depression when the object is below.

Thus, if B be the point of observation (Fig. 71) and A the given object, the angle made by the line joining B to A, with

the horizontal line BC, is called the **angle of elevation.**

In a similar manner if A be the point of observation and B an object, then the angle between the horizontal line (DA, drawn through A) and the line AB is called the **angle of depression.**

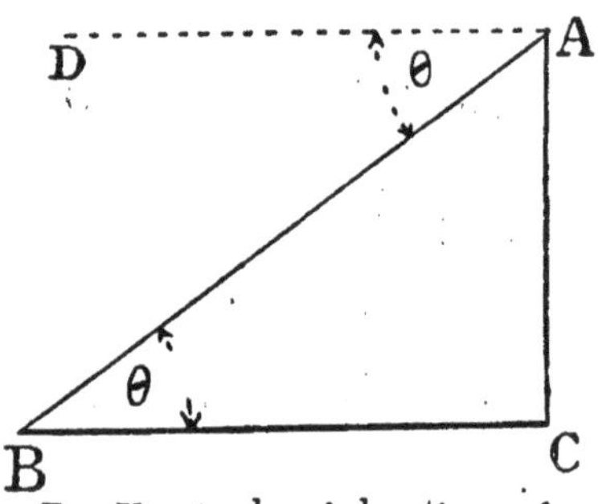

FIG. 71.—Angles of elevation and depression.

The following problems will show the methods adopted in working examples involving these angles :

Ex. 1. At a distance of 100 feet from the foot of a tower the angle of elevation of top of tower is found to be 60°. Find the height of the tower.

*To any convenient scale make AB (Fig. 72) the base of a right-angled triangle equal 100 units.

Draw the line AC, making an angle of 60° with AB and inter-secting BC at C. Then BC is the required height.

$$\frac{BC}{AB} = \tan 60° \; ; \text{ but } \tan 60° = \sqrt{3} \; ;$$

$$\therefore \; BC = AB \tan 60° = 100\sqrt{3} = 173...\text{ft}.$$

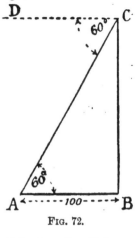

Fig. 72.

Ex. 2.   From the top of a tower, the height of which is $100\sqrt{3}$ ft., the angle of depression of an object on a straight level road on a line with the base of the tower is found to be 60°.   Find the distance of the object from the tower.

In this case, draw CD a horizontal line (Fig. 72) through C, the point of observa-tion, making $CB = 100\sqrt{3}$ ft. on any scale, and BA at right angles to CB.   Then the point at which a line CA, drawn at an angle of 60° to the horizontal line DC, meets BA gives the distance BA required $= 100$ ft.

EXERCISES   XXXI.

1. ABC is a triangle with a right angle at C, CB is 30 feet long, and BAC is 20°.   If CB be produced to a point P, such that PAC is 55°, find the length of CP.

2. The angular height of a tower is observed from two points A and B 1000 feet apart in the same horizontal line as the base of the tower.   If the angle at A is 20° and at B 55°, find the height of the tower.

3. The angle of elevation of the top of a steeple is 30°.   If I walk 50 yards nearer, the angle of elevation becomes 60°.   What is the height of the steeple ?

4. The elevation of a tower from a point A due N. of it is observed to be 45°, and from a point B due E. of it to be 30°.   If $AB = 240$ feet, find the height of the tower.

5. If from a point at the foot of the mountain, at which the elevation of the observatory on the top of Ben Nevis is 60°, a man walks 1900 feet up a slope of 30°, and then finds that the elevation of the observatory is 75°, show that the height of Ben Nevis is nearly 4500 feet.

6. The angle of elevation of a balloon from a station due south of it is 60° ; and from another station due west of the former, and distant a mile from it, the angle of elevation is 45°.   Find the height of the balloon.

7. Two ships leave the harbour together, one sailing N.E. at the rate of 7½ miles an hour, and the other sailing north at the rate of 10 miles an hour; find the shortest distance between the ships an hour and a half after starting.

8. A and B are two points on one bank of a straight river and C a point on the opposite bank; the angle BAC is 30°, the angle ABC is 60°, and the distance AB is 400 feet; find the breadth of the river.

9. From the top of a cliff 1000 feet high, the angles of depression of two ships at sea are observed to be 45° and 30° respectively: if the line joining the ships points directly to the foot of the cliff, find the distance between the ships.

10. Two knots on a plumb-line at heights of 7 feet and 2 feet above the floor cast shadows at distances of 11·4 feet and 1·65 feet respectively from the point where the line produced meets the floor. Find the height of the source of light above the floor.

11. From a ship at sea the angle subtended by two forts A and B is 35°. The ship sails 4·26 miles towards A and the angle is then 51°; find the distance of B from the ship at the second point of observation.

12. A tower stands at the foot of an inclined plane whose inclination to the horizon is 9°; a line is measured up the incline from the foot of the tower of 100 feet in length. At the upper extremity of this line the tower subtends an angle of 54°. Find the height of the tower.

13. From two stations A and B on shore 3742 yards apart a ship C is observed at sea. The angles BAC, ABC are simultaneously observed to be 73° and 82° respectively. Find the distance of A from the ship.

14. Three vertical posts are placed at intervals of one mile along a straight canal, each rising to the same height above the surface of the water. The visual line joining the tops of the two extreme posts cuts the middle post at a point 8 inches below the top; find to the nearest mile the radius of the earth.

15. The altitude of a certain rock is observed to be 47°, and after walking 1000 feet towards the rock, up a slope inclined at an angle of 32° to the horizon, the observer finds that the altitude is 77°. Find the vertical height of the rock above the first point of observation.

16. A balloon is ascending uniformly, and when it is one mile high the angle of depression of an object on the ground is found to be 35° 20′, 20 minutes later the angle of depression is found to be 55° 40′; find the rate of ascent of the balloon.

CHAPTER XVII.

USE OF SQUARED PAPER. EQUATION OF A LINE.

Use of squared paper.—Two quantities, the results of a number of observations or experiments, which are so related that any alteration in one produces a corresponding change in the other, can be best represented by a **graphic method**, in which it is possible to ascertain by inspection the relation that one variable quantity bears to the other.

For this purpose *squared paper*—or paper having equidistant vertical and horizontal lines $\frac{1}{10}''$, $\frac{1}{4}''$, 1 mm., etc., apart—is employed; these cover the surface of the paper with little squares (Fig. 73).

Commencing near the lowest left-hand corner of the paper, one of the lower horizontal lines may be taken as the axis of x and a vertical line near the left edge of the paper as the axis of y.

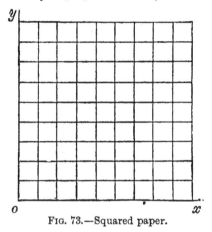

FIG. 73.—Squared paper.

The two lines ox and oy at right angles are called **axes**. The horizontal line ox is called the axis of **abscissae**; oy is the axis of **ordinates**.

One or more sides of the squares, measured along the line ox, is taken as the unit of measurement in one set of the observations; and, in a similar way, one or more sides of the squares bordered by the line oy is taken as the unit of the other set of observations.

Then for any pair of observations two points are obtained, one on *ox* and the other on *oy*.

If a vertical line were drawn upon the squared paper, from the point on *ox*, and a horizontal line from the point on *oy*, the two lines would intersect. It is not necessary to draw such lines ; they are furnished by the lines on the squared paper By dealing with a number of pairs of observations a series of points may be obtained upon the squared paper ; these are usually marked with a small cross or circle.

In this manner a sufficient number of points separated by short distances may be obtained, and the points joined by a **straight** or **curved** continuous line called a **graph** or a **locus**. From such a line intermediate values may be read off.

When the plotted points are obtained by calculation from a given formula, the curve is made to pass through each point ; but when the points denote experimental numbers, and therefore include errors of observation, any attempt to draw a continuous line through the points would give a curve consisting of a series of angles or sharp bends. As such an irregular curve would be for many purposes practically useless it is better to use a piece of thin wood ; this when bent may be used to draw a fairly even curve lying evenly among the points, probably passing through a few of the plotted points, above some, and below others. The curve also furnishes a check on the plotted values and gives a fair idea not only as to the numbers which are in error, but also in each case their probable amount.

In the case of experimental results, the points can often be arranged to lie as nearly as possible on a straight line, and the line which most nearly agrees with the results may be obtained by using a piece of black thread. This thread is stretched and placed on the paper, moved about as required until a good average position lying most evenly among the points is obtained.

A better plan is to use a **strip of celluloid** or **tracing paper** on which a line has been drawn, the transparency of the strip allowing the points underneath it to be clearly seen. When the position of the line is determined, two points are marked at its extremities, and the line is inserted by using a straight edge or the edge of a set square.

The word *curve* is often used to denote *any line, whether straight or curved*, representing the relation between two quantities.

Ex. 1. The following table gives the number of centimetres in a given number of inches. Plot these on squared paper and find from the curves the number of centimetres in 1¼ inches, and the number of inches in 8 centimetres.

Inches.	1	2	2½	3½
Centimetres.	2·54	5·08	6·4	8·8

Use the vertical axis *oy* to denote inches and the horizontal axis *ox* to denote centimetres (Fig. 74). Read off 2·54 on the horizontal and 1 on the vertical axes ; these two values receive various names,

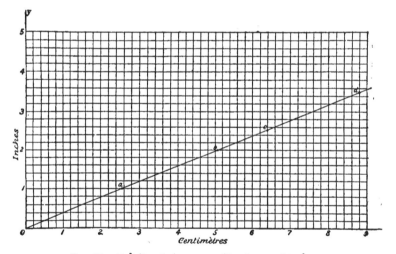

FIG. 74.—Relation between centimetres and inches.

with all of which it is important to become acquainted. Thus, the two values of the point *a* may be called the *x-coordinate* and the *y-coordinate* of point *a* ; or simply the point (*x*, *y*) ; this when the given values are substituted becomes the point (2·54, 1).

The next point *b* is the point (5·08, 2), *i.e.* its abscissa is 5·08, and its ordinate 2.

The two remaining points *c* and *d* are obtained in like manner, and finally a fine line is drawn through the plotted points.

In a similar manner the relation between square inches and square centimetres, or pounds and kilograms, may be obtained.

Interpolation.—When a series of values have been plotted on squared paper, and a line drawn connecting the plotted points, any intermediate value can be read off by what is called interpolation. Thus, at the point, the ordinate of which is 1½ inches, we read off 3·8 centimetres. Similarly, we find 3·15 inches correspond to 8 centimetres.

Positive and negative coordinates.—To denote negative quantities conventional methods similar to those already referred to in measuring angles (p. 159) are adopted. All distances measured above the axis of x are positive, and all distances below negative ; distances on the right of the line oy are positive, and those on the left negative. This may also be expressed by the statement that all abscissae measured to the right of the origin are positive, all to the left negative. All ordinates above the axis of x are positive, and all below are negative.

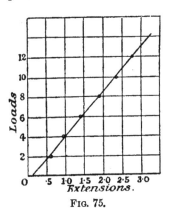

FIG. 75.

Ex. 2. In an experiment on a *spiral spring* the following values for loads and corresponding *extensions* were observed :

Original length of spring 17·2 centimetres.

(*a*) Plot on squared paper the given values of extension and load and find the law connecting them ; (*b*) also find E the modulus of elasticity.

Load on spring in pounds.	2	4	6	8	10	12	14
Extension in centimetres.	·60	·96	1·42	1·88	2·35	2·82	3·28

We can use the vertical axis for loads and the horizontal axis for extensions ; as the coordinates of the first point are 0·6 and 2, mark off the first point as in Fig. 75.

The next point (·96, 4) is marked in like manner, proceeding with the remaining observations a series of plotted points are obtained. These are observed numbers, and therefore contain errors of observation; consequently the line is made to lie evenly among the points instead of being drawn through them.

To obtain E; at one point we find a load of 10·8 lbs. produces an extension of 2·5 cm.

Hence, an extension of 1 cm. would require $\dfrac{10\cdot8}{2\cdot5}$,

∴ to double the length, *i.e.* to increase the length of the spring by 17·2 cm. requires

$$\frac{17\cdot2\times10\cdot8}{2\cdot5}; \ \therefore E=74\cdot13 \text{ lbs.}$$

Plotting a line.—When the relation between two variables is given in the form of an equation, or formula, then assuming values for one, corresponding values of the other variable can be found and the line plotted.

Ex. 3.   Let $y=x+2$ be a given equation.

When $x=0,\ y=2$; $x=1,\ y=3$; $x=2,\ y=4$.

Corresponding values of x and y may be tabulated in two columns thus :

Values of x	0	1	2	3	4	5	6
Values of y	2	3	4	5	6	7	8

When $x=0$, $y=2$; there is no distance to measure off on the axis of x, and as $y=2$ we measure 2 units upwards, and, as in Fig. 76, this gives one point in the line.   When $x=2$, $y=4$; at the intersection of the lines through 2 and 4, make a cross or dot.   Proceeding in this manner, any number of points lying on the line are obtained, and the points joined by a fine line, which will be found to be a straight line.   The line plotted may be written $y=ax+b$, where $a=1$, $b=2$.

Conversely, assuming that the straight line represents a series of plotted results of two variables such as E and R.   To find the law connecting the two it is only necessary to substitute in the equation $E=aR+b$, the simultaneous known values of two points in the line. This will give two equations from which a and b can be determined as in the previous example.

Let the values along oy represent values of E, and those along ox values of R ; $\therefore$ from the line (Fig. 76),

when E is 3, R is 1,
when E is 8, R is 6.

Substituting these values in the equation

$$\therefore \quad 3 = a \times 1 + b \dots\dots\dots\dots\dots\dots\text{(i)}$$
$$8 = a \times 6 + b \dots\dots\dots\dots\dots\dots\text{(ii)}$$

Subtracting $\quad\quad 5 = 5a \quad\quad \therefore a = 1.$

Hence from (i) $\quad b = 3 - 1 = 2.$

Substituting these values for a and b, the required equation is

$$E = R + 2.$$

It will be noticed that the value of the term b gives the point in the axis of y from which the line is drawn. By altering the value of b, the term a remaining constant a series of parallel lines are obtained. Thus, let $b = 0$, then equation (i) becomes $y = x$. $\therefore$ when $x = 0$, $y = 0$, and the result obtained by plotting values of x and corresponding values of y is a line parallel to the preceding, but passing through the origin.

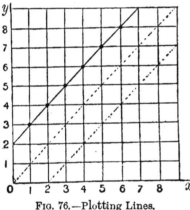

FIG. 76.—Plotting Lines.

Again, let $b = -2$. $\therefore$ when $x = 0$, $y = -2$, and we obtain a line parallel to the preceding, intersecting the axis of y at a distance -2, or 2 units below the origin, the equation is now $y = x - 2$. The three lines are shown in Fig. 76.

When the term b remains unchanged, but the magnitude of a is altered, then when the values of a and b are plotted a series of lines are obtained, all drawn from the same point, but each at a different inclination to the axis of x, or better, the slope of each line is different from that of the rest.

Equation of a line.—When as in Ex. 1, the relation between two variable quantities can be represented by a straight line, the equation of the line is of the form

$$y = ax + b \dots\dots\dots\dots\dots\dots\dots\text{(i)}$$

where a and b are constants. Then, if in (i) simultaneous values of x and y are inserted, the values of a and b can be found.

Thus, in Fig. 74, when $y=1$, $x=2\cdot54$;
also when $y=2$, $x=5\cdot08$.

Substituting these values in (i) we obtain

$$2=a\times5\cdot08+b \dots\dots\dots\dots\dots\dots\text{(ii)}$$
$$1=a\times2\cdot54+b \dots\dots\dots\dots\dots\dots\text{(iii)}$$

By subtraction $\quad\quad 1=a\times2\cdot54$

$$\therefore\ a=\frac{1}{2\cdot54}=\cdot39.$$

Also substituting this value for a in (ii) or (iii) we find $b=0$.

Hence the equation of the line is $y=\cdot39x$, and the line passes through the origin.

Again, in **Ex. 2** the equation of the line as before is of the form $y=ax+b$ (i), if simultaneous values of x and y be inserted for two points the values of a and b can be found.

Referring to Fig. 75 we see that when $x=1$, $y=4$; when $x=3$, $y=13$.

Substituting these values in (i) we obtain

$$13=3a+b \dots\dots\dots\dots\dots\dots\dots\text{(ii)}$$
$$4=\ a+b \dots\dots\dots\dots\dots\dots\text{(iii)}$$

By subtraction $\quad\quad 9=2a$

$$\therefore\ a=4\cdot5.$$

Also substituting this value for a in (ii) or (iii) we find $b=-\cdot5$.

Hence the required equation, or **straight line** graph as it is called, is $y=4\cdot5\ x-\cdot5$ (iv).

At the point where the line cuts the axis of y the value of x is 0. Substituting this value of x in (iv) we get $y=-\cdot5$, *i.e.* the line intersects the axis of y at a point $\cdot5$ below the origin. The point where the line cuts the axis of x is in like manner obtained by making $y=0$.

$$\therefore\ 0=4\cdot5x-\cdot5 \text{ or } x=\frac{\cdot5}{4\cdot5}=\cdot111.$$

Line passing through two points.—From the preceding example it will be obvious that we can readily find the equation to a straight line passing through two given points, and if necessary from the equation proceed to plot the line.

Ex. 1. Find the equation of the straight line passing through the points (1, 4), (3, 13).

The equation of a straight line is $\qquad y = ax + b$ .... (i)

Substituting the coordinates of the first point; $\therefore 4 = a \times 1 + b$ .... (ii)

,, ,, ,, second ,, $13 = a \times 3 + b$ ....(iii)

Subtracting (ii) from (iii) $\qquad\qquad\qquad 9 = 2a$

from either (ii) or (iii) we find $b = -\cdot 5$ $\qquad$ or $a = 4\cdot 5$.

Hence the required equation is $y = 4\cdot 5x - \cdot 5\cdot$

The line can now be plotted, and is shown in Fig. 75.

As the line passing through two given points can be obtained, two such lines will give at their point of intersection simultaneous values of x and y, and this may be made to give the solution of a simultaneous equation.

Simultaneous equations.—Two general methods of solving simultaneous equations have been described on p. 91; another, which may be called a graphical method, is furnished by using squared paper. The method applied to the solution of a simultaneous equation containing two unknown quantities, consists in plotting the two lines given by the two equations. When this is done the point of intersection of the two lines is obtained. This is a point common to both lines, and as the co-ordinates of the point obviously satisfy both equations, it is the solution required.

Ex. 1. Solve the simultaneous equations

(i) $3x + 4y = 18$, (ii) $4x - 2y = 2$.

From (i) $\qquad\qquad y = \dfrac{18 - 3x}{4}.$ ....(iii)

From (ii) $\qquad\qquad y = 2x - 1.$ ....(iv)

To plot the lines it is sufficient to obtain two points in each and join the points by straight lines.

In (iii), when $x = 0$, $y = 4\cdot 5$ and when $x = 5$, $y = \cdot 75$, the first gives point a (Fig. 77), the second point b. Join a and b and the line corresponding to Eq. (iii) is obtained.

In Eq. (iv), when $x = 1$, $y = 1$ and when $x = 3$, $y = 5$, the first gives point c, the second gives d. Join c and d and the two lines are seen to cross at f. This point of intersection is a point common to both lines, its coordinates are seen to be $x = 2$, $y = 3$, and these values satisfy the given simultaneous equations.

In the preceding examples it has been possible to draw a straight line through the plotted points, and to obtain an equation connecting the two variables. When, however, the plotted points lie on a curve, it would be difficult, if not impossible, to express the relation between the two variables by means of a law or an equation. In such cases a straight line may often be obtained by plotting instead of one of the variables, quantities derivable from it, such as *the logarithms, the reciprocals, or the squares, etc., of the given numbers.*

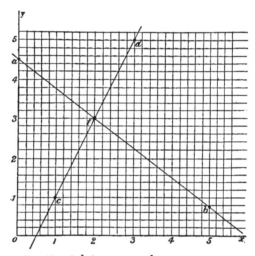

Fig. 77.—Solution of simultaneous equation.

Thus, when a cord is passed round a fixed cylinder and a force N is applied at one end and a force M at the other, the cord remains at rest not only when N is equal to M, but also when N is increased. If the increase in N is made gradually a value is obtained at which the cord just begins to slip on the cylinder. The amount by which N must be greater than M when slipping occurs is readily found by experiment, and depends not only on the surfaces in contact, but also on the fractional part of the circumference of the cylinder embraced by the cord.

Ex. 3. Denoting by n the fractional part of the circumference of a cylinder embraced by a cord, then the following table gives a series of values of n and corresponding values of N. Find the relation between n and N.

n	·25	·5	·75	1	1·25	1·5	1·75	2	2·25	2·5	2·75
N	150	195	295	375	515	615	755	1045	1435	1735	2335
Log N	2·1761	2·29	2 4698	2·5740	2·7118	2·7887	2·8779	3·0191	3·1568	3·2392	3·3683

When simultaneous values of n and N are plotted, a curve lying evenly among the points (Fig. 78) can be found; but by plotting n and log N the points lie approximately on a straight line. The relation between n and log N may be expressed in the form

$$n = a \log N + b.$$

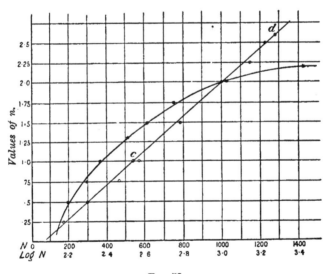

FIG. 78.

To find the numerical values of the constants a and b it is only necessary to substitute for two points on the line simultaneous values of n and log N, thus obtaining two equations from which a and b can be obtained.

Thus at c (Fig 78), $n=1$,   log $N=2·54$,

and at d, $n=2·6$, log $N=3·28$.

Hénce
$$2 \cdot 6 = a \times 3 \cdot 28 + b \ \dots\dots\dots\dots\dots\dots\dots\text{(i)}$$
$$1 = a \times 2 \cdot 54 + b \ \dots\ \dots\dots\dots\dots\dots\dots\text{(ii)}$$

By subtraction,
$$1 \cdot 6 = a \times \cdot 74$$
$$\therefore \ a = \frac{1 \cdot 6}{\cdot 74} = 2 \cdot 162.$$

And substituting this value tor a in (i) we have
$$b = 2 \cdot 6 - 2 \cdot 162 \times 3 \cdot 28 = -4 \cdot 49.$$

Hence the relation between the variables is expressed by
$$n = 2 \cdot 162 \log N - 4 \cdot 49.$$

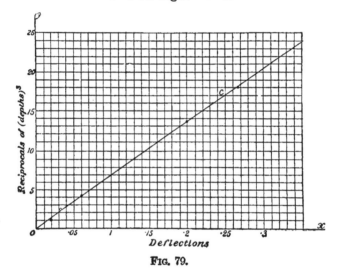

FIG. 79.

Ex. 4. The depths d and deflections δ, when loaded with the same load, of a series of beams of varying depths and constant breadths are given in the annexed table. Find the equation connecting d and δ.

d	1	$\cdot 75$	$\cdot 625$	$\cdot 5$	$\cdot 375$	$\cdot 25$
δ	$\cdot 02$	$\cdot 033$	$\cdot 06$	$\cdot 118$	$\cdot 27$	$\cdot 934$
$\dfrac{1}{d^3}$	1	$2 \cdot 38$	$4 \cdot 1$	8	$18 \cdot 9$	64

When the variables d and δ are plotted a curve is obtained ; but by plotting δ and $\dfrac{1}{d^3}$ a straight line, lying evenly among the points, can be drawn as in Fig. 79.   The line passes through the origin, and its equation may be written $\delta = a \times \dfrac{1}{d^3}$.

From Fig. 79 at c, $\delta = \cdot 25$ and $\dfrac{1}{d^3} = 17$.

Hence $\cdot 25 = a \times 17$; $\therefore$ $a = \dfrac{\cdot 25}{17} = \cdot 0147$.

The relation is therefore $\delta = \cdot 0147 \dfrac{1}{d^3}$.

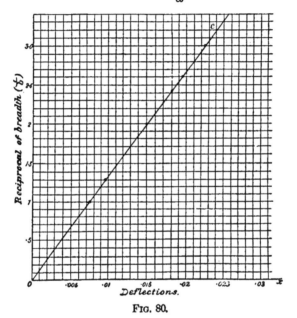

Fig. 80.

Ex. 5.   The following table gives a series of values of the breadths b and deflections δ of a series of beams of constant depths and vari. able breadths ; find the equation connecting b and δ.

b	$\cdot 25$	$\cdot 375$	$\cdot 5$	$\cdot 625$	$\cdot 7$	1
δ	$\cdot 03$	$\cdot 017$	$\cdot 014$	$\cdot 011$	$\cdot 009$	$\cdot 007$
$\dfrac{1}{b}$	4	$2 \cdot 67$	2	$1 \cdot 6$	$1 \cdot 33$	1

If the first two columns are plotted the points lie on a curve, but by plotting the second and last columns δ and $\frac{1}{b}$ (Fig. 80), a straight line through the points and passing through the origin may be drawn. At the point c, $\delta = \cdot024$ and $\frac{1}{b} = 3\cdot2$. Substituting these values in the equation $\delta = a \times \frac{1}{b}$, the relation between the variables is found to be $\delta = \cdot007 \times \frac{1}{b}$.

Slope of a line.—The ratio of increase of one variable quantity relatively to that of one another is of fundamental importance. Simple cases are furnished as in the preceding examples, when on plotting two variable quantities on squared paper a straight line connecting the plotted points can be obtained. The rate of increase which is constant is denoted by the inclination or *slope of the line.* Care should be taken to clearly distinguish between the usual meaning attached to the term "slope of a line" and the meaning given to the same words in Mathematics.

What is usually meant by the statement that a hill rises 1 in 20 is that

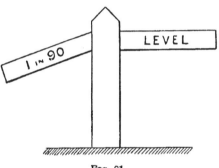

FIG. 81.

for every 20 feet along the hill there is a vertical rise of 1 foot. To indicate the slope of a railway line a post may be placed along the side of the line and a projecting arm indicates roughly the slope by the angle which it makes with the horizontal, and in addition the actual amount is marked on it. As in Fig. 81, the termination of the slope and the commencement of a level line may in like manner be denoted by a horizontal arm with the word *level* on it. This so-called slope of a line which is largely used by engineers and others is not the slope used in Mathematics. The slope of a line such as AB (Fig. 82) is denoted by drawing a horizontal line at any convenient point A and at any other point B a perpendicular BC meeting the former line in C. The slope of the line is then measured by the ratio of $\frac{CB}{AC}$.

If at any point in AB a point B' be taken and a perpendicular $B'C'$ be drawn the ratio remains unaltered;

i.e.
$$\frac{C'B'}{AC'} = \frac{CB}{AC} \text{ (p. 156)}.$$

Denoting the coordinates of the point A by (x, y), then, if B' is a point near to A, the distance AC' may be called the *increment of x*, and the distance $C'B'$ the *increment of y*. Instead of using the word increment it is better to introduce a symbol for it, this is usually the symbol δ; hence, δx is read as "increment of x," and does not mean $\delta \times x$. Similarly increment of y is written δy.

$\dfrac{CB}{AC} = \dfrac{\delta y}{\delta x}$ is the *tangent* of the angle BAC, or the tangent of the angle of slope. It will be obvious that the former method would give the *sine* of the angle of slope.

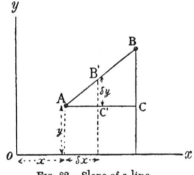

FIG. 82.—Slope of a line.

Rate of increase.—To find the rate at which a quantity is increasing at any given point we find the rate of increase of y compared with the increase of x at the point.

Let the equation of the line AB (Fig. 82) be $y = ax + b$ (i) and let A be the point (x, y), then the coordinates of B' a point near to A may be written as $x + \delta x$, and $y + \delta y$, substituting these values in (i) then we obtain

$$y + \delta y = a\,(x + \delta x) + b \text{ (ii)}.$$

Subtract (i) from (ii), $\therefore \delta y = a\delta x$, or $\dfrac{\delta y}{\delta x} = a$.

Thus we find, as on p. 176, that the slope or inclination of the line depends on the term a.

In Ex. 3, the line $y = x + 2$ has been plotted; proceeding as in the preceding example we find that $\dfrac{\delta y}{\delta x} = 1$. As the slope or tangent of the angle made by the line is 1, we know that the inclination of the line to the axis of x is 45°.

EXERCISES. XXXII.

1. The following numbers refer to the test of a crane.

Resistance just overcome, R lbs.	100	200	300	400	500	600	700	800
Effort just able to overcome resistance, E lbs.	8·5	12·8	17·0	21·4	25·6	29·9	34·2	38·5

Try whether the relation between E and R is fairly well represented by the equation

$$E = aR + b,$$

and if so, find the best values of a and b. What effort would be required to lift a ton with this crane?

2. In the following examples a series of observed values of E, R, and F are given. In each case they are known to follow laws approximately represented by $E = aR + b$, $F = cR + d$; but there are errors of observation. Plot the given values on squared paper, and determine in each case the most probable values of a, b, c, and d.

(i)

E	3·5	5	6·75	8·25	9·75	11·5	13·25	14·78
R	14	28	42	·56	70	84	98	112
F	2·86	3·83	5·00	5·92	6·83	8·00	9·17	10·1

(ii)

E	·5	1	$1\frac{1}{2}$	2	2·5	3	3·5	4
R	4	15	28	40	52	64	76	88
F	32	57	80	104	128	162	176	200

(iii)

E	3·25	4·25	5	5·75	6·75	7·5	8·5	9·25	10
R	14	21	28	35	42	49	56	63	70
F	2·68	3·39	3·86	4·32	5·04	5·5	6·22	6·68	7·14

3. A series of observed values of n and N are given. Find the relation in each case between n and $\log N$.

(i)

n	·25	·5	·75	1	1·25	1·5	1·75	2	2·25	2·5	2·75
N	154	180	265	375	485	635	835	1135	1535	1835	2435

(ii)

n	·25	·5	·75	1	1·25	1·5	1·75	2·0	2·5	3·0
N	145	186	235	296	385	495	558	683	1115	1515

(iii)

n	·25	·5	·75	1	1·25	1·5	1·75	2·0
N	115	145	185	235	300	385	490	605

4. An electric light station when making its maximum output of 600 kilowatts uses 1920 lbs. of coal per hour. When its load factor is 30 per cent. (that is, when its output is $600 \times 30 \div 100$) it uses 1026 lbs. of coal per hour. What will be the probable consumption of coal per hour when the load factor is 12 per cent.?

5. Plot on squared paper the following observed values of A and B, and determine the most probable law connecting A and B. Find the percentage error in the observed value of B when A is 150.

A	0	50	100	150	200	250	300	350	400
B	6·2	7·4	8·3	9·5	10·3	11·6	12·4	13·6	14·5

6. The following observed values of M and N are supposed to be related by a linear law $M = a + bN$, but there are errors of observation. Find by plotting the values of M and N the most probable values of a and b.

N	2·5	3·5	4·4	5·8	7·5	9·6	12·0	15·1	18·3
M	13·6	17·6	22·2	28·0	35·5	47·4	56·1	74·6	84·9

7. (i) The following values, which we may call x and y, were measured. Thus when x was found to be 1, y was found to be ·223.

x	1	1·8	2·8	3·9	5·1	6·0
y	·223	·327	·525	·730	·910	1·095

It is known that there is a law like—
$$y = a + bx$$
connecting these quantities, but the observed values are slightly wrong. Plot the values of x and y on squared paper, find the most likely values of a and b, and write down the law of the line.

(ii)

x	0·5	1·7	3·0	4·7	5·7	7·1	8·7	9·9	10·6	11·8
y	148	186	265	326	388	436	529	562	611	652

State the probable error in the measured value of y when $x = 8\cdot7$.

8. In the annexed table, values of L, the length of a liquid column, and T, its time of vibration, are given. The relation between L and T^2 is given by $L = aT^2 + b$; find a and b.

L	2·4	2·8	3·0	3·2	3·4	3·6
T	1·06	1·23	1·29	1·34	1·38	1·42

9. It is known that the following values of x and y are connected by an equation of the form $xy = ax + by$, but there are slight errors in the given values. Determine the most probable values of a and b.

x	18	28	54	133	− 455	− 111	− 65
y	5	6	7	8	9	10	11

10. The following measurements were made at an Electric Light Station under steady conditions of output:

W is the weight in pounds of feed water per hour, and P the electric power, in kilowatts, given out by the station. When P was 50, W was found to be 3800; and when P was 100, W was found to be 5100.

If it is known that the following law is nearly true—
$$W = a + bP,$$
find a and b, also find W when P is 70 kilowatts.
State the value of $W \div P$ in each of the three cases.

11. Some particulars of riveted lap joints are given in the following table :

t = Thickness of plate,	$\frac{3}{8}$	$\frac{1}{2}$	$\frac{5}{8}$	$\frac{3}{4}$	$\frac{7}{8}$	$1''$
d = Diameter of rivet,	$\frac{3}{4}$	$\frac{7}{8}$	$\frac{15}{16}$	$\frac{17}{16}$	$\frac{9}{8}$	$\frac{5}{4}$
p_1 = Pitch of Rivets (*single riveted*), -	2·06	2·25	2·3	2·37	2·40	2·63
p_2 = Pitch of Rivets (*double riveted*), -	3·33	3·58	3·60	3·63	3·63	3·95

(i) Plot d and t and obtain the values of the constants a and b in the relation $d = at + b$ for plates from $\frac{3}{8}''$ to $\frac{7}{8}''$ thick.

(ii) Plot d and $\sqrt{t}$ and obtain for the whole series of values given in the table the value of c in the relation $d = c\sqrt{t}$.

(iii) Find values of d when t is $\frac{5}{16}$, $\frac{7}{16}$, and $\frac{9}{16}$.

(iv) Plot d and p_1 and d and p_2 and obtain the constants in the relations $p_1 = d + b$; $p_2 = d + c$.

12. The following table gives some standard sizes of Whitworth bolts and nuts. All the dimensions being in inches.

d = Diameter of bolt,	$\frac{1}{4}$	$\frac{3}{8}$	$\frac{1}{2}$	$\frac{3}{4}$	1	$1\frac{1}{4}$	$1\frac{1}{2}$	2	$2\frac{1}{2}$
W = Width of nut across the corners,	·605	·818	1·06	1·50	1·93	2·36	2·77	3·63	4·50
A = Area of bolt at bottom of thread,	·027	·068	·112	·304	·554	·894	1·30	2·31	3·73

(i) Plot d and W and obtain a relation in the form $W = ad + b$.

(ii) Plot A and d^2 and obtain a relation in the form $A = ad^2$.

(iii) Obtain a more accurate relation in the form $A = ad^2 + b$ for bolts from $\frac{1}{2}''$ to $1\frac{1}{2}''$ diameter.

13. In the following table a series of values of the pull P lbs. necessary to tow a canal-boat at speeds V miles per hour are given. If the relation between P and V can be expressed in the form $P = CV^n$, what is the numerical values of the constants C and n ?

P	1·0	1·82	2·77	3·73	4·4
V	1·82	2·53	3·24	3·86	4·27

CHAPTER XVIII.

PLOTTING FUNCTIONS.

In the preceding chapter the student will have noticed that when the numerical values of two variables are obtained from a simple formula, the curve passes *through* the plotted points. When, however, the given numerical values are experimental numbers involving errors of observation the curve is made to lie evenly among the points, in this manner errors of experiment or observation may be corrected, and by interpolation any intermediate value can be obtained.

The applications of squared paper are so numerous and varied that it becomes a difficult matter to make a suitable selection. The following examples may serve to illustrate some of the uses to which squared paper can be applied.

Ex. 1. In a price list of oil engines the prices for engines of a given brake horse power are as follows :

Brake horse power,	$1\frac{1}{2}$	$3\frac{1}{2}$	$6\frac{1}{2}$	$9\frac{1}{2}$	$12\frac{1}{2}$	16
Price in pounds (£),	75	110	160	200	225	250

Plot the given values on squared paper and find the probable prices of engines of 5 and of 8 horse power.

In Fig. 83 the given values are plotted and a curve is drawn, passing through the points. The coordinates of any point on the curve shows the horse power and probable price of an engine. Corresponding to sizes 5 and 8, we obtain the probable prices as £135 and £180 respectively.

The calculation of logarithms.—The following method described by Prof. Perry, and also devised independently by Mr. E. Edser, may be used to calculate a table of logarithms to three or four significant figures. The square root of 10 or $10^{\frac{1}{2}} = 3\cdot162$.

Referring to Table III. it will be found that $\log 3\cdot162 = \cdot5000$.

Again $\qquad \sqrt{3\cdot162}$ or $10^{\frac{1}{4}} = 1\cdot778$,

and $\qquad\qquad \log 1\cdot778 = \cdot2500$.

Now $\quad 10^{0\cdot5} \times 10^{0\cdot25} = 3\cdot162 \times 1\cdot778 = 5\cdot623$;

$$\therefore \ 10^{0\cdot75} = 5\cdot623.$$

In a similar manner we obtain $10^{\frac{1}{8}} = 1\cdot336$, $10^{\frac{1}{16}} = 1\cdot1548$, and $10^{\frac{1}{32}} = 1\cdot0746$;

$$\therefore \ 10^{\frac{1}{2}} \times 10^{\frac{1}{32}} = 10^{\frac{17}{32}} = 3\cdot162 \times 1\cdot0746 = 3\cdot398.$$

Also $\qquad\qquad 10^{\frac{3}{2}} = 1000^{\frac{1}{2}} = 31\cdot62$,

and $\qquad\qquad 10^{1\frac{3}{4}} = 10^{\frac{7}{4}} \quad = 56\cdot23.$

$$10^{0} = 1 \ ; \quad \therefore \ \log 1 = 0.$$

$$10^{1} = 10 \ ; \ \therefore \ \log 10 = 1.$$

When a series of values have been obtained by calculation the logarithms may be plotted on squared paper as ordinates and the numbers as abscissae. By drawing the logarithmic curve through the plotted points any intermediate value can be read off. Even with the cheapest squared paper, tables of logarithms and antilogarithms can be made fairly accurate in this manner. Using better paper and with care, a table of logarithms

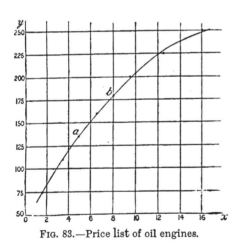

FIG. 83.—Price list of oil engines.

accurately giving logarithms to four figures can be obtained.

Ex. 2. By means of squared paper shew the values of the sine, cosine, tangent, and radian measure of all angles from $0°$ to $90°$.

Find from the curves the values of the sines, cosines, and tangents of 15°, 30°, 45°, 75°.

Here as in Fig. 84 we may denote degrees as abscissa and numerical values as ordinates; these are obtained from Table V. Having drawn curves through the plotted points the values for 15°, etc., can be read off. Notice carefully that when the angle is 45° the sine and cosine curve cross, *i.e.* the values of the sine and cosine are equal and the curve denoting values of the tangent has an ordinate unity at this point.

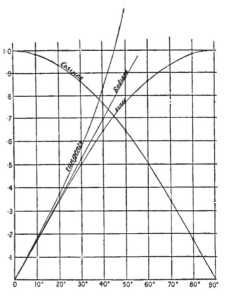

FIG. 84.—Values of the sine, cosine. tangent and radian measure of angles from 0° to 90°.

Again it will be obvious that for small angles not exceeding 20° the values of the sine, radian and tangent are approximately the same.

Ex. 3. In the following table some population statistics of a certain country are given. Let P denote the population and t the time in years. Show the relation between P and t by a curve, and find from the curve the probable population in 1845 and in 1877.

t, year.	1821	1831	1841	1851	1861	1871	1881	1891	1901
P, in millions.	10·2	12·8	15·4	18·4	21·6	25·6	30·0	38·0	

When as in Fig. 85 the given values are plotted and a curve drawn, the probable populations in 1845 at a, and in 1877 at b, can be read off, and are found to be 16·5 millions and 28·2 millions respectively.

Equations.—On p. 75 a method has been indicated by which in a given expression such as $x^2 - 4x + 3$ the factors $(x-1)(x-3)$ can be obtained by substitution; these values $x=1$, $x=3$ are

called the roots of the given equation. In equations which are
more complicated such a method may become very troublesome

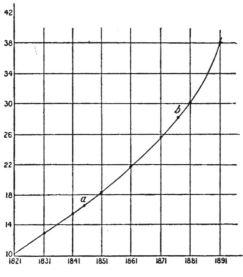

FIG. 85.

and laborious; the roots of an equation, or better, the *solution of
an equation*, which would be difficult by algebraical methods, may
in many cases be obtained by the use of squared paper. To
gain confidence the method may be applied to any simple
equation such as the one above.

Ex. 4. To solve the equation $x^2 - 5x + 5\cdot25 = 0$, let

$$y = x^2 - 5x + 5\cdot25.$$

Substitute values 0, 1, 2, etc., for x, and find corresponding values
of y. Thus, when $x=0$, $y=5\cdot25$; when $x=1$, $y=1-5+5\cdot25=1\cdot25$;
the values so obtained may be tabulated as follows :

x	0	1	2	3	4	5
y	$5\cdot25$	$1\cdot25$	$-\cdot75$	$-\cdot75$	$1\cdot25$	$5\cdot25$

Plotting these values on squared paper a curve of the form shown
in Fig. 86 is obtained. The curve crosses the axis of x in two
points, A and B; the two values of x given by $0A$ and $0B$ make
$y=0$, and therefore are the two roots required; $0A$ is $1\cdot5$ and

OB is 3·5. When these values are substituted they are found to satisfy the given equation. Hence $x = 1\cdot5$ and $x = 3\cdot5$ are the two roots required.

Ex. 5. Find the roots of the equation $x^3 - 3x - 1 = 0$. Let
$$y = x^3 - 3x - 1.$$

As before, put $x = 0$, 1, 2, etc., and calculate corresponding values of y as follows:

x	-2	-1	0	1	2	3
y	-3	1	-1	-3	1	17

Plotting these values as in Fig. 87 the curve cuts the axis of x in three points, C, B, and A. At each of these points the value of x

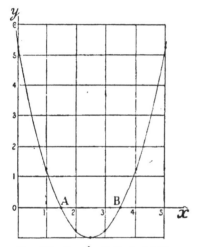

FIG. 86.— Graph of $x^2 - 5x + 5\cdot25 = 0$.

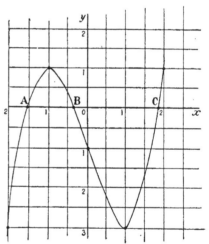

FIG. 87.—Graph of $x^3 - 3x - 1 = 0$.

makes $y = 0$, and hence is a solution of the given equation. At A the value of x is seen to be between $-1\cdot5$ and $-1\cdot7$, and at B between $-\cdot3$ and $-\cdot5$, by plotting this part of the curve to a larger scale, the more accurate values are found to be $-1\cdot532$, and $-\cdot347$. In a similar manner the value at C is found to be $1\cdot879$.

Probably a simpler method than the one described, and which may be shown by an example, is as follows:

Ex. 6. Solve the equation $x^3 - 6x + 4 = 0$.

Write the given equation in the form of two equations.
$$y = x^3 \text{ (i)}, \quad y = 6x - 4 \text{ (ii)}.$$

From (i) we shall by plotting obtain a curve, and from (ii) a line. The points of intersection of two lines, as in p. 179, give values which satisfy the equations, and in like manner the points of intersection of the line and curve will give the required values of x.

Thus in (i), by giving x various values 0, 1, 2, etc., we can calculate corresponding values of y as follows :

x	0	1	2	3	4	5
y	0	1	8	27	64	125

By plotting these values we obtain the curve shown in Fig. 88.

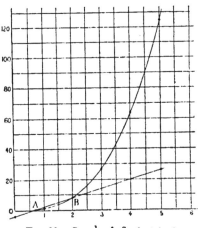

FIG. 88.—Graph of $x^3 - 6x + 4 = 0$.

Positive values of x have been assumed, but if negative values are used the values of y are of the same magnitude but with altered sign. Hence the corresponding part of the curve, below the axis of x, can be obtained.

In (ii), if $x=0$, $y=-4$, and if $x=5$, $y=26$, the line drawn through these plotted points will give at their points of intersection A and B the required values. As all equations of this kind can be reduced to the forms shown at (i) and (ii) the curve indicated by (i) may be used for all equations of this form.

Plotting of functions.—Functions of the form $y=ax^n$, $y=ae^{bx}$, $y=\sin ax$, where a, b, and n may have all sorts of values, are easily dealt with by using squared paper.

Thus in the equation $y=ax^n$, when a and n are known, for various values of x corresponding values of y can be obtained.

Ex. 7.   Let $a=\cdot25$ and $n=2$.   The equation $y=ax^n$ becomes $y=0\cdot25x^2$.

By giving a series of values to x, 1, 2, 3, etc., we can obtain from Eq. (i) corresponding values of y.

Thus, when $\qquad\qquad x=0$, $y=0$,

also when $\qquad\qquad x=1$, $y=\cdot25\cdot$

It will be convenient to arrange the two sets of values of x and y as follows :

Values of x,	0	1	2	3	4	5	...	..	...
Corresponding values of y,	0	·25	1	2·25	4	6·25	...	...	..

As y is 0 when x is 0, the curve passes through the origin (or point of intersection of the axes). Plotting the values of x and y from the two columns, as shown in Fig. 89, a series of points are obtained. The figure obtained is a parabola.

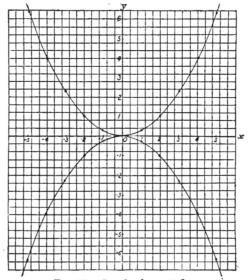

Fig. 89.—Graph of $y = ·25x^2$.

It is sometimes difficult to draw a fairly uniform curve through plotted points, but when a curve has been drawn improvements may be made, or faults detected, by simply holding the paper on which the curve is drawn at the level of the eye, and looking along the curve. Some such simple device should always be used.

As the square of either a positive or a negative number is necessarily positive, it follows that two values of x, equal in magnitude but opposite in sign, correspond to each value found for y. By using positive values of x, the curve shown on the right of the line oy is obtained. The negative values give the corresponding curve on the left.

If the constant a be negative, its numerical value remaining the same, then the equation becomes $y = -\cdot25x^2$; this when plotted will be found to be another parabola below the axis of x (Fig. 89).

The equation $\mathbf{y = ax^n}$ becomes when $a = 1$, $y = x^n$. Giving various values 2, 3, $\frac{1}{2}$, $\frac{1}{3}$, 1, etc., to the index n then functions of the form $y = x^3$, $y = x^{\frac{1}{3}}$, etc., are obtained. Assuming values 0, 1, 2... for x corresponding values of y can be found. The curves can be plotted, and are shown in Fig. 90. It will be seen that the curves $y = x^3$, $y = x^{\frac{1}{3}}$, and the straight line $y = x$ all intersect at the same point.

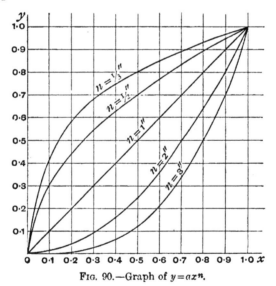

Fig. 90.—Graph of $y = ax^n$.

The hyperbolic curve is of great importance, more especially to an engineer, and is obtained from the general equation $y = ax^n$ by making $n = -1$; the equation then becomes $y = ax^{-1}$ or $y = \dfrac{a}{x}$

$$\therefore \quad xy = a \dots\dots\dots\dots\dots\dots\dots\dots\dots(i)$$

The curve is shown in Fig. 91, and should be carefully plotted. The rectangular hyperbola is the curve of expansion for a gas such as air, at constant temperature, and is often taken to represent the curve of expansion of superheated or saturated steam.

If p and v denote the *pressure* and *volume* respectively of a gas, instead of the form shown by (i), the equation is usually

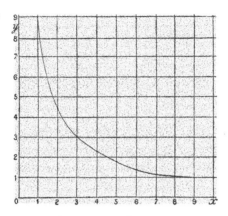

Fig. 91.—Graph of $xy=9$.

written, $pv=$constant$=c$, and is known as **Boyle's Law**; c is a constant, this is either given, or may be obtained from simultaneous values of p and v.

Ex. 8.   Plot the curve $xy=9$.

$$\therefore \ y=\frac{9}{x}. \ \dots\dots\dots\dots\dots\dots\text{(ii)}$$

From (ii), when $\qquad x=1, \ y=9.$

$\qquad\qquad\quad x=2, \ y=4\cdot5.$

$\qquad\qquad\quad x=\tfrac{1}{1000}, \ y=9000.$

$\therefore$ when x is very small y is very great.

Thus let

$$x=\tfrac{1}{1000000}, \text{ then } y=9000000.$$

When $x=0$, then $y=\tfrac{9}{0}$, or is infinite in value. In other words the curve gets nearer and nearer to the axis oy as the value of x is diminished, but does not reach the axis at any finite distance from the origin. This is expressed by the symbols $y=\infty$ when $x=0$.

As Eq. (ii) can be written $x=\dfrac{9}{y}$ it follows as before that when $y=0$,

$x=\infty$.

The two lines or axes ox and oy are called **asymptotes** and are said to meet (or touch) the curve at an infinite distance.

Arranging in two columns a series of values of x and corresponding values of y obtained from Eq. (ii) we obtain.

Values of x,	0	1	2	3	4	5	6	7	8	9
Corresponding values of y,	∞	9	4·5	3	2·25	1·8	1·33	1·3	1·13	1·

Plotting these values of x and y on squared paper then the curve or graph passing through the plotted points is a hyperbola as in Fig. 91.

One of the most important curves with which an engineer is concerned is given by the equation $pv^n = c$, where p denotes the pressure and v the volume of a given quantity of gas.

The constant c and index n depend upon the substance used ; *i.e.* steam, air, etc.

When, as in the preceding example, the values of c and n are known, for various values of one variable, corresponding values of the other can be obtained, and plotted. The converse problem would be, given various simultaneous values of p and v calculate the numerical values of c and n.

To do this it is necessary to write the equation $pv^n = c$ in the form $\log p + n \log v = \log c$.

Plotting $\log p$ and $\log v$ a straight line may be drawn lying evenly among the plotted points, and from two simultaneous values of p and v the values of c and n may be found.

EXERCISES. XXXIII.

1. A man sells kettles. He has only made them of three sizes as yet, and he has fixed on the following as fair list prices :

12 pint kettle, price 68 pence.
6　,,　　　,, 50　,,
2　,,　　　,, 22　,,

He knows that other sizes will be wanted, and he wishes to publish at once a price list for many sizes. State the probably correct list prices of his 4 and 8 pint kettles.

2. Plot the following values of D and θ, and determine
     (i) The value of D when θ is 0.
     (ii) The value of θ when D is 0.
     (iii) The maximum value of D.

θ	$-45°$	$-15°$	$15°$	$45°$	$75°$	$105°$
D inches	-0.25	$\cdot98$	$1\cdot80$	$2\cdot24$	$2\cdot05$	$1\cdot32$

3. Plot the corresponding values of x and y given below, and determine the mean value of y.

x	0	11·5	25	40·5	58·5	75	96·4	109	120
y	7·6	10·2	12·6	14·4	15·6	16	15·2	13·8	11·2

4. Plot the curve given by the equation $y = 0\cdot1e^x$ where $e = 2\cdot718$.

5. The population of a country is as follows:

Year.	1830	1840	1850	1860	1870	1880	1890
Population (millions),	20	23·5	29·0	34·2	41·0	49·4	57·7

Find by plotting the probable population in 1835, 1865, and in 1895. Find the probable population at the beginning of 1848 and the rate of increase of population then.

6. A manufacturer finds that to make a certain type of cast-iron pump the cost is

     45 shillings for a pump of 3 inches diameter, and
     115 shillings for a pump of 6 inches diameter.

Estimate the probable cost of pumps of 4 inches and 5 inches diameters so far as you can from these data.

If the actual cost of a 5-inch pump when made is found to be 82 shillings, what would now be the estimate of the probable cost of a 4-inch pump?

Solve the following equations:

7. $x^2 - 5\cdot45x + 7\cdot181 = 0.$ **8.** $0\cdot24x^2 - 4\cdot37x - 8\cdot97 = 0.$

9. $2\cdot3x^2 - 6\cdot72x - 13\cdot6 = 0.$ **10.** $x^3 - 7x^2 + 14x - 8 = 0.$

Find in each of the following a value of x which satisfies the equations:

11. $2x^{3\cdot1} - 3x - 16 = 0.$ **12.** $2\cdot42x^3 - 3\cdot15\log_e x - 20\ 5 = 0.$

13. $e^x - e^{-x} + 0\cdot 4x - 10 = 0$.

The answer to be given correctly to *three* significant figures.

14. The following values of p and u, the pressure and specific volume of water-steam, are taken from steam tables:

p	15	20	30	40	50	65	80	100
u	25·87	19·72	13·48	10·29	8·34	6·52	5·37	4·36

Find by plotting $\log p$ and $\log u$ whether an equation of the form $pu^n = \text{constant}$ represents the law connecting p and u, and if so, find the best average value of the index n for the range of values given.

15. Find a value of x which satisfies each of the equations

$$\text{(i) } x^3 + 4\cdot 73x - 1\cdot 746 = 0.$$
$$\text{(ii) } x^3 + 9x - 16 = 0.$$

16. Find a value for x for which $\tan x = 2\cdot 75x$.

17. Given $y = 1 - 4\cdot 818x + 7\cdot 514x^3$, calculate and enter the values of y in the following table:

x	0	·1	·2	·3	·4	·5	·6	·7	·8	·9	1·0
y											

Plot the curve and find one root.

Solve the equations:

18. $x^3 - 13x - 12 = 0$. **19.** $x^3 - 237x - 884 = 0$. **20.** $x^3 - 27x - 46 = 0$.

Slope of a curve.—*The slope of a curve* at any point is that of the *tangent to the curve* at the point. The tangent to the curve is the straight line which touches the curve at the point. If, in Fig. 92, the tangent at P makes an angle 42° with the axis of x, then slope of curve at $P = \tan 42° = 0\cdot 90$.

It is an easy matter to draw a line touching a given curve at a point when the inclination of the line is known, but if the direction is not known, then at any point P several lines apparently touching the curve could be drawn, but it would be difficult by mere inspection to draw a tangent at the point. Before this can be done it is necessary to know the direction of the line with some approach to accuracy. This may be effected by taking the values of x and y at a given point and the values x' y' of another point close to the former; from these values

$x' - x$, and $y' - y$ can be obtained, the former may be denoted by δx and the latter by δy.

The ratio $\frac{\delta y}{\delta x}$ gives the *average* rate of increase of one variable compared with the other, and also approximately the slope of the tangent at P.

If Q and P (Fig. 92) be two points on a curve, the former the point (x', y') the latter the point (x, y), then $x' - x$ or δx is $7 - 5 = 2$, and $y' - y$ or δy is $5 - 1\cdot5 = 3\cdot5$.

$$\therefore \frac{\delta y}{\delta x} = \frac{3\cdot5}{2} = 1\cdot75.$$

Draw QM and PM parallel to oy and ox respectively, then if points P and Q be joined by a straight line

$$\frac{\delta y}{\delta x} = \tan QPM = 63\cdot3°.$$

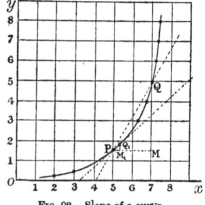

FIG. 92.—Slope of a curve.

This is obviously only a very rough approximation, a better result is obtained by using point Q_1 (Fig. 92), its co-ordinates $(5\cdot4, 1\cdot88)$. The increments δy and δx become $0\cdot38$ and $0\cdot4$ respectively.

$$\therefore \frac{\delta y}{\delta x} = \frac{\cdot38}{\cdot4} = \cdot9499 ; \quad \therefore \text{ slope of line} = 43\cdot3°.$$

If the numbers on the vertical axis denote distances in feet, and along the horizontal axis time in seconds, then the slope of the curve at any point gives rate of change of position or velocity at the point. Thus at P the velocity would be $\cdot9$ ft. per sec. If the ordinates denote velocities and the abscissae times, then the slope of the curve at any point gives rate of change of velocity, or the acceleration at the point.

When the increments δy and δx are made smaller and smaller the slope given by $\frac{\delta y}{\delta x}$ becomes more and more nearly the actual slope at the point. Finally, when each increment is made indefinitely small the ratio is written $\frac{dy}{dx}$ and is the tangent or the actual rate at the point.

The slope of a curve at a given point may be indicated by a simple example as follows :

Ex. 10.   Plot the curve $y=x^2$ and find the slope of the curve at the point $x=2$.

The square of a negative and a positive quantity are alike positive, so that for each value of y there are two values of x.   Thus, when $x=2$ or -2, or $x=\pm2$, $y=4$, etc.   By substituting values 0, 1, 2, 3, etc., for x corresponding values of y are obtained as in the following table :

$x\pm$	0	1	2	3	4	5
y	0	1	4	9	16	25

To obtain the slope of the curve at the point 2, if we take the two points $x=2$ and $x'=3$, then $y=4$, $y'=9$;

$$\therefore \frac{y'-y}{x'-x}=\frac{5}{1}=5.$$

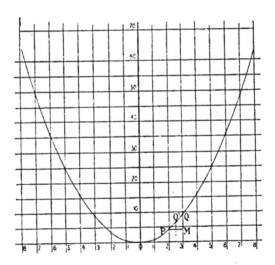

Fig. 93.—Graph of $y=x^2$.

This is obviously only a rough approximation.   The line PQ (Fig. 93) joining the two points cuts the curve, and the slope of the line is

given by $$\frac{y'-y}{x'-x}=\frac{5}{1}.$$

Assuming a point Q' nearer to P, then a better approximation is obtained. Thus if $x = 2\cdot5$, then $y' = (2\cdot5)^2 = 6\cdot25$;

$$\therefore\ x' - x = \cdot5,\ y' - y = 2\cdot25.$$

Hence $\qquad \dfrac{y' - y}{x' - x} = \dfrac{2\cdot25}{\cdot5} = 4\cdot5;\ \therefore\ \dfrac{\delta y}{\delta x} = \dfrac{Q'M'}{PM'} = \dfrac{4\cdot5}{1}.$

As the magnitude of $x' - x$ is diminished the corresponding values obtained approach nearer and nearer to the actual value.

Thus when $\qquad\qquad x' = 2\cdot05,\ y' = 4\cdot2025,$

$$\therefore\ \frac{\delta y}{\delta x} = \frac{\cdot2025}{\cdot05} = 4\cdot05.$$

When $\qquad\qquad x' = 2\cdot005,\ y' = 4\cdot020025,$

$$\therefore\ \frac{\delta y}{\delta x} = \frac{\cdot020025}{\cdot005} = 4\cdot005.$$

In each case we obtain the **average rate of increase** at the point P, the average rate approaching nearer and nearer to the actual rate as the increments get smaller and smaller.

The **actual rate** at P will be 4 when the increments δy and δx are made small enough, and it is easy to show this by using algebraic symbols as follows :

If the equation to a curve be

$$y = x^2 \dotfill \text{(i)}$$

and (x, y) the coordinates of a point on the curve, the coordinates of a point close to the former may be written $x + \delta x$ and $y + \delta y$ (p. 184). Substituting these values in (i) we get

$$y + \delta y = (x + \delta x)^2 = x^2 + 2x\delta x + (\delta x)^2 \dotfill \text{(ii)}$$

Subtracting (i) from (ii),

$$\therefore\ \delta y = 2x\delta x + (\delta x)^2.$$

Dividing both sides by δx,

$$\frac{\delta y}{\delta x} = 2x + \delta x \dotfill \text{(iii)}$$

Equation (iii) is true whatever value is given to δx, and values of $\dfrac{\delta y}{\delta x}$ corresponding to values for δx of $\cdot5$, $\cdot05$, etc., have been obtained, in each case giving an approximation to the tangent at the point and also giving the *average rate of increase* of y with respect to x.

If we imagine the increments δy and δx to get smaller and smaller without limit, then the ratio $\frac{\delta y}{\delta x}$ is denoted by $\frac{dy}{dx}$, and equation (iii) becomes $\frac{dy}{dx}=2x$. This is the *actual rate* at the point P, or in other words is the tangent at P. Hence the **slope** of a curve at a given point is represented by **the tangent of the angle which the tangent to the curve makes with the axis of x.**

The symbol $\frac{dy}{dx}$ is read as the **differential coefficient** of y with respect to x, and simply denotes a *rate of increase*. Its numerical value can be ascertained when the law or the relation connecting two variables x and y is known.

The beginner should notice that the differential of a variable quantity denoting the difference between two consecutive values is an indefinitely small quantity, and is expressed by writing the letter d before the variable x or y. When this is clearly understood the symbol dx (which is read as the differential of x) will not be taken to mean $d \times x$, nor dy as $d \times y$.

It is obvious that there cannot be any rate of change of a constant quantity, hence **the differential of a constant quantity is zero.**

Ex. 11.   Find the slope of the curve $y=x^3$ at the point $x=2\cdot5$.

As before, we may write $y+\delta y=(x+\delta x)^3$;

$$\therefore \quad y+\delta y=x^3+3x^2\delta x+3x(\delta x)^2+(\delta x)^3.$$

Subtracting $\qquad y=x^3,$

$$\delta y=3x^2\delta x+3x(\delta x)^2+(\delta x)^3,$$

or $\qquad\qquad \frac{\delta y}{\delta x}=3x^2+3x\delta x+(\delta x)^2.$

When the increments become indefinitely small the ratio on the left is $\frac{dy}{dx}$, and on the right all terms involving δx disappear.

Hence $\qquad\qquad \frac{dy}{dx}=3x^2.$

The slope of the curve at the point $x=2\cdot5$ is

$$3\times2\cdot5^2=18\cdot75.$$

On p. 176 we have found that the equation $y=ax+b$ represents the equation to a straight line in which the slope or inclination of the line to the axis of x depends on a.

Let $$y = ax + b, \dots\dots\dots\dots\dots\dots\text{(i)}$$
then $$y + \delta y = a(x + \delta x) + b = ax + a\delta x + b, \dots\dots\dots\text{(ii)}$$
Subtracting (i) from (ii),

$$\therefore \quad \delta y = a\delta x \; ;$$

$$\therefore \quad \frac{\delta y}{\delta x} \text{ or } \frac{dy}{dx} = a. \dots\dots\dots\dots\dots\dots\text{(iii)}$$

Hence a is the slope, or the tangent of the angle which the line makes with the axis of x.

Generally if

$$y = ax^n \text{ then } \frac{dy}{dx} = nax^{n-1}, \dots\dots\dots\dots\dots\text{(iv)}$$

where a is a constant and n is any number positive or negative.

Simple differentiation.—The process of finding the value of $\frac{dy}{dx}$, the rate of change from a given expression, is called differentiation, and in simple cases, which are all that are required at the present stage, it is only necessary to apply the rule given by Eq. (iv).

Any constant which is a multiplier or divisor of a term will be a multiplier or divisor after differentiation, but as the differential of a constant is zero, any constant connected to a variable by the signs + or − disappears during differentiation.

The process may be seen from the following examples :

Ex. 12. $y = 3x^2$.
$$\frac{dy}{dx} = 2 \times 3x^{(2-1)} = 6x.$$

Ex. 13. $y = 5x^4$.
$$\frac{dy}{dx} = 4 \times 5x^{(4-1)} = 20x^3.$$

Ex. 14. $y = 4x^3 + 3x^2 + 2x + 3$.
$$\frac{dy}{dx} = 12x^2 + 6x + 2.$$

As the differential of a constant is zero the constant 3 connected to the variables by the sign + disappears during differentiation.

Ex. 15. $y = 3x^{\frac{3}{2}}$.
$$\frac{dy}{dx} = \frac{3}{2} \times 3x^{\left(\frac{3}{2}-1\right)} = \frac{9}{2}x^{\frac{1}{2}}.$$

Ex. 16. $y = \dfrac{x^{\frac{1}{4}}}{2}$.

$$\frac{dy}{dx} = \frac{1}{4} \times \frac{x^{(\frac{1}{4}-1)}}{2} = \frac{x^{-\frac{3}{4}}}{8}$$

Ex. 17. $y = 2x^{-\frac{2}{3}}$.

$$\frac{dy}{dx} = -\frac{2}{3} \times 2x^{(-\frac{2}{3}-1)} = -\frac{4}{3}x^{-\frac{5}{3}}.$$

The process of finding the differential coefficient of a given expression, *i.e.* the value of $\dfrac{dy}{dx}$ is of the utmost importance, and the operations involved in many cases consist of simple algebraic processes which may be easily carried out as in the preceding and in the following examples.

Ex. 18.   Graph the curve $y = 0\cdot1x^{-\cdot25}$. Find the slope of the curve at the point $x = 0\cdot4$.

The equation $y = 0\cdot1x^{-\cdot25}$ is obtained from the general equation $y = ax^n$ by writing $0\cdot1$ for a and $-\cdot25$ for n.

To plot the curve we may assume values 0, $\cdot1$, $\cdot2$, etc., for x, and find corresponding values of y.

Thus when $\qquad x = 0, \quad y = 0.$

When $\qquad\qquad x = \cdot3, \quad y = 0\cdot1 \times 0\cdot3^{-\cdot25}.$

$$\therefore \ \log y = \log \cdot1 - \cdot25 \log 0\cdot3. = \overline{1}\cdot0000 - \tfrac{1}{4}(\overline{1}\cdot4771).$$

$$\therefore \ y = \cdot1351.$$

In a similar manner other values of y can be obtained and tabulated as follows :

x	0	$\cdot1$	$\cdot2$	$\cdot3$	$\cdot4$	$\cdot5$	$\cdot6$	$\cdot7$	$\cdot8$	$\cdot9$	$1\cdot0$
y	0	$\cdot1778$	$\cdot1495$	$\cdot1351$	$\cdot1257$	$\cdot1190$	$\cdot1136$	$\cdot1093$	$\cdot1057$	$\cdot1027$	$\cdot1$

To find the slope of the curve at the point $0\cdot4$ we may find the value of y when $x = \cdot42$;

$$\therefore \ \delta x = \cdot42 - \cdot4 = \cdot02 ;$$

$$\therefore \ y = \frac{\cdot1}{(\cdot42)^{\cdot25}} = \cdot1243 ;$$

$$\therefore \ \delta y = \cdot1243 - \cdot1257 = -\cdot0014.$$

$$\text{Slope of curve} = \frac{\delta y}{\delta x} = \frac{-\cdot0014}{\cdot02} = -\cdot07.$$

This gives approximately the slope of the curve at the point $x = \cdot 4$. The approximation becoming closer and closer to the actual value as the increments are diminished, and, when they become indefinitely small, the slope is that of the tangent at the point or $\frac{dy}{dx}$.

If
$$y = \cdot 1x^{-\cdot 25},$$

then $\quad \frac{dy}{dx} = -\cdot 25 \times \cdot 1x^{(-\cdot 25 - 1)} = \cdot 025x^{-1 \cdot 25}$ or $\cdot 025x^{-\frac{5}{4}}.$

From this we can obtain the value of the tangent at any point by substituting the value of x. Thus when $x = \cdot 4$, we get

$$\frac{dy}{dx} = \cdot 025 (\cdot 4)^{-\frac{5}{4}}.$$

The numerical value of $\cdot 025 (\cdot 4)^{-\frac{5}{4}}$ is readily obtained by logs.

Thus $\quad \log \cdot 025 - \frac{5}{4} \log \cdot 4 = \overline{2} \cdot 3979 - \overline{1} \cdot 5026$

$$= \overline{2} \cdot 8953;$$

$$\therefore \cdot 025 (\cdot 4)^{-\frac{5}{4}} = \cdot 07857.$$

Compound interest law.—The curve $y = ae^{bx}$, where, a, b, and x may have all sorts of values, is known as the *compound interest law*; e is the base of the Napierian logarithms $= 2 \cdot 718$. When definite numerical values are assigned to a and b, the curve can be plotted.

Let $a = \cdot 53$, $b = \cdot 26$, then the equation becomes $y = \cdot 53e^{0 \cdot 26x}$

Ex. 19.   Plot the curve $y = \cdot 53e^{0 \cdot 26x}$.

Calculate the average value of y from $x = 0$ to $x = 8$. Also determine the slope of the curve at the point where $x = 3$.

Assuming values 0, 1, 2, 3, etc., for x corresponding values of y can be obtained.   Thus when $x = 0$,

$$y = \cdot 53e^{0} = \cdot 53.$$

When $\quad\quad x = 2, \ y = 53e^{\cdot 26 \times 2} = \cdot 53e^{\cdot 52};$

$$\therefore \ \log y = \log \cdot 53 + \cdot 52 \log 2 \cdot 718$$

$$= \overline{1} \cdot 7243 + \cdot 52 \times \cdot 4343$$

$$= \overline{1} \cdot 9501.$$

$$\therefore \ y = \cdot 8915 \cdot$$

Other values of x can be assumed and values of y calculated as in the following table :

Values of x.	0	1	2	3	4	5	6	7	8
Corresponding values of y.	·53	·6874	·8915	1·156	1·500	1·945	2·522	3·271	4·242

To obtain the average value of y from $x=0$ to $x=8$ we can apply *Simpson's Rule*, p. 233. Thus

$$\text{Sum of end ordinates} = 4\cdot772,$$
$$\text{,, even ,,} = 7\cdot0594,$$
$$\text{,, odd ,,} = 4\cdot9135.$$

Area of curve from $x=0$ to $x=8$ is
$$4\cdot772 + 7\cdot0594 \times 4 + 4\cdot9135 \times 2 = 42\cdot8366 ;$$
$$\therefore \text{ area of curve} = \frac{42\cdot8366}{3}.$$

But average value of y multiplied by length of base = area ;
$$\therefore \text{ average value of } y = \frac{42\cdot8366}{3 \times 8} = 1\cdot785.$$

Proceeding as in preceding problems,

If $\qquad y = ae^{bx}, \dfrac{dy}{dx} = abe^{bx} ;$

$$\therefore \frac{dy}{dx} = \cdot53 \times \cdot26 \times 2\cdot718^{\cdot78} = \cdot3006.$$

$\therefore$ slope of curve at point $x=3$ is $\cdot3006$.

Maxima and minima.—If a quantity varies in such a way that its value increases to a certain point and then diminishes, the decrease continuing until another point is reached, after which it begins to increase ; then the former point is called a **maximum** and the latter a **minimum** value of the quantity.

Ex. 20.   Given $2x^3 - 15x^2 + 24x + 25 = 0$; determine values of x so that the expression may be a maximum or a minimum.

Denoting by y the value of the left-hand side of the equation, and substituting various values for x, corresponding values of y can be obtained, as shown in the annexed table:

x	0	·5	1	2	3	4	5
y	25	33·6	36	29	16	9	20

Plotting these values on squared paper and joining the points, the curve $ABCD$ (Fig. 94) is obtained. From the tabulated values

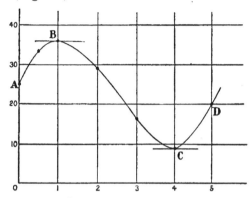

FIG. 94.—Curve showing maximum and minimum values.

the expression seems to be a maximum when $x=1$, and a minimum when $x=4$; this is confirmed by the curve, the former value being located at B, the latter at C. Also, as explained on p. 205, we have,

$$y = 2x^3 - 15x^2 + 24x + 25, \ \frac{dy}{dx} = 6x^2 - 30x + 24 \ ;$$

this gives the *slope* at any point, or the angle which the tangent to the curve at any point makes with the axis of x. At the points B and C the angle is zero, *i.e.* the tangent is horizontal, hence

$$\frac{dy}{dx} = 0 \ ;$$

$$\therefore \quad 6x^2 - 30x + 24 = 0,$$

or $$x^2 - 5x + 4 = (x-1)(x-4),$$

and the values of x which satisfy this equation will give the points at which the tangent to the curve is horizontal.

The required values are therefore $x=1$, $x=4$.

It will be seen from the above example that the values of x corresponding to $\frac{dy}{dx} = 0$ may give either a maximum or minimum value; in many practical questions the conditions of the question will at once suggest whether the value obtained is a maximum or minimum. From Fig. 94, for instance, it is obvious that at point C, where y is a minimum, an increase in x gives an increase in y, i.e. $\frac{dy}{dx}$ is positive, at B where y is a maximum, for an increase in x we obtain a decrease in y, hence $\frac{dy}{dx}$ is negative.

As there may be several loops in a curve similar to those indicated, the terms *maximum* and *minimum* may in each case be taken to denote a point such that a point near to, and on the left of it, has a different slope to a similar point near to, and on the right of it.

Ex. 21. Divide the number 8 into two parts, such that their product is a maximum.

Let x denote one part, then $8 - x$ is the other ; $x(8 - x)$, or $8x - x^2$ is the product ; denote this by y. Substitute values 0, 1, 2, 3, 4, 5, 6 for x, and find the corresponding values 0, 7, 12, 15, 16, 15, 12 for y. Plot on squared paper, and at $x = 4$ a point corresponding to B (Fig. 94) is obtained ; or let $y = 8x - x^2$, then $\dfrac{dy}{dy} = 8 - 2x$;

$$\therefore \ 8 - 2x = 0, \text{ for a maximum, gives } x = 4.$$

Average velocity.—Probably every one has a more or less clear idea of what is meant by saying that a railway train, which may be continually varying its speed, is at any given instant moving at the rate of so many miles per hour.

Suppose that in t hours the train has gone over a distance s miles ; then if the rate were uniform, the rate $\dfrac{s}{t}$ would denote the number of miles per hour.

As a simple numerical example suppose the distance between two places to be 150 miles and the time taken by a railway train from one place to the other is 5 hours ; the uniform rate per hour or average velocity would be $\dfrac{150}{5} = 30$ miles per hour. Such an average would include all the variations of speed, including stoppages on the journey, and is clearly *not* what is meant by the statement that at a given moment the train is going at one particular speed. To obtain the numerical value of such a speed it is necessary to recognise that as the speed is variable the value of $\dfrac{s}{t}$ is continually changing, and can only give a good approximate value of the *average velocity* when the time interval is very short, *i.e.* when t and therefore s are both small quantities. Such small intervals or increments may be denoted by δs and δt,

$$\therefore \ \text{Average velocity} = \frac{\delta s}{\delta t}.$$

Ex. 22. Suppose a body to fall from rest according to the law

$$s = 16t^2 \dots\dots\dots\dots\dots\dots\dots\dots\dots\dots(i)$$

where s is the space in feet and t the time in seconds. Find the actual velocity of the body when t is one second.

In this example, if s and t are plotted the curve is of the form shown in Fig. 93. To find the velocity at time 1, we can, from the given equation, find the space described in a fractional part of a second ; by dividing the space described by the time, the average velocity is obtained. We may take values of t such as 1 and 1·25, 1 and 1·1, and 1 and 1·01, the approximation becoming closer and closer to the actual value as the interval is diminished. When the points are 1 and 1·1, then from (i)

$$s = 16 \times (1{\cdot}1)^2 = 19{\cdot}36 \text{ feet} ;$$

$$\therefore \text{ space in ·1 second} = 16\{(1{\cdot}1)^2 - 1^2\} = 3{\cdot}36 ;$$

$$\therefore \text{ average velocity during ·1 second} = \frac{3{\cdot}36}{\frac{1}{10}} = 33{\cdot}6.$$

Hence the average velocity obtained is too great, and its inaccuracy becomes greater as the interval of time is increased.

Thus space is one quarter of a second $= 16\{(1\frac{1}{4})^2 - 1^2\}$

$$= 16(\tfrac{25}{16} - 1) = 9 ;$$

$$\therefore \text{ average velocity during ·25 second} = \frac{9}{{\cdot}25} = 36.$$

If the interval be from 1 to 1·01,

$$\text{space} = 16(1{\cdot}01^2 - 1^2)$$

$$= 16(1{\cdot}0201 - 1) = {\cdot}3216 ;$$

$$\therefore \text{ average velocity during ·01 second} = \frac{{\cdot}3216}{{\cdot}01} = 32{\cdot}16{\cdot}$$

Other values for t may be assumed, the average value obtained becoming closer and closer to the actual value as the interval of time is diminished. Thus, the intervals of time may be ·001, ·0001 of a second, etc. These small intervals of time and corresponding small space described may be indicated in a convenient manner by the symbols δs and δt. The actual value is obtained when the increments are made indefinitely small, and $\frac{\delta t}{\delta s}$ becomes $\frac{ds}{dt}$.

The preceding results are easily obtained by means of Algebra. The coordinates of any point on the curve $s = 16t^2 \dots$ (i) may be denoted by (s, t), and those of a point near to it by $s + \delta s$ and $t + \delta t$.

Substituting these values in (i) we get

$$s + \delta s = 16(t + \delta t)^2 = 16\{t^2 + 2t(\delta t) + (\delta t)^2\}. \ldots\ldots\ldots\ldots(ii)$$

Subtracting (i) from (ii) we have

$$\delta s = 32t(\delta t) + 16(\delta t)^2.$$

Dividing by δt,

$$\therefore \frac{\delta s}{\delta t} = 32t + 16\delta t. \ldots\ldots\ldots\ldots\ldots\ldots(iii)$$

When δt is made smaller and smaller without limit, then the last term $16\delta t$ is zero and (iii) becomes

$$\frac{ds}{dt} = 32t.$$

Hence the actual value when t is 1 is 32.

Ex. 23. At the end of a time t it is observed that a body has passed over a distance s.

Given that $\qquad s = 10 + 16t + 7t^2, \ldots\ldots\ldots\ldots\ldots\ldots\ldots\ldots(i)$ find s when t is 5. Taking a slightly greater value for t, say $t = 5\cdot01$, calculate the new value of s and find the *average velocity* during the $\cdot01$ second. Also find the exact velocity at the instant when t is 5.

Assuming values 0, 1, 2 ... for t values for s can be found from (i) as follows :

t	0	1	2	3	4	5
s	10	33	70	121	186	265

When t is 5 ; $\qquad s = 10 + 80 + 7 \times 25 = 265.$

,, t is $5\cdot01$; $\qquad s = 10 + 16 \times 5\cdot01 + 7 \times (5\cdot01)^2 = 265\cdot8607.$

$\therefore$ space in $\cdot01$ sec. $= 265\cdot8607 - 265 = \cdot8607.$

Average velocity or $\dfrac{\delta s}{\delta t} = \dfrac{\cdot8607}{\cdot01} = 86\cdot07.$

When t is $5\cdot001$, proceeding as before $\delta s = \cdot086007.$

$$\therefore \frac{\delta s}{\delta t} = \frac{\cdot086007}{\cdot001} = 86\cdot007.$$

Again when t is $5\cdot0001$, $\delta s = \cdot008600007.$

$$\therefore \frac{\delta s}{\delta t} = \frac{\cdot008600007}{\cdot0001} = 86\cdot00007.$$

It will be seen that the average velocities approach a certain value during smaller and smaller intervals of time, and the limiting value is the actual velocity at the point; or, by differentiation,

$$s = 10 + 16t + 7t^2 ;$$

$$\therefore \ \frac{ds}{dt} = 16 + 14t ;$$

and when $t = 5$ this gives 86 as the actual velocity.

Acceleration.—When the ordinates of a curve denote the space or distance passed over by a moving body and the abscissae the time, the slope of the curve at any point gives the velocity at that point. If the ordinates are made to represent the velocity and the abscissae, time, then the slope of the curve, or the tangent to the curve at any point, gives the rate at which the velocity of the moving body is increasing or diminishing. In the former case *the rate of increase* is called **acceleration**, if the latter then *the rate of decrease* is called the **retardation**. Thus the velocity of a body falling freely is known to be g feet per sec. at the end of one second (where g denotes 32·2 ft. per sec.), the velocity at the end of the next second would be $2g$. Hence if we proceed to plot velocities and times we should obtain a straight line, indicating that the "slope" is constant. The body is said to move with uniform acceleration.

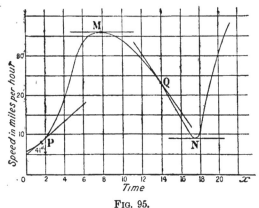

Fig. 95.

The slope of a curve $PQMN$ (Fig. 95) at a point such as P, gives the **rate of increase** of the velocity, or acceleration at the point. At M the tangent to the curve is horizontal, the slope is O, the acceleration is zero, and the body is moving with uniform velocity. At a point such as Q the slope of the curve gives the rate of decrease or retardation, and at N the body is again moving with uniform velocity. The points M and N correspond to maximum and minimum.

EXERCISES. XXXIV.

1. What is meant by the slope of a curve *at a point* on the curve? How is this measured? If the co-ordinates of points on the curve represent two varying quantities, say, distance and time, what does the slope of the curve at any place represent? Obtain an expression for the slope if the distance s and time t are connected by the equation $s = 5t + 2\cdot1t^2$ and give the numerical value at the instant when t is 5.

2. At the end of a time t it is observed that a body has passed over a distance s reckoned from some starting point. If it is known that $s = 20 + 12t + 7t^2$, find s when t is 5, and by taking a slightly greater value of t, say $5\cdot001$, calculate the new value of s and find the average velocity during the $\cdot001$ second. How would you proceed to find the exact velocity at the instant when t is 5, and how much is this velocity?

3. A body is first observed at the instant when it is passing a point A. The time t hours (measured from this instant) and the distance s miles (measured from A) are connected by the equation $s = 20t^2$: find the average speed of the body during the interval between $t = 2$ and $t = 2\cdot1$, between $t = 2$ and $t = 2\cdot001$, and between $t = 2$ and $t = 2\cdot00001$. Deduce the actual speed at the instant when t is exactly 2. How could you otherwise determine this speed, and what symbol is used to denote it?

4. How do we measure (1) the slope of a straight line, (2) the slope of a curve at any point on it?

There are two quantities denoted by v and r which vary in such a way that $v = 4\cdot2r^3$.

Explain what is meant by "the rate of increase of v relatively to the increase of r." How may the *value* of this rate of increase be exhibited graphically for any value of r? Calculate its value when $r = 0\cdot5$.

5. A body weighing 1610 lbs. was lifted vertically by a rope, there being a damped spring balance to indicate the pulling force F lbs. of the rope. When the body had been lifted x feet from its position of rest, the pulling force was automatically recorded as follows:

x	0	11	20	34	45	55	66	76
F	4010	3915	3763	3532	3366	3208	3100	3007

Find approximately the work done on the body when it has risen 70 feet.

6. A body is observed at the instant when it is passing a point P. From subsequent observations it is found that in any time t seconds, measured from this instant, the body has described s feet (measured from P), where s and ι are connected by the equation $s = 2t + 4t^2$. Find the average speed of the body between the interval $t = 1$ and $t = 1 \cdot 1$, between $t = 1$ and $t = 1 \cdot 001$, and between $t = 1$ and $t = 1 \cdot 0001$, and deduce the actual speed when t is exactly 1.

7. Plot the curve $y = \dfrac{1 \cdot 5x}{1 + 0 \cdot 5x}$. Determine the average value of y between $x = 0$ and $x = 10$.

Solve the equations:

8. $0 \cdot 35x^2 - 5 \cdot 23x - 7 \cdot 86 = 0$.   **9.** $2 \cdot 065^{-0 \cdot 48x} = 0 \cdot 826$.

10. Find, correctly to three significant figures, a value of x which will satisfy this equation:

$$9x^3 - 41x^{0 \cdot 8} + 0 \cdot 5e^{2x} - 92 = 0.$$

11. Divide the number 12 into two parts so that the square on one part together with twice the square on the other shall be a minimum.

12. Plot the curve $y = x^2 - 5 \cdot 45x + 7 \cdot 181$ between the points $x = 0$ and $x = 4$, and determine the average value of y between the points $x = 3 \cdot 25$ and $x = 4$.

Plot the following curves from $x = 0$ to $x = 8$.

13. $y = 4x^{0 \cdot 70}$.   **14.** $y = 2 \cdot 3 \sin\left(\cdot 2618x + \dfrac{\pi}{6} \right)$.   **15.** $y = 0 \cdot 53e^{0 \cdot 26x}$.

In each case find the rate of increase of y with regard to x where $x = 3$; also find the average value of y from $x = 0$ to $x = 8$.

16. Given (i) $y = ax^n$, (ii) $y = ax^5 + bx^{\frac{1}{3}} + cx^p + dx^{-\frac{1}{q}}$, write down the value of $\dfrac{dy}{dx}$.

17. By using squared paper, or by any other method, divide the number 420 into two parts such that their product is a maximum. Describe your method.

18. A certain quantity y depends upon x in such a way that

$$y = a + bx + cx^2,$$

where a, b, and c are given constant numbers. Prove that the rate of increase of y with regard to x is $b + 2cx$.

19. Divide the number 20 into two parts, such that the square of one, together with three times the square of the other, shall be a minimum. Use any method you please.

CHAPTER XIX.

MENSURATION. AREA OF PARALLELOGRAM. TRI-
ANGLE. CIRCUMFERENCE OF CIRCLE. AREA OF A
CIRCLE.

Areas of plane figures.—When the numbers of units of length in two lines at right angles to each other are multiplied together, the product obtained is said to be a quantity of two dimensions, and is referred to as so many square inches, square feet, square centimetres, etc., depending upon the units in which the measures of the lengths are taken. The result of the multiplication gives what is called an area. Or, briefly, the area of a surface is the number of square units (square inches, etc.) contained in the surface.

It is obvious that although square inches or units of area are derived from, and calculated by means of, linear measure, those quantities only which are of the same dimensions can be added, subtracted, or equated to each other. Thus, we cannot add or subtract a line and an area, or an area and a volume. Results obtained in such cases would be meaningless.

It must be observed, too, that the two lengths multiplied together to obtain the area are *perpendicular to each other*. This applies to the calculation of all areas.

Area of a rectangle.—*The number of units of area in a rectangular figure is found by multiplying together the numbers of units of length in two adjacent sides.*

Thus, if AB and BC (Fig. 96) are two adjacent sides of a rectangle, its area is the product of the number of units of length in AB and the number of units of length in BC.

Let a be the number of units of length in the lowest line of the figure AB usually called its *base*, or length, and the units in line BC, perpendicular to this, its *altitude*, or breadth b.

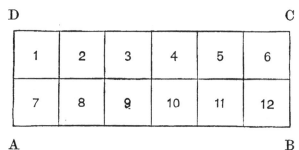

D   C

| 1 | 2 | 3 | 4 | 5 | 6 |
| 7 | 8 | 9 | 10 | 11 | 12 |

A   B

FIG. 96.—Area of a rectangle.

Hence **area = base × altitude,**

or $\qquad$ $= \text{length} \times \text{breadth} = ab.$

Ex. 1. If the base AB and height BC are 6 and 2 units of length respectively, the area is 12 square units. If AB be divided into 6 equal parts and BC into 2, then by drawing lines through the points of division parallel to AB and BC, the rectangle is seen to be divided into 12 equal squares (Fig. 96).

The area is obtained in a similar manner when the two given numbers denoting the lengths of the sides are not whole numbers.

Ex. 2. · Obtain the area of a rectangle when the two adjacent sides are 5 ft. 9 in. and 2 ft. 6 in. in length respectively.

We may reduce to inches before multiplying.

Thus $\qquad$ 5 ft. 9 in. = 69 inches

and $\qquad$ 2 ft. 6 in. = 30 inches ;

∴ area of rectangle = 69 × 30 square inches

$\qquad$ = 2070 square inches = 14·375 square feet ;

Or, instead of first reducing the feet to inches and afterwards multiplying, we may proceed as follows :

$\qquad$ 5 ft. 9 in. = $5\frac{3}{4}$ feet and 2 ft. 6 in. = $2\frac{1}{2}$ feet ;

∴ area of rectangle = $5\frac{3}{4} \times 2\frac{1}{2}$ square feet

$\qquad$ = $\frac{23}{4} \times \frac{5}{2}$ square feet = $14\frac{3}{8}$ square feet.

If a rectangle is divided into three, four, or more rectangles, the area of the whole is equal to the sum of the areas of the several parts. In Fig. 97 the rectangle $ABCD$ is divided into four rectangles.

Area of $AEFK = 3 \times 4 = 12$ sq. in. ; area of $BEFG = 6$ sq. in. ; area of $GFHC = 2$ sq. in. ; area of $HFKD = 4$ sq. in. $\therefore$ Total area is $12 + 6 + 2 + 4 = 24$ sq. in., and this is equal to the area of $ABCD$. Using the letters a, b, c, d to denote the respective sides of the four rectangles, we have a verification of the formula

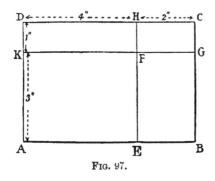

FIG. 97.

$$(a+b)(c+d)$$
$$= ac + bc + ad + bd.$$

Area of a parallelogram. —The rectangle is a particular case of the parallelogram, and

area of parallelogram = base × altitude.

This may also be shown as follows :

Let $ABCD$ (Fig. 98) be the given parallelogram length of sides a and b respectively.

Draw AF and BE perpendicular to AB, and meeting CD at E, and CD produced at F, then the rectangle $ABFE$ is equal in area to the parallelogram $ABCD$.

Hence, area of parallelogram

$$= \text{base} \times \text{altitude} = ab \text{ ;}$$

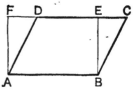

FIG. 98.—Area of a parallelogram.

or, *the vertical distance between a pair of parallel sides multiplied by one of them.*

As $b = AD \sin EDA$

the area $= AB \times AD \sin EDA = ab \sin \theta$;

or, the **product of two** adjacent sides **and the sine of the included** angle gives the area of a parallelogram.

As $\sin 90° = 1$, this formula immediately reduces to that given for a rectangle when the included angle is $90°$.

Of the three terms, area, base, and altitude, any two being given, the remaining term may be found. Similarly, if the area, one side, and included angle be given, the remaining side can be found.

Ex. 1. If the altitude be $1\frac{1}{2}$ ft. and the area 6 sq. ft., then the base is

$$\frac{6}{1\frac{1}{2}} = 4 \text{ ft.}$$

Ex. 2. The area of a parallelogram is 12 sq. ft., one side is 6 ft. and included angle is 30°. Find the remaining side.

Let *a* denote the side. Then we have

$$a \times 6 \sin 30° = 12 ; \quad \therefore \quad a = 4 \text{ ft.}$$

EXERCISES. XXXV.

1. The length of a rectangle is 6·25 ft., its breadth 1·74 ft. Find its area.

2. The length of a room, the sides of which are at right angles, is $31\frac{1}{2}$ ft. and the area 46 sq. yds. What is the breadth?

3. The length and width of a rectangular enclosure are 386 and 300 ft. respectively. Find the length of the diagonal.

4. Show by a figure that the area of a rectangle 8 in. long and 2 in. broad is the same as that of 16 squares each of them measuring one inch in the side.

5. Show that any parallelogram in which two opposite sides are each 15 in. long, while the shortest distance between them is 3 in. has an area of 15 sq. in. Write down an expression for the area of a parallelogram whose base is *a* inches, and altitude *b* inches.

6. The foot of a ladder is at a distance of 36 ft. from a vertical wall, the top is 48 ft. from the ground. Find the length of the ladder.

7. The side of a square is 24 ft. 6 in. What is its area?

8. The sides of a rectangle are as 4 : 3 and the difference between the longer sides and the diagonal is 2. Find the sides.

9. Two sides of a parallelogram are 4·5 ft. and 5·6 ft. respectively, the included angle is 60°. Find the area.

10. How many persons can stand on a bridge measuring 90 ft. in length by 18 ft. in width, assuming each person to require a space of 27 in. by 18 in. ?

11. What will it cost to cover with gravel a court 31 ft. 6 in. long and 18 ft. 9 in. broad at the rate of 1*s.* 4*d.* per square yard ?

12. The length of a rectangular board is 12 ft. 6 in., its area is 18·75 sq. ft. Find its width.

13. The diagonal of a square is 3362 ft. Find the length of a side of the square.

Area of a triangle.—In Fig. 99 the rectangle $ABCD$ and the parallelogram $ABEF$ on the same base AB and of the same altitude, are equal in area.

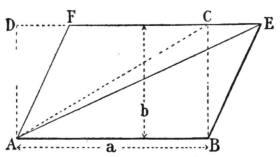

FIG. 99.—Area of a triangle.

When A is joined to C and E it is easily seen that the triangle ACB is half the rectangle $ABCD$, and the triangle AEB is half the parallelogram $ABEF$.

Hence, the two triangles are equal in area, and the area in each case is equal to *half the product of the base and the altitude.*

∴ **area of a triangle** = ½ (base × altitude) = ½ab.

From this rule the area of any triangle can be obtained by measuring the length of any side assumed as a base and the perpendicular on it from the opposite angular point. The area is one-half the product of the base and perpendicular; or, as the perpendicular is the product of the sine of the angle opposite and an adjacent side, the following rule is obtained. **Multiply half the product of two sides by the sine of the included angle.**

This result may be shown graphically by drawing a rectangle on the same, or an equal base, and half the height of the triangle ABC.

Ex. 1.   The base of a triangle is 3·5 in., the height 6·25.   Find the area.

$$\text{Area} = \tfrac{1}{2} \times 3\text{·}5 \times 6\text{·}25 = 10\text{·}94 \text{ sq. in.}$$

Ex. 2.   The sides of an equilateral triangle are 10 ft. in length. Find the area.

As the included angle is 60° the area is given by

$$\tfrac{1}{2} \times 10 \times 10 \sin 60° = \frac{50\sqrt{3}}{2} = 43\text{·}29 \text{ sq. ft.}$$

When the three sides of a triangle are given :

If a, b, c be the three sides and $s = \dfrac{a+b+c}{2}$; or $s =$ half the sum of the three sides, then the area of the triangle is given by the formula

$$\text{area} = \sqrt{s(s-a)(s-b)(s-c)},$$

or, find half the sum of the length of the sides, subtract from this half sum the length of each side separately; multiply the three remainders and the half sum together; the square root of the product is the area of the triangle required.

Ex. 3. We may use this rule to find the area of a right-angled triangle, sides 3, 4, and 5 units respectively. The area can be determined by the method used in Ex. 1.

Here $\qquad s = \frac{1}{2}(3+4+5) = 6.$

Subtract from this the length of each side separately, *i.e.*

$$6-3=3, \quad 6-4=2, \quad 6-5=1.$$

$\therefore$ Area of triangle $= \sqrt{6 \times 3 \times 2 \times 1} = \sqrt{36} = 6$ sq. units.

Ex. 4. Find the area of a triangle, the lengths of the three sides being 3·27, 4·36, and 5·45 respectively.

$$s = \tfrac{1}{2}(3\cdot27 + 4\cdot36 + 5\cdot45) = 6\cdot54.$$

$\therefore$ Area $= \sqrt{6\cdot54(6\cdot54 - 3\cdot27)(6\cdot54 - 4\cdot36)(6\cdot54 - 5\cdot45)}$

$\qquad = \sqrt{6\cdot54 \times 3\cdot27 \times 2\cdot18 \times 1\cdot09} = \sqrt{50\cdot8169}$

$\qquad = 7\cdot129$ sq. ft.

Ex. 5. The sides of a triangular field are 500, 600, and 700 links respectively. Find its area.

$$\tfrac{1}{2}(500 + 600 + 700) = 900,$$

$$\text{area} = \sqrt{900 \times 400 \times 300 \times 200} = 146969 \text{ sq. links}$$

$$= 1 \text{ ac. } 1 \text{ r. } 35\cdot15 \text{ poles.}$$

The area of any rectilineal figure is the sum of the areas of all the parts into which the figure can be divided. Usually the most convenient method is to divide the figure into a number of triangles, then, as the area of each triangle can be found, the sum of the areas will give the area of the figure.

EXERCISES. XXXVI.

1. The base of a triangle is 4·9 ft. and the height 2·525 ft. Find its area?

2. Find the area of a triangle in which the sides are 13, 14, and 15 ft. respectively.

3. Find the area of a triangle sides 21, 20, and 29 in. respectively.

4. Make an equilateral triangle on a base 3 in. long and construct a parallelogram equal to it in area.

5. On a base of 10 yards a right-angled triangle is formed with one side two yards longer than the other. Find its area.

6. The sides of a triangle are 101·5, 80·5, and 59·4. Find the area.

7. The span of a roof is 40 feet, the rise 15 feet. Find the total area covered by slating if the length of the roof is 60 ft.

8. The sides of a triangular field are 300, 400, and 500 yds. If a belt 50 yds. wide is cut off the field, what are the sides of the interior triangle, and what is the area of the belt?

9. Find the area of a triangle, the sides being 15, 36, and 39 ft. respectively.

10. The sides of a triangle are 1·75, 1·03, and 1·11 ft. respectively. Find the area.

11. Find the area of a triangle whose sides are 25, 20, and 15 chains respectively.

12. In a triangle ABC the angle C is 53°, the sides AC and AB are ·523 and ·942 mile respectively. Find the side CB and the area of the triangle.

13. The sides of a triangle have lengths a, b, c inches. State (1) which of the following relations *are true* for all triangles, (2) which *untrue* for all triangles, (3) which are true for some triangles and untrue for others : ($a > b$ denotes a is greater than b ; $a < b$, that a is less than b) :

$$a > b, \quad a = b, \quad a < b ; \quad a + b > c, \quad a + b = c, \quad a + b < c.$$
$$a^2 + b^2 > c^2, \quad a^2 + b^2 = c^2, \quad a^2 + b^2 < c^2.$$

Circumference of a circle.—The number of times that the length of the circumference of a circle contains the length of the diameter of a circle cannot be expressed exactly, but it is very nearly 3·14159265. The number 3·1416 is used for convenience, and is sufficiently exact for nearly all purposes. This number is denoted by the Greek letter π.

An approximate value of π, sufficiently exact for all practical purposes, and very convenient when four-figure logarithms are used, is $3\frac{1}{7}$ or 3·142. Thus, $\pi = 3\cdot14159265 = \frac{22}{7}$ within $\frac{1}{20}$ per cent.

That the length of the circumference of a circle is πd or $2\pi r$, where d is the length of the diameter and r the radius, may be shown in several ways. Two simple experimental methods will be sufficient in this place.

1. *By rolling a disc of metal or wood on any convenient scale.* Make a mark on the circumference of the disc. Put the mark coincident with a scale division. Slowly roll the disc along the scale until the mark is again coincident with the scale, and note carefully the distance in scale divisions moved through. Then by applying the scale to the disc obtain the diameter.

Simple division will then show that the length of the circumference is $3\frac{1}{7}$ times that of the diameter.

2. Or, *wrap a piece of thin paper round the disc*, and mark, by two points, the line along which the edges overlap; unroll the paper, and its length when measured will be found to be $3\frac{1}{7}$ times that of the diameter.

To obtain a good average value the mean of several readings should be taken.

EXERCISES. XXXVII.

1. Find the diameters of circles, the circumferences in inches being 1·57, 2·3562, 4·712, 11·78, 17·28, 128·02.

2. Find the circumferences of circles, the diameters of which are 1·75, 2·5, 4·75, 8, 30·5, 67·5.

3. The minute hand of a clock is 6 ft. long. What distance will its extremity move over in 36 minutes?

4. A carriage wheel is 2 ft. $7\frac{1}{2}$ in. diameter. How many turns does it make in a distance of 7 miles 1332 yards?

5. The circumference of a wheel is 20 ft. How many turns will it make in rolling over 100 miles? Find the diameter of the wheel.

6. A rope is wrapped on a roller 1 ft. diameter. How many coils will be required to reach to the bottom of a well 200 ft. deep? What number of coils will be required if the rope is 1 inch thick?

7. The wheel of a locomotive 5 ft. in diameter made 10,000 revolutions in a distance of 24 miles. What distance was lost due to the slipping of the wheels?

8. How many revolutions per minute would a wheel 56 in. diameter have to make in order to travel at 30 miles an hour?

9. The circumferences of two wheels differ by a foot, and one turns as often in going 6 furlongs as the other in going 7 furlongs. Find the diameter of each wheel.

10. The hind and front wheels of a carriage have circumferences 14 and 16 ft. respectively. How far has the carriage advanced when the smaller wheel has made 51 revolutions more than the larger one?

Area of a circle.—If a regular polygon be inscribed in a circle, a series of triangles are formed by joining the angular points of the polygon to the centre of the circle.

The area of each little triangle is one-half the product of its base and the perpendicular let fall from the centre of the circle on the base of the triangle.

The length of the base may be denoted by a; the length of the perpendicular by p; and the radius of the circle by r. The area of the polygon will be $\frac{1}{2}p(a+a+\ldots)$ or $\frac{1}{2}p\Sigma a$. The

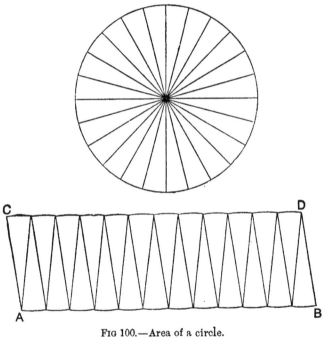

Fig 100.—Area of a circle.

symbol Σ, which denotes "the sum of," is very convenient, and the expression $\frac{1}{2}p\Sigma a$ simply means the product of $\frac{1}{2}p$, and the sum of all the terms each of which is represented by the letter a.

As the number of sides in the polygon is increased, its area becomes nearer and nearer that of the circle, and when the number of sides is indefinitely increased, the perimeter (or sum of the sides) of the polygon becomes equal to the circumference of the circle $=2\pi r$; the perpendicular referred to above also becomes the radius of the circle..

Hence the **area of a circle** $=\frac{1}{2}(2\pi r \times r)=\pi r^2$.

If d is the diameter of a circle, then as $d=2r$ the formula πr^2 becomes $\frac{\pi}{4}d^2$.

By dividing a circle into a large number of sectors, the bases may be made to differ as little as possible from straight lines. Each of the sectors forming the lower half of the circumference could be placed along a horizontal line AB (Fig. 100). A corresponding number of sectors from the upper half of the circumference could be placed along the upper line CD, completing the parallelogram $ABCD$. The length of the base AB will then be half the length of the circumference of the circle and the height of the parallelogram is equal to the radius of the circle, r.

$\therefore$ Area of circle $=AB \times r=r \times \pi r=\pi r^2$.

If a thin circular disc of wood be divided into narrow sectors, and a strip of tape glued to the circumference, then when the tape is straightened the sectors will stand upon it as a series of triangles. By cutting the tape in halves the two portions may be fitted together as in Fig. 100.

Area of sector of a circle.—*To find the area of the sector AE* (Fig. 101) *the angle θ being known.*

As the whole circle consists of 360 degrees, or 2π radians, the area of the sector will be the same fractional part of the whole area that the angle θ is of 360°, or of 2π.

Denoting the angle in degrees by N, then

area of sector $=\dfrac{N}{360}\pi r^2=\dfrac{\theta}{2\pi}\pi r^2=\dfrac{\theta r^2}{2}$.

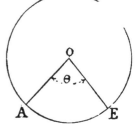

FIG. 101.—Area of sector of a circle.

Thus the area of a sector is given by half the product of the angle and the radius squared.

Ex. 1. Find the area of the sector of a circle containing an angle of 42°, the radius of the circle being 15 feet;

area of circle $=\pi \times 15^2$;

area of sector $=\frac{42}{360}\pi \times 15^2=82\cdot47$ square feet.

Ex. 2. The length of the diameter of a circle is 25 feet. Find the area of a sector in the circle, the length of the arc being 13·09 feet.

The area of the sector will be the same fraction of the whole area that 13·09 is of the circumference ;

$$\therefore \frac{13 \cdot 09}{25 \times \pi} \times \frac{\pi}{4} \times 25^2 = 81 \cdot 79 \text{ square feet.}$$

In a similar manner the length of an arc subtending a given angle θ can be obtained from the relation : length of an arc is the same fractional part of the whole circumference that θ is of 360° or 2π, or if r is radius of circle

$$\therefore \frac{\text{length of arc}}{2\pi r} = \frac{\theta}{360°}$$

Area of an annulus.—If R (Fig. 102) denote the radius of the outer circle, and r the radius of the inner, the area of the annulus is the difference of the two areas ;

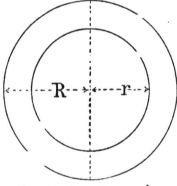

$$= \pi R^2 - \pi r^2$$

$$= \pi(R^2 - r^2) = \pi(R+r)(R-r) ;$$

$\therefore$ **Multiply the sum and difference of the two radii by 3⅟ to obtain the area of an annulus.**

Segment of a circle.—Any chord, not a diameter, such as AB (Fig. 103), divides the circle into two parts, one greater than, and the other less than, a semi-

FIG. 102.—Area of an annulus.

circle. If C is the centre of the circle of which the given arc ADB forms a part, then the area of the segment ADB is equal to the difference between the area of the sector $CADB$ and the triangle ABC.

Length of arc ADB.—To find the length of the arc ADB (Fig. 103), we may proceed to find the centre of the circle of which ADB is a part. Then by joining A and B to C, the angle subtended by the given arc is known and its length can be obtained.

To find the area enclosed by the arc and the chord AB we can find the area of the sector $CADB$ and subtract the area of the triangle *ABC* from it; this gives the area required. To avoid the construction necessary in the preceding cases several

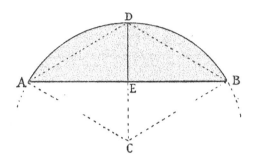

Fig. 103.—Segment of a circle.

approximate rules have been devised. Of these the following give fairly good results :

$$\text{Length of arc } ADB = \frac{8 \cdot AD - AB}{3},$$

or in words, **subtract the chord of the arc from eight times the chord of half the arc and divide the remainder by 3.**

Area of segment.—The area of the segment may be obtained from the rule,

$$\text{area} = \frac{h^3}{2c} + \frac{2}{3}ch,$$

where c denotes the chord AB and h denotes the height ED.

EXERCISES. XXXVIII.

1. Find the diameter of a circle containing 3217 sq. in.
2. The diameter of a circle is 69·75 in. Find its area.
3. The circumference of a circle is 247 in. Find its area.
4. Find the diameter of a circle when the area in square inches is (i) ·7, (ii) ·0000126, (iii) ·00031, (iv) ·0314.

5. Find the area of a circle when its diameter in inches is (i) ·064, (ii) ·109, (iii) 3·3.

6. A pond 25 feet diameter is surrounded by a path 5 feet wide. Find the cost of making the path at 1s. 1½d. per square yard.

7. The perimeter of a circle is the same as that of a triangle the sides of which are 13, 14, and 15 ft. Find the area of the circle.

8. If the two perpendicular sides of a right-angled triangle are 70 and 98 ft. respectively ; find the area of a circle described on the hypothenuse as a diameter.

9. Find the area of the annulus enclosed between two circles, the outer 9 in. and the inner 8 in. diameter.

· **10.** The inner and outer diameters of an annulus are 9½ and 10 in. respectively. Find the area.

11. The area of a piston is 5944·7 square inches. What is the diameter of the air-pump which is one-half that of the piston ?

12. A sector contains 42°, the radius of the circle is 15 ft. Find the area of the sector.

13. The length of the arc of a sector of a given circle is 16 ft. and the angle ⅛ of a right angle. Find the area of the sector ; find also the length of the arc subtending the same angle in a circle whose radius is four times that of the given circle.

14. The diameter of a circle is 5 ft. Find the area of a sector which contains 18°.

15. Find the area of the sector of the end of a boiler supported by a gusset-stay, the radius of the boiler being 42 inches, length of arc 25 inches.

16. A sector of a circle contains 270°. Find its area when the radius of the circle is 25 ft.

· **17.** In an arc of a circle the chord of the arc is 30 ft. and the chord of half the arc 25 ft. Find the length of the arc.

18. The circular arch of a bridge is 50 feet long and the chord of half the arc is 26·9 ft. Find the length of the chord or "span."

19. The length of a circular arc is 136 ft. and the chord of half the arc is 75·5 ft. Find the length of the chord.

20. Find the area of a segment in which the chord is 30 ft. and height 5 ft.

21. Find the area of a segment of a circle when the chord is 120 ft. and height 25 ft.

CHAPTER XX.

AREA OF AN IRREGULAR FIGURE. SIMPSON'S RULE. PLANIMETER.

Area of an irregular figure.—When the periphery of an irregular figure *ABCDE* (Fig. 104) consists of a series of straight lines, the area may be obtained by dividing the figure into a number of triangles, and the area of each triangle may be obtained separately. The sum of the areas of all the triangles into which the figure has been divided will give the area of the figure.

When the ordinates of an irregular figure, in which one or more of the boundaries may consist of curved lines, are given, the area may be obtained by drawing the figure on squared paper and counting the squares enclosed by the periphery. In this method there will

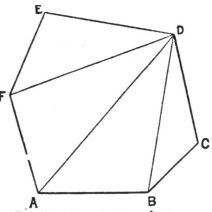

FIG. 104.—Area of an irregular figure.

usually be a number of complete squares enclosed by the periphery and a number cut by it. To estimate the value of any square cut by the outline it is convenient to neglect any square obviously less than one-half and to reckon as a whole square any one cut which is equal to, or greater than, one-half.

One defect of this method is that large errors are likely to occur when portions of the periphery are nearly parallel to the lines of ruling.

To avoid the errors likely to be introduced in the preceding method, other methods depending upon calculation are preferable.

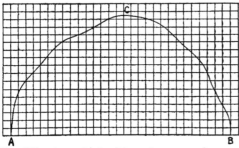

FIG. 105.—Area obtained by using squared paper.

Of these the two in general use are known as the **Mid-ordinate Rule** and **Simpson's Rule**. The latter is usually the more accurate of the two.

Mid-ordinate rule. —A common method of estimating the area of an irregular figure, such as *GFED* (Fig. 106), in which one of the boundaries is a curved line, is to divide the base *GF* into a number of equal parts, and at the centre of each of the equal parts

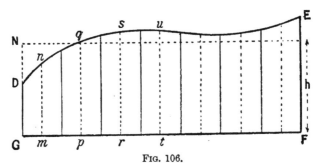

FIG. 106.

to erect ordinates. The length of each ordinate, *mn*, *pq*, *rs*, etc., from the base *GF* to the point where the vertical cuts the curve, is carefully measured, and all these ordinates are added together. The sum so obtained, divided by the number of ordinates, gives approximately the mean height, *h*, or mean ordinate, *GN*.

A convenient method of adding the ordinates is to mark them on a slip of paper, adding one to the end of the other until the total length is obtained.

The degree of approximation depends upon the number of ordinates taken. The approximation more closely approaches the actual value the greater the number of ordinates used.

The product of the mean ordinate and the base is the area required. For comparatively small diagrams, such as an indicator diagram (Fig. 107), ten strips are usually taken. This

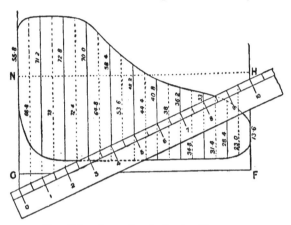

FIG. 107.—Area of an indicator diagram.

number is sufficiently large to give a fair average, and, moreover, dividing by 10 can be effected by merely shifting the decimal point.

The length GF (Fig. 107) may correspond on a reduced scale to the travel of the piston in a cylinder, and the ordinates of the curve represent, to a known scale, the pressure per square inch of the steam in the cylinder at the various points of the stroke.

Hence, the mean height GN indicates the mean pressure P of the steam, in pounds per sq. inch, throughout the stroke (the stroke being the term applied to the distance moved through by the piston in moving from its extreme position at one end of the cylinder to a corresponding position at the other end).

If A denote the area of the piston in square inches, then the total force exerted by the steam on the piston is $P \times A$, and the work done by this force in acting through a length of stroke L is $P \times A \times L$. If N denote the number of strokes per minute, the work done per minute by the steam $= PALN$.

But the unit of power used by engineers, and called a Horse-power, is 33000 ft. lbs. per minute.

Hence Horse-power of the engine $=\dfrac{P \times L \times A \times N}{33000}$.

Ex. 1. In Fig. 107 the indicator card of an engine is shown; the diameter of the piston is $23\frac{1}{8}$ inches, length of stroke 3 ft., and revolutions 100 per minute. Find the mean pressure of the steam, also the horse-power of the engine.

Adding together the ten ordinates shown by dotted lines, we have

$$66\cdot6 + 73\cdot0 + 72\cdot4 + 64\cdot8 + 53\cdot6 + 44\cdot4 + 38\cdot0 + 34\cdot8 + 31\cdot4 + 23\cdot0$$
$$= 502.$$

As there are 10 ordinates,

$$\therefore \text{ mean pressure} = \frac{502}{10}$$
$$= 50\cdot2 \text{ lbs. per sq. inch.}$$

Area of piston $= 420$ sq. inches;
number of strokes per minute $= 200$.

$$\therefore \text{ Horse-power} = \frac{50\cdot18 \times 3 \times 420 \times 200}{33000} = 383\cdot2.$$

Simpson's rule.—By means of what is called Simpson's rule the area of an irregular figure $GFED$ (Fig. 108) can usually be ascertained more accurately than by the mid-ordinate rule.

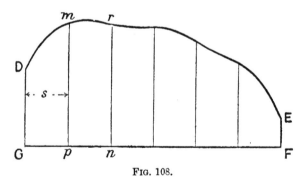

Fig. 108.

The base GF is divided into a number of equal parts. This ensures that the number of ordinates is an odd number, 3, 5, 7, 9, etc. In Fig. 108 the base GF is divided into 6 equal parts, and the number of ordinates is therefore 7.

Denoting, as before, the lengths of the ordinates GD, pm, nr, etc., by h_1, h_2, $h_3 \ldots h_7$; then, if s denotes the common distance or, space between the ordinates, we have

$$\text{Area of } GFED = \frac{s}{3}\{h_1 + h_7 + 4(h_2 + h_4 + h_6) + 2(h_3 + h_5)\}$$

$$= \frac{s}{3}(A + 4B + 2C),$$

where A denotes the sum of the first and last ordinates.

,, B ,,   ,,   even ordinates.

,, C ,,   ,,   odd ordinates.

∴ Add together the extreme ordinates, four times the sum of the even ordinates, and twice the sum of the odd ordinates (omitting the first and the last). Multiply the result by one-third the common interval between two consecutive ordinates.

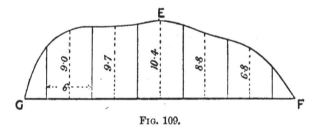

FIG. 109.

The end ordinates at G and F may both be zero, the curve commencing from the line GF (Fig. 109). In this case A is zero, and the formula for the area becomes

$$\frac{s}{3}(0 + 4B + 2C).$$

Or, using the given values in Fig. 109, where the length of the ordinates are expressed in feet, we have

$$\text{Area} = \frac{6}{3}\{0 + 4(9 \cdot 0 + 10 \cdot 4 + 6 \cdot 8) + 2(9 \cdot 7 + 8 \cdot 8)\}$$

$$= 2(104 \cdot 8 + 37) = 283 \cdot 6 \text{ sq. ft.}$$

Comparison of methods.—It will be found a good exercise to compare the various methods of obtaining the area of a plane figure by using them to obtain the area of a figure such as a quadrant of a circle.

Ex. 2. Draw a quadrant of a circle of 4 in. radius and divide the figure into eight strips each ½ in. wide. Measure all the ordinates (including the mid-ordinate) and find the area by (i) the *mid-ordinate* rule, (ii) by *Simpson's* Rule, (iii) the *ordinary* rule $\dfrac{\pi r^2}{4}$.

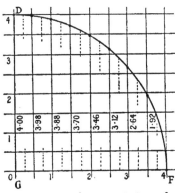

FIG. 110.—Area of quadrant of a circle.

Compare the results and find the percentage errors. Find also the mean ordinate in each case.

In Fig. 110 a quadrant of a circle is shown in which the two radii GF and GD are horizontal and vertical. Divide the base into eight equal parts; then, if the figure is drawn on squared paper, the lengths of the ordinates and also the mid-ordinate can be read off and marked as in Fig. 110. As the distance between each ordinate is $\frac{1}{2}$ in., we have by Simpson's Rule

$$\text{Area } FGD = \frac{\cdot 5}{3}\{0 + 4 + 4(3\cdot98 + 3\cdot7 + 3\cdot12 + 1\cdot92) + 2(3\cdot88 + 3\cdot46 + 2\cdot64)\}$$

$$= 12\cdot61 \text{ sq. in.}$$

Using the formula $\dfrac{\pi r^2}{4}$, area $= 12\cdot57$ when $\pi = \dfrac{22}{7}$,

area $= 12\cdot575$ sq. in. when $\pi = 3\cdot1416$.

Accepting the latter as the more accurate value the difference is $12\cdot61 - 12\cdot575 = \cdot035$;

$\therefore$ Percentage error is $\dfrac{\cdot035 \times 100}{12\cdot57} = \cdot3 \%$.

The mean ordinate is $\dfrac{12\cdot6}{4} = 3\cdot15$ in. or $\dfrac{12\cdot57}{4} = 3\cdot14$ in.

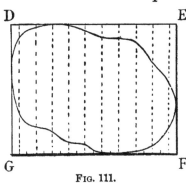

FIG. 111.

In a similar manner the area and percentage error by using the mid-ordinate rule can be obtained.

When the area is *not* symmetrical about a line, and its boundary is an irregular curve, lines are drawn touching the curve; two of these, FG and ED (Fig. 111) may be made parallel to each other and GD, FE drawn perpendicular to the former.

As before, the base GF is divided into a number of equal

parts and the ordinates of the curve measured ; from these values, proceeding as before, the area can be obtained.

As the area of an irregular figure is the product of the length of the base GF and the mean ordinate, it follows that when the area is obtained, the mean ordinate may be found by dividing the area by the length of the base. Thus in Fig. 107, p. 231, where GF is 6 inches, the division into 10 equal parts will give the common distance between each ordinate to be ·6 inch. On p. 232 a rough result for the mean ordinate has been obtained. A more accurate result can be found by Simpson's rule, as follows :

$$\text{Extreme ordinates} = 55\text{·}8 + 13\text{·}6 = 69\text{·}4,$$

$$\text{Even ordinates} = 71\text{·}2 + 70 + 48\text{·}2 + 36\text{·}2 + 28\text{·}4 = 254,$$

$$\text{Odd ordinates} = 72\text{·}8 + 58\text{·}4 + 40\text{·}8 + 33 = 205,$$

$$\therefore \ \text{Area of figure} = \frac{\text{·}6}{3}(69\text{·}4 + 4 \times 254 + 2 \times 205)$$

$$= 299\text{·}08 \text{ sq. in.}$$

$$\therefore \ \text{Mean ordinate} = \frac{299\text{·}08}{6} = 49\text{·}85 \text{ in.}$$

In the preceding examples the given ordinates are equidistant ; when this is not the case, points corresponding to the given ordinates can be plotted on squared paper and a fair curve drawn through the plotted points. The area enclosed by the curve the two end ordinates, and the base, is the area required. This value may be obtained by counting the enclosed squares ; or, better, by dividing the base into an even number of parts and reading off the values of the ordinates at each point of division. The area may then be obtained either by the application of *Simpson's* or *Mid-ordinate rule.*

Ex. 4. The following table gives the values of the ordinates of a curve and their distances from one end. Find the mean ordinate and the area enclosed by the curve.

Distances in feet.	0	2·3	4·5	7·0	12·2	18·0	24·0	30·0
Ordinates.	7	6·3	5·89	5·48	4·67	3·96	3·39	2·9

Plotting the given values on squared paper as in Fig. 112, we obtain a series of points through which a fair curve is drawn. Next, dividing the base into six equal parts, seven ordinates are drawn. The values of these ordinates are shown in Fig. 112.

By Simpson's Rule, as the common distance is 5 feet,

$$\text{Area} = \frac{5}{3}\Big\{ 7 + 2\cdot 9 + 4(5\cdot 8 + 4\cdot 3 + 3\cdot 3) + 2(5\cdot 0 + 3\cdot 75) \Big\} = 135 \text{ sq. ft.}$$

Mean ordinate × 30 = 135 ;

$$\therefore \text{ Mean ordinate} = \frac{135}{30} = 4\cdot 5 \text{ ft.}$$

Other methods of finding the area of an irregular figure, instead of those which have now been studied, are by means of weighing, and by using a planimeter.

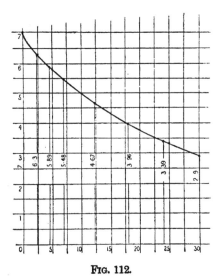

By weighing.—Draw the figure to some convenient scale, or, if possible, full size, on thick paper or cardboard of uniform thickness. Cut it out carefully. Also cut out a rectangular piece from the same sheet; find the weight of the rectangular piece, and hence deduce the weight of a square inch. Then, knowing the weight of the irregular figure and the weight of unit area, the area of the figure can be calculated.

Fig. 112.

Planimeter.—The planimeter is an instrument for estimating the areas of irregular figures. There are many forms of the instrument to which various names—Hatchet, Amsler, etc.—are applied. Of these the more expensive and accurate forms are mostly modifications of the *Amsler planimeter.*

Hatchet planimeter.—A hatchet planimeter in its simplest form may consist of a ∩-shaped piece of metal wire (Fig. 113), one end terminating in a round point, the other in a knife edge. This knife edge is rounded or hatchet-shaped, the distance between the centre of the edge K and the point T may be made

5, 10, or some such convenient number. This length may be denoted by TK.

To determine the area of a figure we proceed as follows :

(*a*) Estimate approximately the centre of area, and through this point draw a straight line across the figure.

(*b*) Set the instrument so that it is roughly at right angles to this line, with the point T at the centre of gravity. When in this position a mark is made on the paper by a knife edge K.

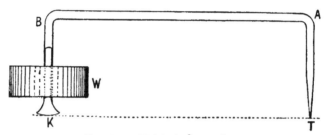

Fig. 113.—Hatchet planimeter.

Holding the instrument in a vertical position, the point T is made to pass from the centre to some point in the periphery of the figure, and then to trace once round the outline of the figure until the point is again reached, thence to the centre again. In this position a mark is again made with the edge K. The distance between the two marks is measured, the product of this length and the constant length TK gives approximately the area of the figure.

To obtain the result more accurately, it is advisable when the point T (after tracing the outline of the figure) arrives at the centre to turn the figure on point T as a pivot through about 180°, and trace the periphery as before, but in the opposite direction. This should, with care, bring the edge K either to the first mark or near to it. The nearness of these marks depends to some extent on the accuracy with which the centre of area has been estimated.

The area of the figure is the product of TK, and the mean distance between the first and third marks.

To prevent the knife edge K from slipping, a small weight W (Fig. 113) is usually threaded on to the arm BK ; the portion of the arm on which the weight is placed is flattened to receive it.

The arm BA is usually adjustable, and this enables the instrument to be used, not only for small, but also for comparatively large diagrams.

Amsler planimeter.—One form of the instrument is shown in Fig. 114 and consists of two arms A and C, pivoted together at a point B. The arm BA is fixed at some convenient point s. The other arm BC carries a tracing point T. This is passed round the outline of the figure, the area of which is required.

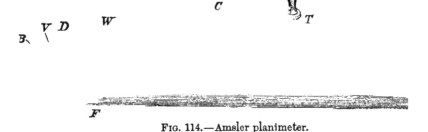

FIG. 114.—Amsler planimeter.

The arm BC carries a wheel D, the rim of which is usually divided into 100 equal parts.

When the instrument is in use the rim of the wheel rests on the paper, and as the point T is carried round the outline of the figure, the wheel, by means of a spindle rotating on pivots at a and b, gives motion to a small worm F, which in turn rotates the dial W.

One rotation of the wheel corresponds to one-tenth of a revolution of the dial. A vernier, V, is fixed to the frame of the instrument, and a distance equal to 9 scale divisions on the rim of the wheel is divided into ten on this vernier. The readings on the dial are indicated by means of a small finger or pointer shown in Fig. 114. If the figures on the dial indicate units, those on the wheel will be $\frac{1}{10}$ths; as each of these is subdivided into 10, the subdivisions indicate $\frac{1}{100}$ths. Finally, the vernier, V, in which $\frac{9}{100}$ of the wheel is divided into 10 parts, enable a reading to be made to three places of decimals.

To obtain the area of a figure, the fixed point s may be set at some convenient point outside the area to be measured, and the point T at some point in the periphery of the figure. Note the reading of the dial and wheel. Carefully follow the outline of the figure until the tracing point T again reaches the starting-point, and again take the reading. The difference between the two readings multiplied by a constant will give the area of the figure, the value of the constant may be found by using the instrument to obtain a known area, such as a square, a rectangle, etc.

EXERCISES. XXXIX.

1. Find the area of a quadrant of a circle of 5 inches radius by ordinary rule and by Simpson's Rule. Find the percentage error.

2. The transverse sections of a vessel are 15 feet apart and their areas in square feet up to the load water line are 4·8, 39·4, 105·4, 159·1, 183·5, 173·3, 127·4, 57·2, and 6·0 respectively. Find the volume of water displaced by the ship between the two end sections given above.

3. The half ordinates of an irregular piece of steel plate of uniform thickness, and weighing 4 lbs. per sq. ft., are 0, 1·5, 2·5, 3, 5, 6·75, 7·25, 9, 8·75, 7, 6, 5·25, 3·5, 2, and 0 ft. respectively, the common distance between the ordinates is 5 ft. Find the weight.

4. The ordinates of an irregular piece of land are 3·5, 4·75, 5·25, 7·5, 8·25, 14·75, 6, 9·5, and 4 yards respectively, the common interval is $1\frac{1}{4}$ yds. Find the area in square yards.

5. The equidistant ordinates of an irregular piece of sheet lead weighing 6 lbs. per sq. ft. are respectively 2, 4, 9, 5, and 3 ft., the length of the base is 8 ft., find the weight.

6. The ordinates of a curve and the distances from one end are given in the following table. Find the area and the mean ordinate.

Distances from one end (in inches).	0	20	35	56	72	95	110	140	156
Ordinates.	405	380	362	340	325	304	287	260	252

7. The girth or circumference of a tree at five equidistant places being 9·43, 7·92, 6·15, 4·74, and 3·16 ft. respectively, the length is $17\frac{1}{4}$ ft. Find the volume, using the mid-ordinate and Simpson's Rule.

8. Find the area of a curved figure when the distances and ordinates both in feet are as follows :

Distances from one end.	0	4	14	26	32
Ordinates.	20	16·5	12	8	7·5

9. Find the area of a half of a ship's water plane of which the curved form is defined by the following equidistant ordinates spaced 12 feet apart :

·1, 5·1, 7·17, 8·75, 10·1, 9·17, 8·05, 6·4, ·1 feet.

10. The half-ordinates of the load water-plane of a vessel are 13 ft. apart, and their lengths are ·4, 3·3, 6·9, 10·5, 13·8, 16·3, 18·3, 19·5, 19·9, 20·0, 19·6, 19·0, 17·8, 15·7, 11·8, 6·0, and ·8 feet respectively.   Calculate the area of the plane.

11. The load water-plane of a ship is 240 ft. long, and its half-ordinates, 20 ft. apart, are of the following lengths ·2, 8·0, 12·4, 14·4, 15·6, 16·0, 16·0, 15·6, 14·2, 12·0, 9·2, 5·0 and ·2 feet.   What is the total area of the water plane ?

12. A river channel is 60 ft. wide, the depth (*y*) of the water at distances *x* ft. from one bank are given in the following table. Find the area of a cross-section and the average depth of the water.

x	0	10	20	30	40	50	60
y	5	7	15	21	30	16	6

13. The work done by force is the product of the force and its displacement in the direction of the force, hence show that the work done by a variable force through a distance AB can be represented graphically by an area.   If the distance AB be divided into two equal parts at C, and the magnitudes of the force at A, B, and C are P, Q, R respectively, show that the work is $AB(P+4R+Q)\div 6$. Given P, Q, and R to be 50, 28, and 24 lbs. respectively, AB to be 12 ft., show that the work done is 372 ft. lbs.

CHAPTER XXI.

MENSURATION. VOLUME AND SURFACE OF A PRISM, CYLINDER, CONE, SPHERE, AND ANCHOR RING. AVERAGE CROSS SECTION AND VOLUME OF AN IRREGULAR SOLID.

A solid figure or **solid** has the three dimensions of length, breadth, and thickness. When the surfaces bounding a solid are plane, they are called faces, and the edges of the solid are the lines of intersection of the planes forming its faces.

What are called the regular solids are five in number, viz., the *cube, tetrahedron, octahedron, dodecahedron,* and *icosahedron.*

The *cube* is a solid having six equal square faces.

The *tetrahedron* has four equal faces, all equilateral triangles.

The *octahedron* has eight faces, all equilateral triangles.

The *dodecahedron* has twelve faces, all pentagons.

The *icosahedron* has twenty faces, all equilateral triangles.

Cylinder.—If a rectangle *ABCD* (Fig. 115) be made to revolve about one side *AB*, as an axis, it will trace out a right cylinder. Thus, a door rotating on its hinges describes a portion of a cylinder. Or, a cylinder is traced by a straight line always moving parallel to itself round the boundary of a curve, called the *guiding curve.*

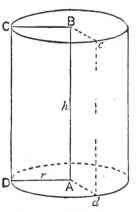

Fig. 115—Cylinder.

Pyramid.—If one end of the line *AB* always passes through a fixed point, and the other end be made to move round the boundary of a curve, a *pyramid* is traced out.

Cone.—If the curve be a circle and the fixed point is in the line passing through the centre of the circle, and at right angles to its plane, a *right cone* is obtained (Fig. 116); the fixed point is called the *vertex* of the cone; an *oblique cone* results when the fixed point is not in a line at right angles to the plane of the base.

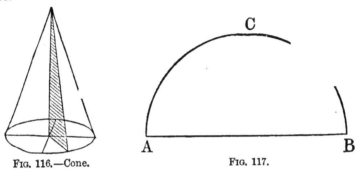

FIG. 116.—Cone. FIG. 117.

Sphere.—If a semicircle ACB (Fig. 117) revolve about a diameter AB, the surface generated is a *sphere*.

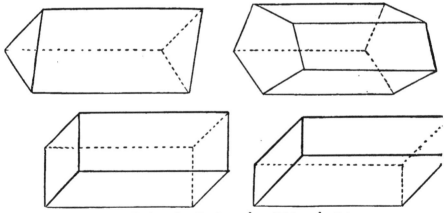

FIG. 118.—Rectangular, Pentagonal, and Triangular Prisms.

Prism.—When the line remains parallel to itself and is made to pass round the boundary of any rectilinear polygon, the solid formed is called a *prism*.

The ends of a prism and the base of a pyramid may be polygons of any number of sides, *i.e.* triangular, rectangular, pentagonal, etc.

A prism is called *rectangular, square, pentagonal, triangular, hexagonal,* etc. (Fig. 118), according as the end or base is one or other of these polygons.

A prism which has six faces all parallelograms is also called a *parallelopiped.*

A right or **rectangular prism** has its side faces perpendicular to its ends. Other prisms are called **oblique.**

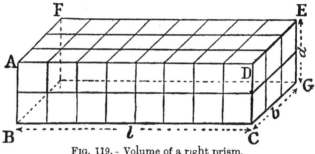

FIG. 119. - Volume of a right prism.

In Fig. 119 a right prism, the ends of which are rectangles, is shown; to find its volume, sometimes called the *content,* or *solidity,* it is necessary to find the area of one end *DCGE,* and multiply it by the length *BC.* Let $v, l, b,$ and d denote the volume, length, breadth, and depth or altitude of the right prism respectively.

Then area of one end $= b \times d$.

And volume of prism $= b \times d \times l$.

As $b \times l =$ area of base; volume = **area of base** × **altitude.**

As $v = bdl$ it follows that if any three of the four terms be given the remaining one can be obtained.

When the volume is obtained, the weight can be found by multiplying the volume by the weight of unit volume.

Ex. 1. The length of a rectangular wrought iron slab is 8 ft., its depth is 3 ft. Find its breadth if its weight is 23040 lbs. (one cubic foot weighs 480 lbs.).

Here volume $= 8 \times 3 \times b,$

Weight $= 23040 = 8 \times 3 \times b \times 480;$

$$\therefore \ b = \frac{23040}{8 \times 3 \times 480} = 3 \text{ ft.}$$

In Fig. 119, the length BC is divided into 8 equal parts, the breadth into 3, and the depth into 2. It is seen easily that there are 6 square units in the end $DCGE$ of the slab, and these are faces of a row of six unit cubes. There are 8 such rows; hence the volume is $8 \times 6 = 48$ cub. ft.

Total surface of a right prism.—The total surface is, from Fig. 119, seen to be twice the area of the face $ABCD$, and twice the area of $ADEF$, together with the area of the two ends :

$$\therefore \ \text{Surface} = 2\,(ld + bl + bd) \ ;$$

or, the **total surface of a right prism is equal to the perimeter of base multiplied by altitude together with areas of the two ends.**

Ex. 2. The internal dimensions of a box without the lid are: length 8 ft., breadth 3 ft., and depth 2 ft. Find the cost of lining it with zinc at 7d. per square foot.

$$\text{Area of base} = 8 \times 3 = 24 \text{ sq. ft.}$$
$$,, \quad \text{sides} = 2(8 \times 2) = 32 \text{ sq. ft.}$$
$$,, \quad \text{ends} = 2(3 \times 2) = 12 \quad ,,$$
$$\therefore \ \text{Total area} = 68 \text{ sq. ft.}$$
$$\therefore \ \text{Cost} = \frac{68 \times 7}{12} = \pounds 1. \ 19\text{s. } 8\text{d.}$$

EXERCISES.   XL.

1. A cistern (without a lid) 6 feet long and 3 feet broad when two-thirds full of water is found to contain 187·5 gallons. Find the depth of the cistern, also the cost of lining it with zinc at $2\frac{1}{2}d.$ per square foot.

2. If the inside edge of a cubical tank is 4 ft., find its volume; also find the number of gallons it will hold when full.

3. The internal dimensions of a rectangular tank are 4 ft. 4 in., 2 ft. 8 in., and 1 ft. $1\frac{1}{2}$ in. Find its volume in cubic feet, the number of gallons it will hold when full, and the weight of the water.

4. A cistern measures 7 ft. in length, 3 ft. 4 in. in width. What is the depth of the water when the tank contains 900 gallons?

.5. A tank is 4 metres long, ·75 metres wide, and 1 metre deep. Find the weight of water it will hold.

6. A metal cistern is 12 ft. long, 8 ft. wide, and 4 ft. deep external measurements. If the average thickness of the metal is $\frac{1}{4}$ in., find the number of gallons of water it will hold.

7. Three edges of a rectangular prism are 3, 2·52, and 1·523 ft. respectively. Find its volume in cubic feet. Find also the cubic space inside a box of the same external dimensions made of wood one-tenth of a foot in thickness.

8. A Dantzic oak plank is 24 ft. long and 3¾ in. thick. It is 7 in. wide at one end and tapers gradually to 5¾ in. at the other. Find its volume and weight, the specific gravity being ·93.

9. A Riga fir deck plank is 22 ft. long and 4 in. thick and tapers in width from 9 in. at one end to 6 in. at the other. If the specific gravity of the timber be ·53, find the volume and weight of the plank.

10. Find what weight of lead will be required to cover a roof 48 ft. long, 32 ft. wide, with lead $\frac{1}{12}$ in. thick, allowing 5 per cent. of weight for roll joints, etc.

11. A reservoir is 25 ft. 4 in. long, 6 ft. 4 in. wide. How many tons of water must be drawn off for the surface to fall 7 ft. 6. in. ?

12. If the surface of a cube be 491·306 square inches, what is the length of its edge ?

13. A cistern is 9 ft. 4 in. long and 7 ft. 6 in. wide and contains 6 tons 5 cwt. of water. Find the depth of the water in the cistern.

14. Find the volume of a rectangular prism 3 ft. 4 in. long, 2 ft. wide, and 10 in. deep. Find also the increase in its volume when each side is increased by 8 in.

15. The internal dimensions of a rectangular tank are : length 2 metres, depth ·75 metres, and width 1 metre. Find the weight of water it contains when full.

Cylinder.—It has been seen that the volume of a prism is equal to the area of the base multiplied by the length.

In the case of a cylinder the base is a circle.

If r denote the radius of the base and l the length of the cylinder (Fig. 120),

$$\text{Area of base} = \pi r^2 ; \quad \therefore \ \text{volume} = \pi r^2 \times l.$$

More accurately a cylinder of this kind in which the axis is perpendicular to the base should be called a *right cylinder*. This distinguishes it from an oblique cylinder in which the axis is not perpendicular, and from cylinders in which the base is not a circle. It is only necessary for practical purposes to consider a right cylinder.

Surface of a cylinder.—The surface of a cylinder consists of two parts, the curved surface and the two ends which are plane circles.

If the cylinder were covered by a piece of thin paper this when unrolled would form a rectangle of length l and base $2\pi r$. Thus, if the curved surface of a cylinder be conceived as unrolled and laid flat, it will form a rectangle of area $2\pi r \times l$ (Fig. 120).

$$\therefore \ \text{Curved surface of cylinder} = 2\pi r l.$$

To obtain the whole surface the areas of the two ends must be added to this.

∴ Total surface of cylinder $= 2\pi r l + 2\pi r^2$

$$= 2\pi r (l + r).$$

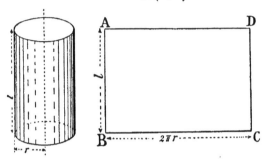

Fig. 120.—Surface of a cylinder.

In any problem in Mensuration it is advisable in all cases to express a rule to be employed as a formula. Thus, if V denote the volume and S the curved surface of a cylinder, then the preceding rules may be briefly written

$$V = \pi r^2 l \; ; \; S = 2\pi r l.$$

Ex. 1.   Find the volume, weight, and surface of a cast-iron cylinder, 18·5 inches diameter, 20 inches length.

Area of base $= \pi \times (9{\cdot}25)^2 = 268{\cdot}8$ sq. in.

Volume $= 268{\cdot}8 \times 20 = 5376$ cub. in.

Weight $= 5376 \times {\cdot}26 = 1397{\cdot}76$ lbs.

$$S = 2\pi \times \frac{18{\cdot}5}{2} \times 20 = 1162{\cdot}4 \text{ sq. in.}$$

Area of each end $= \dfrac{\pi \times (18{\cdot}5)^2}{4} = 268{\cdot}8$;

∴ Total surface $= 1699{\cdot}99$ or 1700 sq. in.

Ex. 2.   Find the effective heating surface of a boiler 6 ft. diameter, 18 ft. long, with 92 tubes $3\frac{1}{2}$ in. diameter, assuming the effective surface of the shell to be one-half the total surface.

Effective heating surface of shell $= \dfrac{6 \times \pi \times 18}{2} = 169{\cdot}6$ sq. ft.

Heating surface of 92 tubes $= \dfrac{3\frac{1}{2} \times \pi \times 18 \times 92}{12} = 1517{\cdot}4$ sq. ft. ;

∴ Effective surface $= 169{\cdot}6 + 1517{\cdot}4 = 1687$ sq. ft.

Cross-section.—The term cross-section should be clearly understood. A section of a right cylinder by any plane perpendicular to the axis of the cylinder is a circle ; any oblique section gives an ellipse. Hence, the term *area of cross-section* is used to indicate the area of a section at right angles to the axis.

Ex. 1. A piece of copper 4 inches long, 2 inches wide, and $\frac{1}{2}$ inch thick is drawn out into a wire of uniform thickness and 100 yards long. Find the diameter of the wire.

$$\text{Volume of copper} = 4 \times 2 \times \tfrac{1}{2} = 4 \text{ cubic inches.}$$
$$\text{Length of wire} = 100 \times 3 \times 12 = 3600 \text{ inches.}$$

Let d denote the diameter of the wire.

Then $\qquad\qquad$ volume of wire $= \dfrac{\pi}{4}d^2 \times 3600.$

Hence $\qquad\qquad \dfrac{\pi}{4}d^2 \times 3600 = 4 ;$

$$\therefore d^2 = \frac{4 \times 4}{\pi \times 3600} = \frac{1}{225 \times \pi} ;$$

$$\therefore d = \cdot 0376 \text{ inches.}$$

Ex. 2. A piece of round steel wire 12 inches long weighs 0·65 lbs. If its specific gravity is 7·8, find the area of cross-section, also the diameter of the wire.

Let a denote the area.

$\qquad$ Volume $= 12 \times a$ cubic inches.

Also from Table I., weight of a cubic inch of water $= \cdot 036$ lbs.

$\qquad$ Weight $= 12 \times a \times 7 \cdot 8 \times \cdot 036,$ but this is equal to ·65 lbs. ;

$\qquad \therefore 12a \times 7\cdot 8 \times \cdot 036 = \cdot 65 \cdot$

$$a = \frac{\cdot 65}{12 \times 7\cdot 8 \times \cdot 036} = \cdot 193 \text{ sq. in.}$$

$$\therefore \frac{\pi}{4}d^2 = \cdot 193.$$

Hence $\qquad\qquad d = \tfrac{1}{2}$ inch nearly.

Hollow cylinder.—The volume, V, is as before equal to area of base multiplied by the altitude.

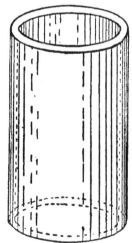

FIG. 121.—Hollow cylinder.

If R and r denote the radii of the outer and inner circles respectively, D and d the corresponding diameters (Fig. 121),

$$\text{Area of base} = \pi R^2 - \pi r^2 = \pi (R^2 - r^2),$$
$$\text{And volume} = \pi (R^2 - r^2)l,$$
$$\therefore \ V = {\cdot}7854(D^2 - d^2)l.$$

To use logarithms, it is better to write this as

$${\cdot}7854(D - d)(D + d)l.$$

Also $W = Vw$, where W represents the weight of the cylinder and w denotes the weight of unit volume of the material.

Ex. 1. The external diameter of a hollow steel shaft is 18 inches, its internal diameter 10 inches. Calculate the weight of the shaft if the length is 30 feet.

$$\text{Area of cross section} = {\cdot}7854(18^2 - 10^2)$$
$$= {\cdot}7854(18 + 10)(18 - 10)$$
$$= {\cdot}7854 \times 28 \times 8,$$
$$\text{volume} = {\cdot}7854 \times 28 \times 8 \times 30 \times 12 \text{ cubic inches,}$$
$$\text{weight} = \frac{{\cdot}7854 \times 28 \times 8 \times 360 \times {\cdot}29}{2240} \text{ tons}$$
$$= 8{\cdot}2 \text{ tons.}$$

EXERCISES. XLI.

1. Let V denote the volume and S the curved surface of a cylinder of radius r and length l.

(i) If $V = 150$ cub. in., $l = 6$ in., find r and S.

(ii) If $V = 100$ cub. in., $r = 3$ in., find l and S.

(iii) Given $r = 4$ in., $l = 10$ in., find V and S.

2. The volume of a cylinder is 1608·5 cub. ft., the height is 8 ft. Find its diameter.

3. The curved surface of a cylinder is 402·124 sq. ft. If the height be 8 ft., what is the radius of the base?

4. 260 feet of round copper wire weighs 3 lb.; find its diameter if a cubic inch of the copper weighs 0·32 lb. If the same weight of the copper is shaped like a hollow cylinder, 1 inch internal diameter and 2 inches long, what is its external diameter?

5. A hollow cylinder is 4·32 inches long; its external and internal diameters are 3·150 and 1·724 inches. Find its volume and the sum of the areas of its two curved surfaces.

6. A portion of a cylindrical steel stern shaft casing is $12\frac{3}{4}$ ft. long, $1\frac{1}{4}$ inches thick, and its external diameter is 14 inches. Find its weight.

7. What is the external curved surface and weight of a cast-iron pipe $1\frac{1}{2}$ ft. internal diameter, 48 ft. long, and $\frac{1}{2}$ in. thick?

8. The outer circumference of a cast-iron cylinder is 127·2 in., and length 3 ft. 6 in. If the weight is 686 lbs., find its internal diameter.

9. If a cube of stone whose edge is 9 in. is immersed in a cylinder of 12 in. diameter half full of water, how far will it raise the surface of the water in the cylinder?

10. Find the length of a coil of steel wire when the diameter is ·025 inch and its weight 49 lbs.

Cone.—Volume of cone $= \frac{1}{3}$ (*area of base × altitude*)

$$= \tfrac{1}{3}\pi r^2 \times h,$$

where $\qquad\qquad\qquad$ $r =$ radius of base

and $\qquad\qquad\qquad$ $h =$ altitude of cone.

Or, **the volume** of a cone **is one-third** that **of a cylinder on the same base and** the same **altitude**.

This result may be checked in a laboratory in many different ways. Thus, if a cone of brass and a cylinder of the same material, of equal heights, and with equal bases, be weighed, the weight of the cylinder will be found to be three times that of the cone.

Or, the cone and cylinder may both be immersed in a graduated glass vessel, and the height to which the water rises measured.

Or, if a cylindrical vessel of the same diameter and height as the cone is filled with water, it will be found, by inserting the cone point downwards, that one-third the water will be displaced by the cone, and will overflow.

Curved surface of a cone.—If the base of the cone be divided into a number of equal parts AB, BC, etc. (Fig. 122), then by joining A, B, C, etc., to the vertex V, the curved surface of the solid is divided into a number of triangles, VAB, VBC, etc.

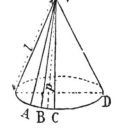

If a line be drawn perpendicular to BC, and passing through V; and its length be p, then

$\qquad$ Area of triangle $VBC = \frac{1}{2}(BC \times p)$.

If n denote the number of triangles into which the base is divided, and a the length BC, then

FIG. 122.—Cone.

$$\text{Curved surface} = \frac{n}{2} \times ap \text{ approximately}.$$

As the number of parts into which the base is divided is increased, the product na becomes more nearly equal to the circumference of the base ; and becomes equal to the circumference when the number of parts is indefinitely increased, also p becomes at the same time equal to l, the slant height.

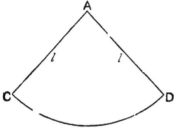

FIG. 123.—Development of a cone.

$\therefore$ Curved surface $=\frac{1}{2} 2\pi r l = \pi r l.$

Or, we may proceed as follows: Cut out a piece of thin paper to exactly cover the lateral surface of a cone. When opened out it will form a sector of a circle of radius l (Fig. 123).

The length of arc $CD=$ circumference of base of cone $=2\pi r.$

But, as we have seen on p. 225, the area of a sector is equal to half the arc multiplied by the radius.

$\therefore$ Curved surface $=\frac{1}{2}(CD \times l) = \frac{1}{2}(2\pi r \times l) = \pi r l,$
the curved surface of a cone equals half perimeter of base multiplied by the slant height.

Thus, if V denote the volume and S the curved surface of a cone, then $V=\frac{1}{3}\pi r^2 h$; $S=\pi r l.$

If h denote the height of the cone, then

$$l=\sqrt{(h^2+r^2)}.$$

Ex. 1. Find the volume and curved surface of a right cone, diameter of base 67 in., height 30 in.

$$\text{Area of base} = 67^2 \times \frac{\pi}{4} = 3525\cdot66 \text{ sq. in.}$$
$$\text{Volume of cone} = \frac{1}{3}(3525\cdot66 \times 30) = 35256\cdot6 \text{ cub. in.}$$
$$\text{Slant height} = \sqrt{33\cdot5^2 + 30^2} = 44\cdot98.$$
$$\therefore \text{ Surface} = \frac{1}{2}(\pi \times 67 \times 44\cdot98) = 4733\cdot85 \text{ sq. in.}$$

EXERCISES. XLII.

1. From the two formulae $V=\frac{1}{3}\pi r^2 h$ and $S=\pi r l$ the volume and curved surface of a right cone can be obtained.
    (i) Given $V=200$ cub. in., $h=8$ in., find r.
    (ii) If $V=200$ $r=6$ in., find h.
    (iii) If $r=6$ in., $h=8$ in., find V and S.

2. The circumference of the base of a cone is 9 ft. Find the height when the volume of the cone is 22·5 cub. ft.

3. Find the volume and weight of a cast-iron cone, diameter of base 4 in., height 12 in.

4. Find the volume and surface of a cone, radius of base 3 in., height 5 in.

5. If the weight of petroleum, specific gravity ·87, which a conical vessel 8 inches in depth can hold is 3·22 lbs., what is the diameter of the base of the cone?

6. If the volume of a cone 7 ft. high with a base whose radius is 3 ft. be 66 cubic feet, find that of a cone twice as high standing on a base whose radius is half as large as the former.

7. If the volume of a cone 7 ft. high with a base whose radius is 3 ft. be 66 cubic feet, find that of a cone half as high standing on a base whose radius is twice as large as the other one.

8. A right circular cone was measured. The method of measurement was such that we only know that the diameter of base is not less than 6·22 nor more than 6·24 inches, and the slant side is not less than 9·42 nor more than 9·44 inches. Find the slant area of the cone, taking (1) the lesser dimensions, (2) the greater dimensions. Express half the difference of the two answers as a percentage of the mean of the two.

In calculating the area, if a man gives 10 significant figures in his answer, how many of these are unnecessary?

The sphere.—A semicircle of radius r, if made to rotate about its diameter as an axis, will trace out a sphere.

Any line such as AB or CD (Fig. 124) passing through the centre and terminated both ways by the surface is a **diameter**, and any line such as OA or OC passing from the centre to the circumference is a **radius**.

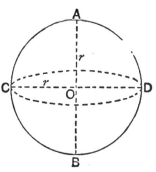

By cutting an orange or a ball of soap it is easy to verify that *any section of a sphere by a plane is a circle*. The section by any plane which passes through the centre of the sphere is called a **great circle**.

Fig. 124.—Sphere.

Surface and volume of a sphere.—The following formulae for the surface and volume of a sphere of radius r should be carefully remembered.

$$\text{Surface of a sphere} = 4\pi r^2 \quad\text{..........................(i)}$$

$$\text{Volume of a sphere} = \tfrac{4}{3}\pi r^3 \quad\text{.............. ....(ii)}$$

The formula for the area of the curved surface may be easily remembered as follows :

The area of a great circle CD (Fig. 124) is πr^2, where r is the radius of the sphere.

The area of the curved surface or hemisphere DAC is twice that of the plane surface, and is therefore $2\pi r^2$.

Hence the area of the surface of the sphere is $2 \times 2\pi^2 = 4\pi r^2$.

The area of the surface of a sphere is equal to that **of the circumscribing cylinder.**

Thus, in Fig. 125, the circumscribing cylinder or the cylinder which just encloses a sphere of radius r is shown. The curved

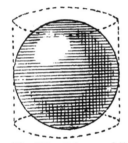

surface of the cylinder will be the circumference of the base $2\pi r$ multiplied by the height $2r$;

∴ Curved surface of cylinder
$$= 2\pi r \times 2r = 4\pi r^2.$$

The volume of the sphere is two-thirds that of the circumscribing cylinder.

Thus, area of base of cylinder $= \pi r^2$,

Fig. 125.—Sphere and its and   height of cylinder $= 2r$;
circumscribing cylinder.

∴ volume of cylinder $= 2\pi r^3$;

two-thirds of $2\pi r^3$ is $\frac{4}{3}\pi r^3$, and this is equal to the volume of the sphere.

The formulae for the surface and volume of a sphere assume a much more convenient form when expressed in terms of the diameter of the sphere.

Let d denote the diameter, then $r = \dfrac{d}{2}$.

$$Surface\ of\ a\ sphere = 4\pi\left(\frac{d}{2}\right)^2 = \pi d^2.$$

$$Volume\ of\ a\ sphere = \frac{4}{3}\pi\left(\frac{d}{2}\right)^3 = \frac{\pi}{6}d^3$$

$$= \cdot 5236 d^3. \quad\dots\dots\dots\dots\dots\text{(iii)}$$

From Eq. (iii) (as ·5 is one-half), the *approximate* method of quickly obtaining the volume of a sphere is seen to be, for **the volume of a sphere, take half the volume of the cube on the diameter and add 5 per cent. to it.**

Ex. 1. Find the surface, volume, and weight of a cast-iron ball; radius 6·25 in.

$$Surface = \pi \times 12 \cdot 5^2 \text{ sq. in.}$$

2 log 12·5 = 2·1938
log π = ·4972   antilog 6910 = 4909
         2·6910   ∴ Surface = 490·9 sq. in.

$$Volume = \cdot 5236 d^3 \text{ cub. in.}$$

3 log 12·5 = 3·2907
log ·5236 = $\bar{1}$·7190   antilog ·0097 = 1023.
         3·0097   ∴ Volume = 1023 cub. in.

Weight of ball = (volume) × (weight of unit volume)
            = 1023 × ·26 lbs. = 266 lbs.

Hollow sphere.—If the external and internal diameters of a hollow sphere be denoted by r_2 and r_1 respectively, then the volume of the material forming the sphere would be

$$\tfrac{4}{3}\pi r_2^3 - \tfrac{4}{3}\pi r_1^3, \text{ or } \tfrac{4}{3}\pi(r_2^3 - r_1^3).$$

This may be replaced by its equivalent

$$\cdot 5236 \, (d_1^3 - d_2^3).$$

Ex. 1. Find the weight of a cast-iron ball, external diameter 9 inches, internal diameter 4 inches.

$$Volume = \cdot 5236(9^3 - 4^3) = \cdot 5236(729 - 64) = \cdot 5236 \times 665.$$

Weight of ball = ·5236 × 665 × ·26 = 90·53 lbs.

EXERCISES. XLIII.

1. The external diameter of a cast-iron shell is 12 in. and its weight 150 lbs. Find the internal diameter; also find the external surface of the sphere.

2 What is the weight of a hollow cast-iron sphere, internal diameter 18 in. and thickness 2 in.?

3. Find the weight of a cast-iron sphere 8 in. diameter, coated with a uniform layer of lead 7 in. thick.

4. Determine (i) the radius of a sphere whose volume is 1 cub. ft., (ii) of a sphere whose surface is 1 sq. ft.

5. A sphere, whose diameter is 1 ft., is cut out of a cubic foot of lead, and the remainder is melted down into the form of another sphere. Find the diameter.

6. A leaden sphere one inch diameter is beaten out into a circular sheet of uniform thickness of $\frac{1}{100}$ inch. Find the radius of the sheet.

7. Find the weight of a hollow cast-iron sphere, internal diameter 2 in., thickness $\frac{1}{6}$ of an inch.

8. The diameter of a cylindrical boiler is 4 ft., the ends are hemispherical, and the total length of the boiler is 8 ft. Find the weight of water which will fill the boiler.

9. The volume of a spherical balloon is 17974 cub. ft. Find its radius.

10. A solid metal sphere, 6 in. diameter, is formed into a tube 10 in. external diameter and 4 in. long. Find the thickness of the tube.

11. Two models of terrestrial globes are 2·35 ft. and 3·35 ft. in diameter respectively. If the area of a country is 20 sq. in. on the smaller globe, what will it be on the larger?

Solid ring.—If a circle, with centre C, rotate about an axis such as AB (Fig. 126), the solid described is called a *solid circular ring*, or simply a solid ring. By bending a length of round solid indiarubber, a ring such as that shown in Fig. 127 may be ob-

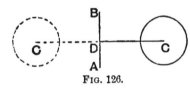

FIG. 126.

tained. The length of such a piece of rubber is the distance DC from the axis multiplied by 2π.

Examples of solid rings are found in curtain rings, in anchor rings, etc. Any cross-section of such a ring will be a circle. The ring may be considered as a cylinder, bent round in a circular arc until the ends meet. *The mean length of the cylinder will be equal to $2\pi CD$, or the circumference of a circle which passes through the centres of area of all the cross-sections.*

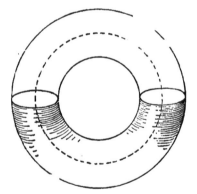

FIG. 127.—Solid ring.

Area of a ring.—The curved surface of a ring is equal to the circumference or perimeter of a cross-section multiplied by the mean length of the ring.

If r denote the radius of the cylinder from which the ring may be imagined to be formed, R the mean radius of the ring, and A the area of the ring, then

Perimeter or circumference of cross-section $= 2\pi r$.

Mean length $= 2\pi R$.

Area of ring $= 2\pi r \times 2\pi R$ ....(i)

$\therefore$ **A** $= 4\pi^2 Rr$ ....(ii)

Eq. (i) will probably be easier to remember than Eq. (ii).

Volume of a ring.—**The volume of a ring is the area of a cross-section multiplied by the mean length.**

Area of cross-section $= \pi r^2$.

Mean length $= 2\pi R$.

Volume $= \pi r^2 \times 2\pi R$.

$\therefore$ $V = 2\pi^2 R r^2$ ....(iii)

In a similar manner the volume may be obtained when the cross-section is a rectangle (Fig. 128), by considering the ring to form a short hollow cylinder.

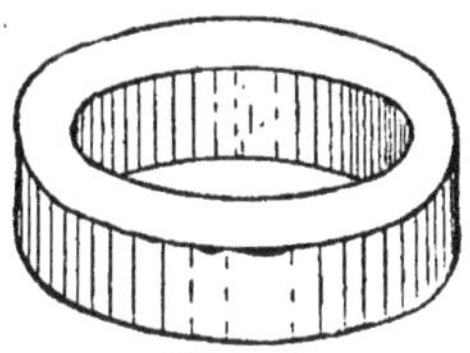

FIG. 128.

Dividing (iii) by (ii) we get

$$\frac{V}{A} = \frac{2\pi^2 R r^2}{4\pi^2 Rr} = \frac{r}{2},$$

from which when V and A are known r can be found, and by substitution in (ii) or (iii) the value of R can be obtained.

Ex. 1. The cross-section of a solid wrought-iron ring, such as an anchor ring, is a circle of 5 inches radius, the inner radius of the ring is 3 ft. Find (*a*) the area of the curved surface, (*b*) the volume of the ring, (*c*) its weight.

(*a*) Here $r = 5$; $R = 36 + 5 = 41$.

Area of curved surface $= 4\pi^2 \times 41 \times 5$ sq. in.

$$= \frac{\pi^2 \times 20 \times 41}{144} \text{ sq. ft.} = 56\cdot2 \text{ sq. ft.}$$

Volume.—Area of cross-section $= \pi \times 5^2$.

Mean length $= 2\pi \times 41$.

$\therefore$ Volume $= \dfrac{2\pi^2 \times 5^2 \times 41}{1728}$ cub. ft. $= 11\cdot71$ cub. ft.

Ex. 2. The cross-section of the rim of a cast-iron fly wheel is a square of 5 inches side. If the inner diameter of the ring is 5 ft., find (*a*) the area, (*b*) the volume, (*c*) the weight of the rim.

As the inner diameter is 60 inches, the outer diameter will be 70.

$\therefore$ Mean diameter $= \frac{1}{2}(60 \times 70) = 65$ inches.

The rim may be considered as a square prism, side of base 5 inches, length $\pi \times 65$.

(a) Perimeter of square $=4 \times 5 = 20$ inches.

$$\therefore \text{ Total surface} = 20 \times \pi \times 65 \text{ sq. in.}$$
$$= 1300\pi \text{ sq. in.}$$

(b) Volume $=$ (area of base) $\times$ (length) $= 5^2 \times \pi \times 65 = 1625\pi$ cub. in.

(c) Weight $= 1625\pi \times \cdot 26$ lb.

EXERCISES. XLIV.

1. The inner diameter of a wrought-iron anchor ring is 12 inches, the cross-section is a circle 4 inches diameter. Find the surface, volume, and weight of the ring.

2. The cross-section of the rim of a cast-iron fly wheel is a rectangle 8 in. by 10 in. If the mean diameter is 10 ft., find the weight of the rim.

3. The volume of a solid ring is 741·125 cub. in. and inner diameter 21 in. Find the diameter of the cross-section.

4. The outer diameter of a solid ring is 12·6 in. if the volume is 54·2 cub. in. Find the inner diameter of the ring.

5. Find the volume of a cylindrical ring whose thickness is 27 in. and inner diameter 96 in.

6. The section of the rim of a fly wheel is a rectangle 6 in. wide and 4 in. deep, the inner radius of the rim is 3 ft. 6 in. Find the volume and weight of the rim, the material being cast iron.

7. In a cast-iron wheel the inner diameter of the rim is 2 ft. and the cross-section of the rim is a circle of 6 in. radius. Find the weight of the rim.

8. Let V denote the volume and A the area of a ring.

(i) If $R=6$, $r=1$, find V and A.

(ii) If $A = 200$ sq. in. and $V = 100$ cub. in., find the dimensions.

(iii) If $V = 200$ cub. in., $R = 12$ in., find r.

9. A circular anchor ring has a volume 930 cub. in. and an area 620 sq. in. Find its dimensions.

10. The cross-section of the rim of the fly wheel of a small gas engine is a rectangle 2·33 in. by 2·5 in. If the mean diameter is 38·4 in., find the volume of the rim in cubic inches and its weight, the material being cast iron.

Similar solids.—Solids which have the same form or shape, but the dimensions not necessarily the same, are called *similar solids*.

All *spheres* and all *cubes* are similar solids.

As a simple case we may consider two right prisms; in one the length, breadth, and depth are 8, 3, and 4 respectively; and in the other, 16, 6, and 8,—*i.e* every linear dimension of the first is doubled in the second. These are similar solids. Further, if a drawing of the first is made to any scale it would answer for the second prism by simply using a *scale* twice the former. In other words two solids are similar when of the same shape or form but made to different scales. It will be seen that the area of any face of the second solid (as each linear dimension is doubled) is four times that of the first, and the volume of the second is 8 times that of the first. If a denote the area of the first and s the scale, then area of second is $s^2 \times a$ and volume $s^3 \times a$.

Ex. 1. The lengths of the edges of two cubes are 2 in. and 4 in. respectively. Compare the surfaces and volumes of the two solids. If the first cube weighs 2 lbs., what is the weight of the second?

The area of each face of a cube of 2 in. edge, is 2^2. As there are 6 similar faces the surface is $6 \times 2^2 = 24$ sq. in.

In a similar manner the surface of the second cube is $6 \times 4^2 = 96$ sq. in.

Thus the surface of the second is 4 times that of the first.

The volume of the first cube is $2^3 = 8$.

The volume of the second cube is $4^3 = 64$.

Hence the volume of the second is 8 times that of the first.

As the weight of the first is 2 lbs., the weight of the second is $8 \times 2 = 16$ lbs.

The definition that two solids are similar when a drawing of one to any convenient scale may by a mere alteration of the scale represent the other, will be found to be a serviceable practical definition of similarity.

And such a definition can be easily applied to cones, cylinders, and pyramids.

Ex. 2. An engine and a small model are both made to the same drawings, but to different scales. If each linear dimension of the engine is 8 times that of the model, find its weight if the weight of the model is 100 lbs. If 1 lb. of paint is required to cover the surface of the model, what amount will probably be required for the engine?

Here volume of engine is 8^3 times that of model;

$$\therefore \text{ weight} = 512 \times 100 = 51200 \text{ lbs.}$$

Area of surface is 8^2 times that of model;

$$\therefore \text{ amount of paint required} = 64 \times 1 = 64 \text{ lbs.}$$

Irregular solids.—When, as is often the case, the given cross-sections are not equidistant, as in Fig. 129, squared paper may be used with advantage. The given distances are set off along a horizontal axis, and the areas are plotted as ordinates. A series of plotted points are thus obtained.

When a curve is drawn through the plotted points the distance between the two end ordinates is divided into an even number of parts, and from the known values of the equidistant ordinates so obtained the area of the curve may be determined by any of the previous rules.

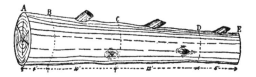

FIG. 129.

Ex. 2. The trunk of a tree (Fig. 129) 32 ft. long has a straight axis and has the following cross-sectional areas at the given distances from one end. Find its volume.

Distances (in feet) from one end.	0	4	14	26	32
Areas of cross-section.	20	16·5	12	8	7·5

Plotting the given values on squared paper a series of points are obtained, and through these points a curve is drawn as in

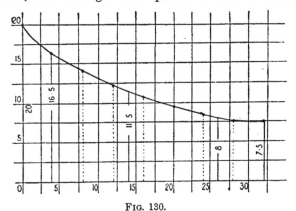

FIG. 130.

Fig. 130. Dividing the base into eight equal parts we obtain nine equidistant ordinates, the values of which can be read off. These are shown in Fig. 130. The common distance between the ordinates is 4 ft.

$$\text{Area} = \tfrac{4}{3}\{20 + 7\cdot5 + 4(16\cdot5 + 12\cdot05 + 9\cdot4 + 7\cdot6) + 2(14 + 10\cdot5 + 8\cdot4)\}$$
$$= 367\cdot3 \text{ cub. ft.}$$

Practical methods of finding volumes and weights.—In many cases a quick method of finding the volume (or weight) of a body is required. For example, if a casting has to be made from a wooden pattern, the weight of metal in the casting may be found approximately by multiplying the weight of the wooden pattern by the ratio of the weight of unit volume of the metal to unit volume of the wood. There are, however, many sources of error in such a calculation. Nails, screws, etc., which are used in the construction of the pattern have a different density to the wood.

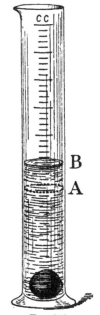

FIG. 131.

Another method is to obtain a volume of water equal to that of the given body. When the body is of small size it may be placed in a graduated cylinder (Fig. 131), and the height A before and B after immersion noted, then from the difference of the readings the volume is at once found. When the body is of comparatively large size it may be placed in a bath of water and the amount of water displaced by the body obtained, from this the volume is obtained. The volume of the *clearance space* in a steam or gas engine cylinder may be determined by noting the quantity of water required to fill it.

EXERCISES. XLV.

1. Find the cubical contents of a body 30 ft. long, the cross-sectional areas at intervals of 5 ft. being respectively 7·5, 5·08, 3·54, 2·52, 1·86, 1·34, 0·92 sq. ft. Find also what the volume would be if only the areas of the two ends and the middle were given.

2. Values of A the area of the cross-section of a body at distances x from one end are given in the following table.   Find the average value of A and the volume of the body.

A Square inches.	53	75	84	94·5	123	139	134	106	76	45
x Inches.	0	9	22	41	62	78	97	114	128	144

3. A body like the trunk of a tree, 13 feet long, its axis being straight, has the following cross-sectional areas of A square inches at the following distances, x inches from its end.   Find its volume, using squared paper.
The following table gives the value of A for each value of x :

x.	0	20	35	56	72	95	110	140	156
A.	405	380	362	340	325	304	287	260	252

4. The length of a tree is 16 ft., its mean girth at five equidistant places is 9·43, 7·92, 6·15, 4·74, and 3·16 ft. respectively.   Find the volume.

5. The areas of the cross-sections of a tree 30 ft. long are as follows :

Distance from one end in feet.	0	2·3	4·5	7	12·2	18	24	30
Area of cross-section in square feet.	7	6·3	5·8	5·2	4·8	4·0	3·8	2·9

Find the volume.

6. In excavating a canal the areas of the transverse sections are in square feet 687·6, 822·2, 735·8, 809·5, 509·5, the common distance between the sections 30 ft.   Find the volume in cubic yards.

7. The transverse sections of an embankment are trapeziums, the distance between each section is 25 ft. ; the perpendicular distances between the parallel sides and the lengths of the parallel sides are given in the following table.   Find the volume of the embankment.

Parallel sides in feet.	22 46	21·9 46·9	21·6 47·6	21·6 50·2	21·8 52·4	21·6 55·2	22 62
Perpendicular distance.	6	6·3	6·6	7·2	7·8	8·4	10

8. The height in feet of the atmospheric surface of the water in a reservoir above the lowest point of the bottom is h; A is the area of the surface in square feet.

When the reservoir is filled to various heights the areas are measured and found to be :

Values of h.	0	13	23	33	47	62	78	91	104	120
Values of A.	0	21000	27500	33600	39200	44700	50400	54700	60800	69300

How many cubic feet of water leave the reservoir when h alters from 113 to 65?

9. A pond with irregular sides when filled to the following heights above a datum level has the surface of the water of the following areas : at datum level the area is 3·16 sq. ft ; 4 ft. above datum, 4·74 sq. ft.; 8 ft., 6·15 sq. ft. ; 12 ft., 7·92 sq. ft. ; 16 ft., 9·43 sq. ft. What is the volume of water in the pond above datum level?

10. Find the cubical contents of a reservoir 42 feet deep, the sectional areas A (sq. ft.) at heights h (ft.) above the bottom being as follows :

h.	0	5	10	17	21	25	29	33	38	42
A.	0	2100	8200	13100	15500	19500	25400	32400	47100	52000

11. A log of timber, 20 feet long, has the following cross-sections at the given distances from one end. Find the average cross-section and the volume in cubic feet.

Distance from one end in feet.	0	2·6	5	7·4	10	12	15	17·6	20
Area in square feet.	5·0	4·3	4·0	3·8	3·46	3·5	3·26	3·1	3·0

CHAPTER XXII.

POSITION OF A POINT OR LINE IN SPACE.

Lines.—Lines may be straight or curved, or straight in one part of their length and curved in another.

Straight line.—A straight line may be defined for practical purposes as the shortest distance between two points ; or as that line which lies evenly between its extreme points.

Planes.—A plane is a surface such that the straight line joining any two points on it lies wholly in that surface.

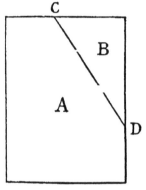

FIG. 132.—A plane surface.

Perhaps a clear notion of what this definition implies may be obtained by using a flat sheet of paper, as in Fig. 132. If any two points, *A* and *B*, on the surface of the paper be selected, it will be seen that the line joining them lies in the surface. Now bend or crease the paper between the points, as along *CD*. The surface no longer remains in one plane, and the shortest, or straight line, joining the two points *A* and *B*, does not lie in the surface.

The intersection of two planes is a straight line, because the straight line joining any two points in their line of intersection must lie in both planes.

Projections of a line.—The projection of a line *AB* on a plane *MN* (Fig. 133) is obtained as follows :

From *A* and *B* let fall perpendiculars (as shown by the dotted lines) on the plane *MN*. The line joining the points where these dotted lines meet the plane is the projection required.

The angle between a line and plane, or the inclination of a line to a plane, is the angle between the line and its projection on the plane. Thus, if *BA* produced meets the plane *NM* (Fig. 133), the inclination of the line to the plane is the angle between the line and its projection on the plane.

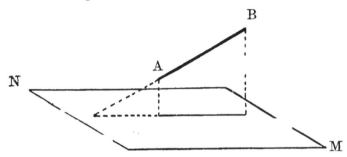

FIG. 133.—Angle between a line and a plane.

Three co-ordinate planes of projections.—A point or points in space may be represented by means of the projections on three intersecting, or co-ordinate planes, as they are called; these projections determine the distances of the point from the three planes, and hence the position of the point is known. The planes are usually mutually at right angles to each other, such as the corner of a cube, or roughly, the corner of a room.

The floor may represent the horizontal plane, sometimes spoken of as the plane *xy*, one vertical wall the plane *xz*, and the other vertical wall at right angles to *xz* the plane *zy*.

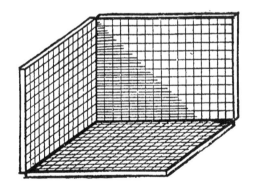

FIG. 134.—Model of the three co-ordinate planes of projection.

A model to illustrate this may consist of a piece of flat board (Fig. 134) and two other pieces mutually at right angles to each other. It is advisable to have the latter two boards hinged. This enables the two sides to be rotated until all three planes lie in one plane. If the three pieces of wood are painted black they may be ruled into squares, or squared paper may be

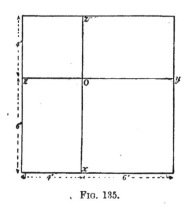

, FIG. 135.

fastened on them. By means of hat pins many problems can be effectively illustrated.

If preferred, a model can be easily made from drawing paper or cardboard. Draw a square of 9 or 10 inches side (Fig. 135). Along two of its sides mark off distances of 4″ and 6″ and letter as shown. Cut through one of the lines OZ, and fold the paper so that the two points marked Z coincide.

To fix the position of a point in space, imagine such a point P; from P let fall a perpendicular on the horizontal plane and meeting it in p (Fig. 136). pP is the distance of the point P from the plane xy, or is the *z co-ordinate* of P. In a similar

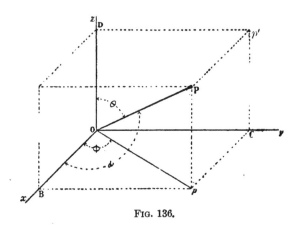

FIG. 136.

manner a perpendicular let fall on the plane yz, meeting it in p', will give the distance from the plane yz or the x co-ordinate of the point. And the distance Pp'' the y co-ordinate of the point is the distance of the point from the plane zx.

The three projections of a point on three intersecting planes definitely determine the distance of a point from these planes.

In the example we have assumed the z co-ordinate to be **above** the plane of xy, but the method applies equally to distances **below** the plane.

Hence, a point or object above or below the earth's surface could be specified if two intersecting vertical planes, such as two walls meeting at right angles, were to be found in the neighbourhood. The distances from the two walls, together with the remaining or z co-ordinate, would completely define the position of the point. Stores of buried treasure may in this manner be located. A person unable to carry away treasure might select a place in which two convenient intersecting walls are to be found in the neighbourhood. If deposited at some depth below the surface the treasure could be recovered at any future time, provided that the respective distances from the two walls and the depth below the surface were known.

It will be found that the problems dealing with the projections of a point, line, or plane, may be solved either by graphic methods, using a fairly accurate scale and protractor, or by calculation. One method should be used as a check on the other.

Ex. 1. Given the x, y, and z co-ordinates of a point as $2''$, $1\cdot5''$, and $2''$ respectively. Draw the three projections of the line OP on the three planes xy, yz, and zx, and in each case measure the length of the projection. Find the distance of P from the origin O, and the angles made by the line OP with the three axes.

Let P (Fig. 136) be the given point and O the origin of co-ordinates. Join OP.

The projection on the axis of x is the line OB; on the axis of y is the line OC; and on the axis of z is the line OD.

$$OB = 2'',\ OC = 1\cdot5'',\ \text{and}\ OD = 2''.$$

Graphic Construction.—The relations of the lines and angles can be seen from the pictorial view (Fig. 136). To measure the lengths of the lines and the magnitudes of the angles, proceed as follows :

Draw the three axes intersecting at O (Fig. 137) and letter as shown. Set off along the axis of z a distance $= 2''$, along the axis of y a distance $= 1\cdot5''$. Draw lines parallel to the axes, and join p_1, their point of intersection, to the origin O. Then Op_1 is the pro-jection of OP on the plane xy; its length is $2\cdot5''$.

In a similar manner the projections, Op_2 on the plane yz and Op_3 on the plane xz are obtained; $Op_2 = 2\cdot5''$ and $Op_3 = 2\cdot83''$.

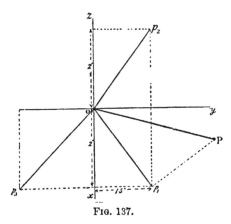

FIG. 137.

The distance of P from the origin, or the length of the line OP, is the hypot-enuse of a right-angled triangle, of which Op_1 is the base and the perpen-dicular p_1P the height of P above the plane of xy, or simply the z co-ordinate of the point. Hence draw p_1P perpendicular to Op_1 and equal to $2''$. Join O to P; OP is the distance required $= 3\cdot2''$.

To obtain the angles with the three axes it is necessary to *rabat* the line into the three planes. This is easily effected by using O as centre and length of $OP = 3\cdot2''$ as radius. Describe a circle cutting the lines passing through p_1, p_2, and p_3 at P_1, P_2, and P_3 respectively. Join O to P_1, P_2, and P_3, then three angles will be found to be $51°\cdot3$, $62°\cdot1$, and $51°\cdot3$.

Ex. 2. Find the distance between the two points $(3, 4, 5\cdot3)$ $(1, 2\cdot5, 3)$ and the angles the line joining them makes with the axes.

The solution of this problem can be made to depend on the pre-ceding example by taking as origin the point $(1, 2\cdot5, 3)$, then the co-ordinates of the remaining point will be $(3-1)$, $(4-2\cdot5)$ and $(5\cdot3-3)$, or $(2, 1\cdot5, 2\cdot3)$. Hence the true length, the projections and the angles made with the axes may be obtained as in Ex. 1.

The manner in which the three axes are lettered should be noticed. It would appear at first sight to be more convenient to letter as the axis of x the line going from the origin O to the right instead of y as in the diagram; but when it becomes necessary to apply mathematics to mechanical or physical prob-lems the notation adopted in Fig. 136 is necessary, and therefore it is advisable to use it from the commencement.

Calculation.—In Fig. 136 let θ denote the angle made by OP

with the axis of z, and ϕ the angle which the projection Op makes with the axis of x; then we have:

$$x = OB = Op \cos \phi,$$

but
$$Op = OP \sin \theta;$$

$$\therefore \quad x = OP \sin \theta \cos \phi. \quad \dots\dots\dots\dots\dots\dots\dots\dots\dots(\mathrm{i})$$

$$y = OC = Op \cos pOC = Op \sin \phi;$$

$$\therefore \quad y = OP \sin \theta \sin \phi, \quad \dots\dots\dots\dots\dots\dots\dots(\mathrm{ii})$$

and
$$z = OP \cos \theta. \quad \dots\dots\dots\dots\dots\dots\dots\dots\dots(\mathrm{iii})$$

Ex. 3.   Let $OP = 100$, $\theta = 25°$, $\phi = 70°$.

Then
$$x = 100 \sin 25° \cos 70°$$
$$= 100 \times \cdot4226 \times \cdot3420 = 14\cdot45;$$
$$y = 100 \sin 25° \sin 70°$$
$$= 100 \times \cdot4226 \times \cdot9397 = 39\cdot71;$$
$$z = 100 \cos 25° = 100 \times \cdot9063 = 90\cdot63.$$

Ex. 4.   Given the co-ordinates of a point $x = 3$, $y = 4$, $z = 5$. Find OP or r, θ, and ϕ.

$$OP^2 = r^2 = Op^2 + pP^2 \text{ also } Op^2 = OB^2 + Bp^2;$$

$$\therefore \quad r^2 = OB^2 + Bp^2 + pP^2$$
$$= x^2 + y^2 + z^2 = 9 + 16 + 25 = 50;$$

$$\therefore \quad r = \sqrt{50} = 7\cdot071.$$

From (iii)
$$z = r \cos \theta = 7\cdot071 \cos \theta;$$

$$\therefore \quad \cos \theta = \frac{5}{7\cdot071} = \cdot7071; \quad \therefore \quad \theta = 45°.$$

From (i)
$$3 = r \sin \theta \cos \phi,$$

or
$$\cos \phi = \frac{3}{r \sin \theta} = \frac{3}{7\cdot071 \times \cdot7071}; \quad \therefore \quad \phi = 53° 6'.$$

Direction-cosines.—We have found that when the x, y, z co-ordinates of a point are given, its distance from the origin may be denoted by r where $r^2 = x^2 + y^2 + z^2$. Hence we can proceed to find the ratios $\dfrac{x}{r}$, $\dfrac{y}{r}$, and $\dfrac{z}{r}$. These are called the direction-cosines of the line.

Thus, if OP (Fig. 136) makes angles a, β, and θ with the axes of x, y, and z respectively, then

$$\cos a = \frac{x}{OP} = \frac{x}{r}.$$

Similarly $\qquad \cos \beta = \frac{y}{r}$ and $\cos \theta = \frac{z}{r}$.

Squaring each ratio and adding, we get

$$\cos^2 a + \cos^2 \beta + \cos^2 \theta = \frac{x^2}{r^2} + \frac{y^2}{r^2} + \frac{z^2}{r^2} = \frac{r^2}{r^2} = 1.$$

The letter l is usually used instead of $\cos a$, and similarly m and n replace $\cos \beta$ and $\cos \theta$ respectively. Thus we get

$$r = \frac{x}{l} = \frac{y}{m} = \frac{z}{n},$$

where l, m, and n denote the direction-cosines of the line.

From the relation $\cos^2 a + \cos^2 \beta + \cos^2 \theta = 1$ or its equivalent $l^2 + m^2 + n^2 = 1$ it is obvious that, if two of the angles which a given line OP makes with the axes are known, the remaining angle can be found. Also, as indicated in Ex. 1, the angles a, β, and θ can be obtained by construction, but more accurately by calculation. We may repeat Ex. 1 thus :

Ex. 5. The co-ordinates of a point P are 2, 1·5, 2. Find the distance of the point from the origin O, and the angles made by the line OP with three axes.

True distance, $OP = \sqrt{2^2 + 1 \cdot 5^2 + 2^2} = 3 \cdot 2$.

Denoting the distance OP by r to find the angles a, β, and θ, we have $\qquad x = OB = r \cos a$;

$$\therefore \cos a = \frac{x}{r} = \frac{2}{3 \cdot 2} = \cdot 6250;$$

$$\therefore a = 51° \ 19'.$$

$$y = OC = OP \cos \beta,$$

$$\cos \beta = \frac{1 \cdot 5}{3 \cdot 2} = \cdot 4688;$$

$$\therefore \beta = 62° \ 3'.$$

$$z = OD = OP \cos \theta,$$

or $\qquad\qquad \cos \theta = \frac{2}{3 \cdot 2} = \cdot 6250;$

$$\therefore \theta = 51° \ 19'.$$

Ex. 6. A line OP makes an angle 60° with one axis, 45° with another. What angle does it make with the third?

Let γ denote the required angle, then as

$$\cos 60° = \tfrac{1}{2} \quad \text{and} \quad \cos 45° = \frac{1}{\sqrt{2}},$$

we have from the relation

$$\cos^2 60° + \cos^2 45° + \cos^2 \theta = 1,$$
$$\tfrac{1}{4} + \tfrac{1}{2} + \cos^2 \theta = 1,$$

or $$\cos^2 \gamma = 1 - \tfrac{3}{4} = \tfrac{1}{4} ;$$
$$\therefore \ \cos \theta = \tfrac{1}{2} \quad \text{and} \quad \theta = 60°.$$

A practical application.—Some of the data we have assumed may perhaps be better expressed by the terms latitude and longitude of a place on the earth's surface. Thus, at regular

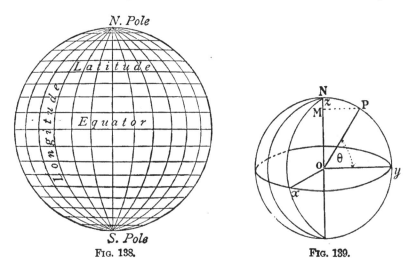

FIG. 138.   FIG. 139.

distances from the two poles a series of parallel circles are drawn (Fig. 138) and are called *Parallels of Latitude.* The parallel of latitude midway between the poles is called the Equator. These parallels are crossed by circles passing through the poles and called meridians of longitude. Selecting one meridian as a standard (the meridian passing through Greenwich), the position of any object on the earth's surface can be accurately determined. This information, together with the depth below the surface or the height above it, determines any point or place.

The plane *xoy* may be taken to represent the equatorial plane of the earth, and OZ the earth's axis. Then the position of a point P (Fig. 139) on the surface of the earth, or that of a point outside the surface moving with the earth, is known when we are given its distance OP (or r) from the centre, its latitude θ, or co-latitude $(90-\theta)$, and its ϕ or east longitude, from some standard meridian plane, such as the plane passing through Greenwich.

Assuming the earth to be a sphere of radius r, then the distance of a point on the surface can be obtained. If P be a point on the surface, the distance of P from the axis is the distance PM, but $PM = r \sin POM = r \cos \theta$.

Ex. 7. A point on the earth's surface is in latitude 40°. Find its distance from the axis, assuming the earth to be a sphere of 4000 miles radius.

Required distance $= 4000 \times \cos 40°$

$= 4000 \times \cdot766 = 3064$ miles.

Having found the distance PM, the speed at which such a point is moving due to the rotation of the earth can be found.

Ex. 8. Assuming the earth to be a sphere of 4000 miles radius, what is the linear velocity of a place in 40° north latitude? The earth makes one revolution in 29·93 hours.

Radius of circle of latitude $= 4000 \times \cos 40°$;

$$\therefore \text{ speed} = \frac{4000 \times \cos 40° \times 2\pi}{29 \cdot 93} = \frac{4000 \times \cdot 766 \times 2\pi}{29 \cdot 93}$$

$= 642 \cdot 77$ miles per hour.

Line passing through two given points.—If the co-ordinates of two given points P and Q be denoted by (x, y, z) and (a, b, c) the equation of the line passing through the two points is

$$\frac{x-a}{l} = \frac{y-b}{m} = \frac{z-c}{n}.$$

Through P draw three lines Pp, Pp', Pp'', parallel to the three axes respectively, and draw the remaining sides of the rectangular block as in Fig. 140. Complete a rectangular block having its sides parallel to the former one and q for an angular point.

$PL = Nq = NR - qR = Pp' - Lp' = x - a.$

$PF = Mq = Md - dq = y - b.$

$PS = Eq = Eq' - qq' = z - c.$

The line Pq is the diagonal of a rectangular block, the edges of which are $x-a$, $y-b$, $z-c$, and therefore to find the length of Pq we have

$$Pq = \sqrt{(x-a)^2 + (y-b)^2 + (z-c)^2}$$

The angle between the line Pq and the axis of Z is the angle between Pq and qE a line parallel to the axis of Z. Hence denoting the angle by θ,

$$n = \cos\theta = \frac{z-c}{Pq} = \frac{z-c}{\sqrt{(x-a)^2 + (y-b)^2 + (z-c)^2}}.$$

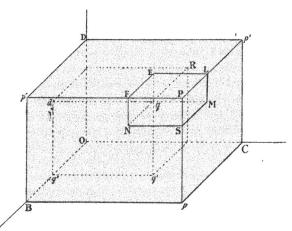

FIG. 140.—Line passing through two points.

Similarly,

$$l = \frac{x-a}{Pq}, \quad m = \frac{y-b}{Pq}.$$

It will be obvious that when the second point is the origin, a, b, and c are each zero, and the equation

$$\frac{x-a}{l} = \frac{y-b}{m} = \frac{z-c}{n}$$

becomes

$$\frac{x}{l} = \frac{y}{m} = \frac{z}{n}.$$

Ex. 9.   If $x=3$, $y=4$, $z=5$, find r, l, m, and n.

We have
$$r^2 = x^2 + y^2 + z^2 = 9 + 16 + 25 = 50;$$
$$\therefore \ r = \sqrt{50} = 7\cdot071,$$
$$l = \frac{x}{r} = \frac{3}{7\cdot071} = \cdot4242,$$
$$m = \frac{y}{r} = \frac{4}{7\cdot071} = \cdot5657,$$
$$n = \frac{z}{r} = \frac{5}{7\cdot071} = \cdot7071.$$

Ex. 10. Find the distance between the two points (3, 4, 5·3) (1, 2·5, 3) and the angles made by the line with the three axes.

$$\text{Distance} = \sqrt{(3-1)^2 + (4-2\cdot5)^2 + (5\cdot3 - 3)^2}$$
$$= \sqrt{2^2 + 1\cdot5^2 + 2\cdot3^2} = 3\cdot4.$$

$$l = \cos a = \frac{3-1}{3\cdot4} = \cdot5882; \qquad \therefore a = 53°\,58'.$$

$$m = \cos \beta = \frac{4-2\cdot5}{3\cdot4} = \cdot4413; \quad \therefore \beta = 63°\,48'.$$

$$n = \cos \gamma = \frac{5\cdot3 - 3}{3\cdot4} = \cdot6764; \quad \therefore \theta = 47°\,24'.$$

When the given point or points are in the plane of x, y, a resulting simplification occurs. Thus, denoting the co-ordinates of two points P and Q by (x, y) and $(a, b$ respectively), and the angles made by the line PQ with the axes of x and y by a and β.

Then, if r be the distance between the points,

$$r = \sqrt{(x-a)^2 + (y-b)^2}.$$

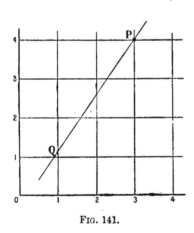

FIG. 141.

Also $\dfrac{x-a}{\cos a} = \dfrac{y-b}{\cos \beta};$

$$\therefore y - b = \frac{\cos \beta}{\cos a}(x - a);$$

but β is the complement of a;

$$\therefore \cos \beta = \sin a.$$

Hence we get

$$y - b = \tan a(x - a),$$

and the equation of the line joining the two points may be written

$$y - b = m'(x - a),$$

where m' is the tangent of the angle made by the line with the axis of x.

Thus, given $x = 3$, $y = 4$, the point P (Fig. 141) is obtained by marking the points of intersection of the lines $x = 3$, $y = 4$.

In a similar manner the point Q (1, 1·34) is obtained. Join P to Q, then PQ is the line through the points (3, 4), (1, 1·34), and

$$PQ = \sqrt{(3-1)^2 + (4 - 1\cdot34)^2} = 3\cdot33.$$

$$y - 1\cdot34 = \tfrac{4}{3}(x-1);$$
$$\therefore \quad y = \tfrac{4}{3}x \text{ or } y = 1\cdot33x.$$

Polar co-ordinates.—If from the point P a line be drawn to the origin, then if the length of OP be denoted by r, and the angle made by OP with the axis of x be θ, when r and θ are known, the position of the point can be determined, and the rectangular co-ordinates can be found.

Conversely, given the x and y of a point, r and θ can be obtained.

Ex. 11. Let $r = 20$, $\theta = 35°$; find the co-ordinates x and y.

Here
$$x = r \cos 35° = 20 \times \cdot8192 = 16\cdot384;$$
$$y = r \sin 35° = 20 \times \cdot5736 = 11\cdot472\cdot$$

Ex. 12. Given the co-ordinates of a point P (4, 3); find r and θ.
$$r^2 = 4^2 + 3^2 = 25; \quad \therefore \quad r = 5;$$
$$\tan \theta = \tfrac{3}{4} = \cdot75, \quad \theta = 36° \; 54'.$$

EXERCISES. XLVI.

1. A point P is situated in a room at a height of 3 ft. above the floor, 4 ft. from a side wall, and 5 ft. from an end wall. Determine the distance of P from the corner where the two walls and the floor meet. Scale, $\tfrac{1}{2}'' = 1'$.

2. Determine the length of a line which joins two opposite corners of a brick $9'' \times 4\tfrac{1}{2}'' \times 3''$. Scale, $\tfrac{1}{3}$.

3. The floor $ABCD$ of a room is rectangular. AB and CD are each 18 feet long, and AD, BC each 24 feet. A small object P in the room is 6 ft. above the floor, 10 ft. from the vertical wall through AB, and 8 ft. from the wall through BC. Find and measure the distances of P from A, B, C, and D, the four corners of the floor. Scale, $0\cdot1'' = 1'$.

4. A small object P is situated in a room at a distance of $17''$ from a side wall, $24''$ from an end wall, and $33''$ above the floor. Find the distance of P from the corner O of the room where these three mutually perpendicular planes meet. If a string were stretched from O to P, find and measure the angles which OP would make with the floor, the end wall, and the side wall respectively. Scale, $\tfrac{1}{10}$.

5. The co-ordinates of two points A and B are $(2\tfrac{1}{2}'', 1\tfrac{1}{2}'', \tfrac{1}{2}'')$ and $(\tfrac{1}{2}'', 4'', 2'')$.
Determine the length of AB and the angles which AB makes with the planes XY, YZ.

6. The co-ordinates of a point are $1\frac{1}{2}''$, $2''$, $1''$. Draw and measure the three projections of the line OP on the planes of xy, yz, and zx. Find the true length of OP.

7. The three rectangular co-ordinates of a point P are $x=1\cdot5$, $y=2\cdot3$, $z=1\cdot8$. Find (1) the length of the line joining P to O the origin, (2) the cosines of the angles which OP makes with the three rectangular axes.

8. The polar co-ordinates of a point are
$$r=20'', \quad \theta=32°, \quad \phi=70°.$$
Find the rectangular co-ordinates.

9. There are three lines OX, OY, and OZ mutually at right angles. The following lengths are set off along these lines:

OA, of length 2 inches, along OX.
OB, ,, 3·4 ,, ,, OY.
OC, ,, 2·95 ,, ,, OZ.

A plane passes through A, B, and C.
Determine and measure the angle between this plane and the plane which contains the lines OX and OY.
Also determine and measure the angle between the plane and the line OZ.

10. Describe any system which you know of that enables us to define exactly the position of a point in space.
The three rectangular co-ordinates of a point P are 3, 4, and 5; determine (i) the length of the line joining P to O, the origin of co-ordinates; (ii) the cosines of the angles which OP makes with the three rectangular axes.

11. The polar co-ordinates of a point are
$$r=3, \quad \theta=65°, \quad \phi=50°.$$
Determine its rectangular co-ordinates.

12. The earth being supposed spherical and of 4000 miles radius, what is the linear velocity in miles per hour of a point in 36° North latitude? The earth makes one revolution in 23·93 hours.

13. A point is in latitude 52°. If the earth be assumed to be a sphere of 3960 miles radius, how far is the point from the axis? Find the length of the circumference of a circle passing through the point called a parallel of latitude. What is the 360th part of this length, and what is it called?

14. If the earth were a sphere of 3960 miles radius, what is the 360th part of a circle called a meridian? What is it called?

15. The polar co-ordinates of a point A are $3''$, 40°, and $50°$ respectively. Find the rectangular co-ordinates.

16. The three rectangular co-ordinates of a point P are 2·5, 3·1, and 4. Find (1) the length of line joining P with O the origin, (2) the cosines of the angles which OP makes with the three axes, and (3) the sum of the squares of the three cosines.

CHAPTER XXIII.

ANGULAR VELOCITY. SCALAR AND VECTOR QUANTITIES.

Angular velocity.—When a point moves in any manner in a plane, the straight line joining it to any fixed point continually changes its direction ; the rate at which such a straight line is rotating is called the *angular velocity* of the moving point about the fixed point. Angular velocity is *uniform* when the straight line connecting the moving and fixed points turns through equal angles in equal times, but variable when unequal angles are described in equal times.

Measurement of angular velocity.—The angular velocity of a rotating body is the angle through which it turns per second, expressed in radians.

One of the most important cases of angular motion is when P (Fig. 142) is a point in a rigid body rotating about a fixed axis O. All points of the body move in circles having their planes perpendicular to and their centres in the axis. Hence, at any instant the angular velocity for all points of the body is the same.

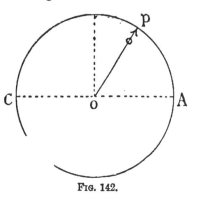

FIG. 142.

If a point P is describing the circle APC, of radius r, with a uniform velocity of v feet per second, then, denoting the angular velocity by ω, the length of arc described in one second v, is the product of the angular velocity and the radius.

$$\therefore \ v = r\omega, \text{ or } \omega = \frac{v}{r}.$$

In one revolution the moving point P describes a distance equal to the circumference of the circle. Hence, if t denote the time (in seconds) of a complete revolution

$$\omega = \frac{2\pi r}{rt} = \frac{2\pi}{t}.$$

In one revolution angle turned through is 2π, and in n revolutions $2\pi n$. From $v = \omega r$,

$$v^2 = \omega^2 r^2 = 2^2 \pi^2 n^2 r^2.$$

If the number of revolutions per minute is given, it is necessary to divide by 60.

Ex. 1. A wheel makes 100 turns a minute, what is its angular velocity? Find the linear speed of a point on the wheel 7 feet from the axis.

In one revolution the angle turned through is 2π radians;

$$\therefore \ \text{angular velocity} = \frac{100 \times 2\pi}{60} = 10\cdot47 \text{ radians per sec.}$$

Linear velocity = angular velocity × radius ;
or, $\qquad\qquad 10\cdot47 \times 7 = 73\cdot29$ feet per sec.

EXERCISES. XLVII.

1. A wheel diameter 5 ft. turns 40 times a minute. Find its angular velocity and the linear velocity of a point on the circumference.

2. Explain what is meant by angular velocity of a rotating body ; knowing the angular velocity, how would you proceed to obtain the linear velocity? P is a point of a body turning uniformly round a fixed axis, and PN is a line drawn from P at right angles to the axis. If PN describes an angle of 375° in 3 sec., what is the angular velocity of the body? If PN is 6 ft. long, what is the linear velocity of P?

3. What is the numerical value of the angular velocity of a body which turns uniformly round a fixed axis 25 times per minute?

4. The radius of a wheel is 14 feet, and it makes 42 revolutions a minute. Find its angular velocity and the linear velocity of the extremity of the radius.

5. A wheel is 5 feet diameter, and a point on its circumference has a speed of 10 feet per second. Express in radians the angle turned through in ¼ second. How many revolutions will the wheel make per minute?

6. Define *angular velocity*. A wheel makes 90 turns per minute. What is its angular velocity in radians per second? If a point on the wheel is 6 feet from the axis, what is its linear speed?

7. The diameter of a wheel is 3½ feet, what is its angular velocity when it makes 120 revolutions per minute? What is the linear speed of a point in the rim of the wheel?

Scalar quantities.—Those quantities which are known when their *magnitudes* (which are simply numbers) are given, such as masses, areas, volumes, etc., are called **scalar quantities.**

Vector quantities.—Quantities which require for their complete specification the enumeration of both magnitude and direction are called **vector quantities, or shortly, vectors.** Thus, forces, velocities, accelerations, displacements, etc., are vectors, and may in each case be represented by a straight line.

To completely specify a vector we require to know

(1) Its *magnitude.*

(2) The *direction* in which it acts, or its line of action.

(3) Its *point of application.*

The term *direction* applied to vector quantities is not sufficiently explicit. For example, in the specification of a vector the direction may be given as vertical, but a vertical direction may be either upward or downward, hence what is called the *sense* of a vector must be known. Thus, if we include sense, four things require to be known before a vector is completely specified.

The properties of a vector quantity may be represented by a straight line; thus, for example, a vector acting at a point A can be fully represented by a straight line.

The length of the line to some convenient scale may represent the *magnitude* of the force. One end of the line A (Fig. 143) will represent the *point of application*, while the direction in which the line is drawn as from O to A will represent the direction or *sense* of the vector.

Direction of a vector.—The direction of a vector is specified when the angle made by it with a fixed line is known. When two, or more, vectors are given the line referred to may be one of the vectors.

In many cases the points of the compass are used. Thus, in Fig. 143, the vectors B, C, D, E, and F make angles of 30°, 45°, 90°, 135°, and 180° respectively with the line OX, or with the vector A.

It is important to remember that all angles are measured in the *opposite* direction to the hands of a clock.

Using the points of the compass A is said to be towards the East, B is 30° N. of E., C is N.E., D is North, E is N.W., and F is W.

The *sense* of a vector is indicated by an arrow-head on the line representing the vector ; the clinure of the line, or the

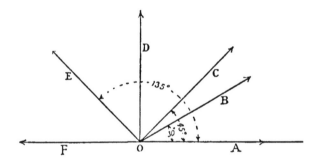

Fig. 143.—Specification of vectors.

direction of the line may be called the *clinure* or *ort* of the vector.

Addition and subtraction of vectors.—If A, B, C (Fig. 144) represent three vectors acting at a point O, to find their resultant, or better, to add them, we make them form consecutive sides of a polygon. Thus, starting from any convenient point a, the line ab is drawn parallel to, and equal in magnitude to, the vector A. In like manner bc is made equal to, and parallel to, B, and cd to C. The last side of the polygon from a to d represents the resultant, or the sum of the three given vectors. The sides of the polygon, a, b, c, d, taken in order, indicate the magnitude and direction of each vector, but arrow-heads on each side of the polygon also indicate the sense of each vector. When taken in order, i.e. a to b, b to c, and c to d, as in Fig. 144, the vectors are said to be *circuital*, hence an arrow-head in a non-cir-cuital direction on the last side of the polygon represents the resultant or the **sum** of the given vectors.

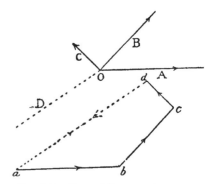

Thus, denoting the sum by D, we have

$$ab + bc + cd = ad,$$

or $\quad A + B + C = D.$

The result obtained is the same if we begin with B or C. In fact,

FIG. 144.—Sum of three vectors.

taking them in quite a different order as the sides of a polygon, the same result follows :

i.e. $\qquad A + B + C - D = 0,$ ....(i)

or $\qquad\qquad A + B = D - C.$

If at O a fourth vector (shown by the dotted line) equal in magnitude and parallel to da, be inserted, the polygon is a closed figure having the arrow-heads on its sides *circuital*. The four vectors acting at O are in equilibrium. Hence we can write Eq. (i) as $A + B + C + D = 0$. Vector quantities may in fact be added or subtracted by the *parallelogram, triangle,* or *polygon law*.

As a simple example consider two displacements A and B. The vector sum is at once obtained by setting off oa and ob (Fig. 145) equal in magnitude to A and B respectively. Completing the parallelogram the diagonal oc is the resultant, or sum, of the given vectors.

A negative sign prefixed to a vector indicates that the vector is to be reversed. Thus, if A and B are represented by oa and ob, then $A - B$ will be represented by oa and the dotted line ob. Hence, one diagonal of the parallelogram gives $A + B$ and the other gives $A - B$.

Ex. 1. There are two vectors in one plane, A of amount 10 in the direction towards the East, B of amount 15 in the direction towards 60° North of East.

(i) Find the vector sum $A + B$.

(ii) The vector difference $A - B$.

(iii) Find $A + B$ when B is in the direction towards the North.

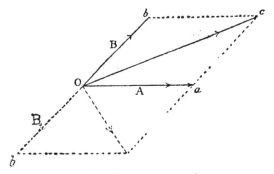

FIG. 145.—Resultant of two displacements.

(i) Starting at any point a (Fig. 146), draw a line ab equal in magnitude to, and parallel to, A. From b draw bc parallel and

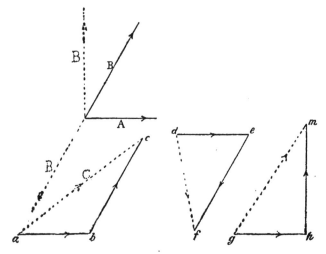

FIG. 146.—Sum and difference of two vectors.

equal to B. Then ac is the magnitude and direction of the vector sum, and its sense is denoted by an arrow-head, non-circuital with the rest; ac measures 21·79 and is directed towards 36° 46′ N. of E,

(ii) Again starting at a point d, draw de as before ; but, from e, draw ef in the opposite direction. Then, df, as before, is the required vector. Its magnitude is 13·2, and its direction 10° 15′ E. of S.

(iii) When the vector is in a direction towards N., then the angle between the two vectors is 90°, and $C = gh^2 + hm^2 = \sqrt{A^2 + B^2} = 18·02$. Its direction is 56·5° N. of E. (*i.e.* $\tan\theta = \frac{15}{10}$).

Ex. 2. A ship at sea is sailing apparently at 8 knots to the East, and there is an ocean current of 3 knots to the South-west. Find the actual velocity of the ship.

We have to find the resultant of a velocity 8 in a direction E., and a velocity 3 in a direction S.W. If a velocity of 1 knot be represented by 1 inch, then 8 inches will represent 8 knots and 3 inches will represent 3 knots.

Make $op = 8$ knots and $oq = 3$ knots (Fig. 147). On the two lines op and oq as sides complete the parallelogram $oprq$. The diagonal

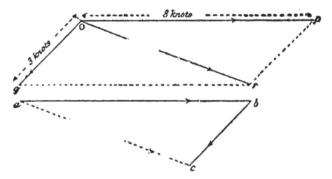

FIG. 147.—Resultant of two velocities.

or is the resultant required. Measuring or we find it to be 6·25 inches, therefore representing 6·25 knots.

Its direction is given by the angle $por = 340°·2$, or it may be written as 19°·8 S. of E.

We may obtain the same result by drawing from any point a the lines ab and bc equal and parallel to op and oq respectively. The resultant is then given in direction and magnitude by the line ac.

Resolution of vectors.—We are able to replace two vectors acting at a point by a single vector which will produce the same effect. Thus, in Fig. 148, the two vectors A and B may be replaced by the vector C.

Conversely, we may replace a single vector by two vectors acting in different directions. The two directions are usually assumed at right angles to each other.

Let OC (Fig. 148) represent in direction and magnitude a vector acting at a point O. If two lines OX and OY at right

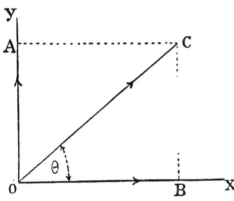

angles to each other be drawn passing through O, and BC and AC be drawn parallel to OY and OX respectively, we obtain two vectors OB and OA, which, acting simultaneously, produce the same effect on the point O as the single vector OC.

The two vectors OA and OB are called the

FIG. 148.—Rectangular components of a vector.

rectangular components of OC, and the process of replacing a vector by its *components* is called *resolution*. The vector OC can be drawn to scale, and the components measured to the same scale. Or, they can, by means of a slide-rule or logarithm tables, be easily calculated as follows :

Denoting the angle BOC by θ.

By definition (p. 155) $\dfrac{OB}{OC} = \cos\theta$; if θ and OC are known, then

$$OB = OC \cos\theta.$$

In a similar manner, $OA = OC \times \cos COA = OC \times \cos(90° - \theta)$

$$= OC \sin\theta.$$

This important relation may be stated as follows : **The resolved part of a vector in any given direction is equal to the magnitude of the vector multiplied by the cosine of the angle made by the vector with the given direction.**

If the vector is a given velocity V, then the *resolved part of the velocity* in any given direction making an angle θ with the direction of the velocity is $V \cos\theta$.

If a body is moving N.E. with a velocity of 10 feet per

second, it has a velocity East of 10 cos 45° and a velocity North of 10 cos 45°.

As the angle made by line OC (Fig. 148) increases, the horizontal component diminishes, and the vertical component increases. When $\theta = 90°$, the vector is vertical and the vertical component is simply the magnitude of the vector, its horizontal component is O. Conversely when the angle is 0° the vertical component is O.

The process just described may be extended to two or more vectors acting at a point. The horizontal and vertical components of each vector are obtained, the sum of all the horizontal components is denoted by X, and the sum of all the vertical components by Y; X and Y are then made to form the base and perpendicular of a right-angled triangle, the hypotenuse of which will be the vector sum required. Denoting the vector sum by R and its inclination to the axis of x by θ, then

$$R = \sqrt{X^2 + Y^2}, \text{ and } \tan\theta = \frac{Y}{X}.$$

By means of Table V. the values of the sine and cosine of any angle can be obtained and the calculations for R and θ are easily made. The results obtained from this and the graphical method may, if necessary, be used to check the result. The application of the rule can best be shown by an example :

Ex. 3. The magnitudes and directions of three vectors in one plane are given in the following table. Find the vector sums and differences (i) $A + B + C$; (ii) $A + B - C$.

	A.	*B.*	*C.*
Magnitude, -	50	30	20
Direction, - -	30° N. of E.	N.	N. W.

Graphically—(i) Show the three vectors acting at a point O (Fig. 149); draw the polygon making ab, bc, and cd to represent the three given vectors. The non-circuital side ad is the sum $A + B + C = 75\cdot2$, its inclination is 67° N. of E.

(ii) In $A + B - C$ the direction of the vector C is reversed; hence, produce CO, and on the line produced put an arrow head indicating a direction opposite to that of the vector C. Also in the polygon produce dc to d', making $cd'=cd$. Join ad'; ad' represents $A + B - C$; the magnitude is 70 and the inclination 36°.

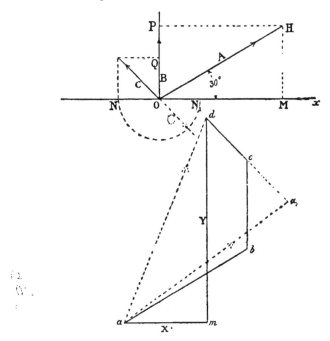

FIG. 149.—Sum of three vectors.

Set off a length OH on the vector A equal to 50 units, draw HM perpendicular to OX. Then OM is the horizontal component of A. Similarly, the horizontal component of C is the line ON. As ON is measured in a negative direction, the sum of the horizontal components is $OM - ON$.

We may measure either OM and ON and subtract one from the other; or, using O as centre and a radius equal to ON, describe an arc of a circle to obtain N'. Then $N'M = OM - ON$, where $N'M$ denotes the sum of the horizontal components;

$$\therefore \ X = N'M = 29 \cdot 16.$$

In a similar manner, projecting on the vertical line OY, OP, is the vertical component of A, and OQ the vertical component of C. Hence, $Y = OP + 30 + OQ = 69 \cdot 14$.

Having found X and Y draw a right-angled triangle in which the base am is 29·16, and the perpendicular md equal to 69·14, then the hypotenuse gives the magnitude and direction of the resultant. It is equal to 75·2, and 67° N. of E.

By calculation,

$$X = 50 \cos 30° - 20 \cos 45°$$
$$= 50 \times ·866 - 20 \times ·7071 = 29·16$$
$$Y = 50 \sin 30° + 30 + 20 \sin 45°$$
$$= 50 \times ·5 + 30 + 20 \times ·7071 = 69·14$$
$$R = A + B + C = \sqrt{29·16^2 + 69·14^2}$$
$$= 75·2.$$

If θ denote the inclination of R, then

$$\tan \theta = \frac{Y}{R} = \frac{69·14}{29·16} = 2·371;$$
$$\therefore \ \theta = 67°.$$

It will be obvious from the figure that the vertical line, or perpendicular, md, represents the sum of the vertical components, and the horizontal line, or base, am, represents the sum of the horizontal components of the polygon $abcd$.

The general case.—In the preceding examples the given vectors have been taken to act in one plane. In the general case, in which the vectors may act in any specified directions in space, the sum or resultant of a number of vectors may be obtained by using, instead of two, the three co-ordinates, x, y, and z. In this manner the resolved parts of each vector may be obtained, and from these the magnitude and direction of the line representing their sum.

The process may be seen from the following example :

Ex. 4. In the following table r denotes the magnitudes of each of three vectors A, B, and C, and α and β the angles made by each vector with the axes of x and y respectively. Find for each vector the values of θ, x, y, and z, and tabulate as shown.

Vector.	r.	α.	β.	θ.	x.	y.	z.
A	50	45°	60°	60°	35·35	25	25
B	20	30°	100°	61° 21′	17·31	−3·472	9·59
C	10	120°	45°	60°	−5	7·071	5

From the given values of a and β the value of θ (where θ denotes the inclination to the axis of z) can be calculated from the relation

$$\cos^2 a + \cos^2 \beta + \cos^2 \theta = 1.$$

Thus, for vector A, we have

$$\cos^2 \theta = 1 - \cos^2 a - \cos^2 \beta = 1 - \tfrac{1}{2} - \tfrac{1}{4} = \tfrac{1}{4} ;$$
$$\therefore \ \cos \theta = \tfrac{1}{2} \text{ and } \theta = 60°.$$

Similarly for B,

$$\cos^2 \theta = 1 - (·866)^2 - (·1736)^2 = ·23 ; \quad \therefore \ \theta = 61° \ 21'.$$

And for C, $\quad \cos^2 \theta = 1 - \tfrac{1}{4} - \tfrac{1}{2} = \tfrac{1}{4} ; \qquad \qquad \therefore \ \theta = 60°.$

To obtain the projections x, y, and z of each vector, we use the relations $\qquad x = r \cos a, \quad y = r \cos \beta, \quad z = r \cos \theta.$

Thus, for vector A,

$$r = 50°, \quad a = 45°, \quad \beta \text{ and } \theta \text{ are each } 60° ;$$
$$\therefore \ x = 50 \cos 45° = 50 \times ·7071 = 35·35,$$
$$y = 50 \cos 60° = 50 \times ·50 = 25,$$
$$z = 50 \cos 60° = 25.$$

For vector B we have

$$x = 20 \cos 30° = 17·31, \quad y = -20 \cos 80° = 3·472,$$
$$z = 20 \cos 61° \ 21' = 9·59.$$

For C, $\qquad x = -10 \cos 60° = -5, \quad y = 10 \cos 45° = 7·071,$
$$z = 10 \cos 60° = 5.$$

Adding all the terms in column x and denoting the sum by Σx,

$$\Sigma x = 35·35 + 17·31 - 5 = 47·66.$$

Similarly, $\qquad \Sigma y = 25 - 3·472 + 7·071 = 28·6,$
$$\Sigma z = 25 + 9·59 + 5 = 39·59.$$

Hence the resultant, or sum of the three vectors, is

$$A + B + C = \sqrt{(47·66)^2 + (28·6)^2 + (39·59)^2} = (68·4).$$

To find the angles made by the resultant vector with the three axes we have

$$\cos a = \frac{47·66}{68·4} = ·6966 ; \quad \therefore \ a = 45° \ 50'.$$

$$\cos \beta = \frac{28·6}{68·4} = ·4181 ; \quad \therefore \ \beta = 65° \ 18'.$$

$$\cos \theta = \frac{39·59}{68·4} = ·5788 ; \quad \therefore \ \theta = 54° \ 38'.$$

Multiplication of vectors.—Addition, subtraction, and multiplication of *scalar* quantities involving magnitude and not direction may be carried out by any simple arithmetical process.

In the case of *vectors*, addition and subtraction are performed by using a parallelogram or a polygon. In multiplication we may write the product of two vectors A and B as A, B, but it must be remembered that the letters indicate, not only magnitude, but also direction. The process may be shown by the product of two vectors such as a displacement and a force.

Ex. 5. The direction of the rails of a tramway is due N., and a force A of 300 lbs. in a direction 60° N. of E. acts on the car. Find the work done by the force during a displacement of 100 ft.

If θ denote the angle between the direction of the force A and the direction of the displacement ON, then the resolved part of A in the direction ON is $A \cos \theta$.

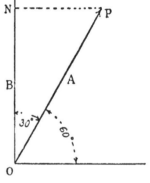

FIG. 150.

The product of a force, or the resolved part of a force, and its displacement, or distance moved through, is the work done by the force. Thus, in Fig. 150, if B denote the displacement of the car, then the work done is

$$A B \cos \theta \quad \text{(i)}.$$

As A is 300, $B = 100$, and $\theta = 30°$.

$$A B \cos 30° = 300 \times 100 \times \cdot 866 = 25980 \text{ ft.-pds.}$$

Observe by way of verification that if θ be 0°, then $\cos 0° = 1$; the force A is acting in the direction ON, and hence

$$\text{work done} = 300 \times 100 = 30,000 \text{ ft.-pds.}$$

When θ is 90°, then $\cos 90° = 0$;

$$\therefore \text{ work done} = 0.$$

This latter result is obvious from the fact that, when the angle is 90°, the force is in a direction at right angles to the direction of motion, and hence no work is done by the force. Again, if the direction of the force were South, then negative work equal to $-300 \times 10 = -3000$ would be done.

From Eq. (i) it follows that the product of two unit vectors such as unit force and unit displacement, is $\cos \theta$. In any diagram, when two vectors are shown acting at a point, care must be taken that the arrow-heads denoting the *sense* of each vector are made to go in a direction outwards from the point. When this is done θ is the angle between the vectors.

EXERCISES. XLVIII.

1. Two vectors A and B act at a point. The magnitude of A is 50, its direction E. B is 100, direction 30° N. of E. Find the resultant or sum $A + B$.

2. Two forces of 8 and 12 units respectively act at a point, the angle between them is 72°. Find their resultant.

3. A ship is sailing apparently to the East, and there is an ocean current of 8·7 knots to the South-west. Find the actual velocity.

4. There are three vectors in one plane :

A, of amount 2, in the direction towards the North-east.
B, of the amount 3, in the direction towards the North.
C, of the amount 2·5, in the direction towards 20° East of South.

By drawing, or any methods of calculation, find the following vector sums and differences :

(i) $A + B + C$, (ii) $B + C - A$, (iii) $A - C$.

5. There are three vectors in a horizontal plane :

A, of amount 1·5, towards the South-east.
B, of amount 3·9, in the direction towards 20° West of South.
C, of amount 2·7, towards the North.

(*a*) Find the vector sums or differences :

$$A + B + C, \quad A - B + C, \quad B - C.$$

(*b*) Find the scalar products AB and AC.

6. Three horizontal vectors are defined as follows :

Vector.	Magnitude.	Direction and Sense.
A	25	Eastward.
B	20	25° North of East.
C	14	80° North of East.

Determine (i) $A + B + C$, and (ii) $A + B - C$, and write down the results. Prove by drawing that $A + B + C = A + C + B$, and $A - (B - C) = A - B + C$.

7. You are given the following three vectors :

	A.	B.	C.
Magnitude, - -	21	15	12
Direction, - · -	0°	75°	120°

Determine and measure the magnitude and direction of the vector sum $A + B + C$.

Also, verify by drawing, that $A - (B - C) = A - B + C$.

8. A force A acts on a tramcar, the direction of the rails being due north. If B denote the velocity of the car, find the vector product $A \times B$ (called the activity or the power).

(i) A is 300 lbs. N. ; B is 20 ft. per sec.

(ii) A is 250 lbs. N.E. ; B is 15 ,,

(iii) A is 200 lbs. E. ; B is 20 ,,

(iv) A is 150 lbs. S.E. ; B is 10 ,,

9. The following five vectors represent displacements :

	A.	B.	C.	D.	E.
Magnitude, -	20	12	6·8	3·3	15·5
Direction, -	0°	75°	310°	225°	120°

Find the vector sums of

(i) $A + B + C + D + E$.   (ii) $A + B + E + D + C$.

(iii) $A + B - C + D - E$.   (iv) $A + B - E + D - C$.

10. A cyclist is travelling at 10 miles per hour in a northerly direction and a south-west wind is blowing at 5 miles an hour. Determine the magnitude and direction of the wind which the rider experiences.

11. Three vectors A, B, and C act at a point. The magnitudes and inclinations of each vector to the axes of x and y are given in the following table. Find in each case the inclination to the axis of z. Also find the sum and the inclination which the line representing the sum makes with the three axes.

	r.	a.	β.	θ.
A	100	30°	120°	
B	50	135°	30°	
C	10	45°	60°	

12. Let A_a denote a vector, where A gives its magnitude, and a its direction.

Find A and a in the following vector equation, that is, add the three given vectors, which are all in the plane of the paper :

$$A_a = 3 \cdot 7_{30°} + 1 \cdot 4_{82°} + 2 \cdot 6_{157°}.$$

Find also B and β from the equation

$$B_\beta = 3 \cdot 7_{30°} - 1 \cdot 4_{82°} + 2 \cdot 6_{157°}.$$

Use a scale of 1 inch to 1 unit.

CHAPTER XXIV.

ALGEBRA (*continued*); SQUARE ROOT; QUADRATIC EQUA-
TIONS; ARITHMETICAL, GEOMETRICAL, AND HAR-
MONICAL PROGRESSIONS.

Square root.—In the process of division advantage is taken
of the results obtained from multiplication. In like manner, the
square root (p. 25) of an algebraical expression can often be
obtained by comparing it with known forms of the squares of
different expressions.

Thus, the square of $(a+b)$, or $(a+b)^2$ is $a^2+2ab+b^2$. Hence,
when any expression of this form is given, its *square root* can
be seen at once and written down, *e.g.* $\sqrt{x^2+2xy+y^2}=x+y$.

We may, from this example, proceed to derive a general rule
for the extraction of the square root.

$$a^2+2ab+b^2\,(\,a+b$$
$$\underline{a^2}$$
$$2a+b\,)\,2ab+b^2$$
$$\underline{2ab+b^2}$$

Thus, arrange the terms according to the dimensions of one
term, as a. The square root of the first term is a; taking its
square from the whole expression, $2ab+b^2$ remains; mentally
dividing $2ab$ by $2a$, the double of the first term of the required
square root, we find that it is contained b times in $2ab$. Hence,
adding b to the $2a$ previously obtained, we obtain the full trial
divisor $2a+b$. Multiply this result by the new term of the
required root, b, and subtract the product from the first remainder. Then, as there is no remainder $a+b$ is the root
required.

Ex. 1. Find the square root of $4x^2 + 24xy + 36y^2$.

Here $2x$ is clearly the root of the first term $4x^2$. Put $2x$ for the first term of the required root ; square it, and subtract its square from the given expression, Bring down the other

$$4x^2 + 24xy + 36y^2 \,(\, 2x + 6y$$
$$\underline{4x^2}$$
$$4x + 6y \,)\, 24xy + 36y^2$$
$$\underline{24xy + 36y^2}$$

two terms $24xy + 36y^2$. Multiply the first term of the root $2x$ by 2, giving $4x$, and using this as a trial divisor, the remaining term of the root is found to be $6y$. Hence, put $6y$ as the second term in the root and multiply $4x + 6y$ by $6y$, giving as a product $24xy + 36y^2$. subtract this from the two remaining terms of the given expression, and there is no remainder. The required root is $2x + 6y$.

Following the steps in the preceding worked out example the next will be readily made out.

Ex. 2. Find the square root of
$$4x^4 + 4x^2y^2 - 12x^2z^2 + y^4 - 6y^2z^2 + 9z^4.$$

$$4x^4 + 4x^2y^2 - 12x^2z^2 + y^4 - 6y^2z^2 + 9z^4 \,(\, 2x^2 + y^2 - 3z^2$$
$$\underline{4x^4}$$
$$\, 4x^2y^2 - 12x^2z^2 + y^4 - 6y^2z^2 + 9z^4$$
$$4x^2 + y^2 \,)\, 4x^2y^2 + y^4$$
$$\underline{\, 4x^2y^2 + y^4}$$
$$4x^2 + 2y^2 - 3z^2 \,)\, -12x^2z^2 - 6y^2z^2 + 9z^4$$
$$\underline{\, -12x^2z^2 - 6y^2z^2 + 9z^4}$$

The expression for the expansion of $(1 + a)^n$ is given on p. 111. When n is $\frac{1}{2}$, and a is small compared with unity, the square root of $(1 + a)$ can be obtained to any desired degree of accuracy.

Ex. 3. Find the first five terms of the square root of $1 + x$, and use them to find the value of $\sqrt{101}$.

$$(1 + x)^n = 1 + \frac{nx}{1} + \frac{n(n-1)}{1 \cdot 2} x^2 + \frac{n(n-1)(n-2)}{1 \cdot 2 \cdot 3} x^3 + \cdots .$$

When $n = \frac{1}{2}$, this becomes

$$1 + \frac{1}{2} x + \frac{\frac{1}{2}\left(\frac{1}{2} - 1\right)}{1 \cdot 2} x^2 + \frac{\frac{1}{2}\left(\frac{1}{2} - 1\right)\left(\frac{1}{2} - 2\right)}{1 \cdot 2 \cdot 3} x^3 + \cdots$$

$$= 1 + \frac{1}{2} x - \frac{1}{8} x^2 + \frac{1}{16} x^3 - \frac{5}{128} x^4 + \cdots .$$

$$\sqrt{101} = \sqrt{(100+1)} = 10\sqrt{\left(1+\frac{1}{100}\right)}$$

$$= 10\left(1+\frac{1}{200}-\frac{1}{80000}+\frac{1}{16000000}-\frac{5}{12800000000}+\text{etc.}\right)$$

$$= 10\cdot04988....$$

EXERCISES. XLIX.

Find the square root of

1. $9x^4 - 42x^3 + 37x^2 + 28x + 4$. 2. $4x^4 + 12x^3 - 11x^2 - 30x + 25$.

3. $x^4 + 4ax^3 + 2a^2x^2 - 4a^3x + a^4$.

4. $x^6 - 22x^4 + 34x^3 + 121x^2 - 347x + 289$.

5. $25x^8 - 60x^6 - 34x^4 + 84x^2 + 49$. 6. $a^4 - 2a^3 + 9a^2 - 8a + 16$.

7. $25x^4 - 30x^3 + 49x^2 - 24x + 16$. 8. $x^4 + 8x^3 - 26x^2 - 168x + 441$.

9. $16x^4 - 40x^3 + 89x^2 - 80x + 64$. 10. $\dfrac{4x^2}{y^2} + \dfrac{4y^2}{x^2} - 8$.

11. $9x^4 - 30x^3y + 31x^2y^2 - 10xy^3 + y^4$.

12. Show that the square of the sum of two quantities together with the square of their difference is double the sum of their squares.

13. Show that the sum of the squares of two quantities is greater than the square of their difference by twice the product of the quantities.

14. Find the square root of the difference of the squares of
$$5x^2 - 8x + 13 \text{ and } 4x^2 + 2x - 12.$$

Quadratic equations.—As already indicated (p. 83) when a given equation expressed in its simplest form involves the *square* of the unknown quantity it is called a **quadratic** equation. Such an equation may contain only the square of the unknown quantity, or it may include both the square and the first power.

Ex. 1. Solve the equation $x^2 - 9 = 0$.
We have $\qquad x^2 = 9 = 3^2$; $\therefore x = \pm 3$.

It is necessary to insert the double sign before the value obtained for x as both $+3$ and -3, when squared give 9.

The solution of a given quadratic equation containing both x^2 and x can be effected by any of the three following methods.

FIRST METHOD.—The method most widely known, and generally used, may be stated as follows :

Bring all the terms containing x^2 and x to the left hand side of the equation, and the remaining terms to the right hand side.

Simplify, if necessary, and make the coefficient of x^2 unity.

Finally, add the square of one-half the coefficient of x to both sides of the equation and the required roots can be readily obtained.

Ex. 2. Solve the equation $x^2 + 4x - 21 = 0$.

We have $\qquad x^2 + 4x = 21.$ ....(i)

Add the square of one-half the coefficient of x to each side ;

$\qquad \therefore\ x^2 + 4x + (2)^2 = 21 + 4 = 25\ ;$

$\qquad i.e.\ (x+2)^2 = 5^2\ ;$

$\qquad \therefore\ x + 2 = \pm 5\ ;$ ....(ii)

$\qquad \therefore\ x = -2 \pm 5 = 3,\ \text{or}\ -7.$

It will be noticed that (ii) may be written

$$x + 2 = +5\ \text{and}\ x + 2 = -5.$$

From these equations the values $x = 3$ and $x = -7$ are at once obtained.

SECOND METHOD.—The second and the third methods of solution are explained on p. 191, but it may be advisable to refer to them again here. Where the given equation can be resolved into factors, then the value of x which makes either of these factors vanish, is a value of x which satisfies the given equation.

Ex. 3. Solve the equation $x^2 + 4x - 21 = 0$.

Since $\qquad x^2 + 4x - 21 = (x - 3)(x + 7)\ ;$

$\qquad\qquad x - 3 = 0,\ \text{when}\ x = 3\ ;$

and $\qquad\qquad x + 7 = 0,\ \text{when}\ x = -7.$

Hence $x = 3$ or $x = -7$ is a solution of the equation and 3 and -7 are the roots of the given equation.

THIRD METHOD.—Let $y = x^2 + 4x - 21.$ Substitute values 1, 2, 3 ... for x and calculate corresponding values of y. Plot the values of x and y on squared paper. The two points of intersection (of the curve passing through the plotted points) with the axis of x are the roots required.

A quadratic equation in its general form may be written

$$ax^2 + bx + c = 0.$$

Then $\qquad\qquad x^2 + \dfrac{b}{a}x = -\dfrac{c}{a}$

adding to each side the square of half the coefficient of x, or $\left(\dfrac{b}{2a}\right)^2$, we have

$$x^2 + \frac{b}{a}x + \left(\frac{b}{2a}\right)^2 = \frac{b^2}{4a^2} - \frac{c}{a} = \frac{b^2 - 4ac}{4a^2}.$$

$$\therefore \quad x = -\frac{b}{2a} \pm \frac{\sqrt{b^2 - 4ac}}{2a}. \quad \ldots\ldots \quad \ldots\ldots\ldots(i)$$

The following important cases occur. If b^2 is greater than $4ac$ i.e. $b^2 > 4ac$, there are two values, or roots, satisfying the given equation. If $b^2 = 4ac$ the two roots are equal; each is $-\frac{b}{2a}$. If $b^2 < 4ac$, there are no real values which satisfy the given equation, and the roots are said to be .imaginary. All these results may be clearly appreciated by using squared paper.

Ex. 4. Solve the equations :
   (i) $2x^2 - 4x + 1 = 0$,
   (ii) $2x^2 - 4x + 2 = 0$,
   (iii) $2x^2 - 4x + 3 = 0$.

(i) Let $y = 2x^2 - 4x + 1$. Assume $x = 0, 1, 2, \ldots$ etc., and find cor-responding values of y.

Thus, when $x = 0$, $y = 1$; when $x = 1$, $y = -1$; when $x = 2$, $y = 1$.

Plot these values on squared paper; then the curve passing through the plotted points intersects the axis of x at points A and B (Fig. 151) for values of $x = \cdot293$, and $1\cdot707$, and these are the roots required.

It will be noticed each time the curve intersects the axis of x the value of y changes sign. Hence we know that one value lies between $x = 0$ and $x = 1$; and between $x = 1$ and $x = 2$.

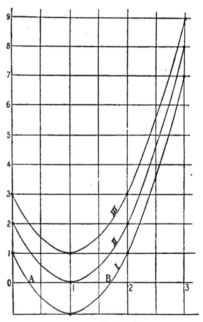

FIG. 151. –To illustrate Ex. 4.

(ii) Let $y = 2x^2 - 4x + 2$. Values of x and corresponding values of y are as follows :

x	0	1	2	3
y	2	0	2	8

Plotting as before the curve (ii) (Fig. 181) is obtained and touches, or better is tangent to, the axis of x at the point $x=1$. Hence the two roots of the equation are equal.

(iii) Proceed as before and obtain the following values:

x	0	1	2	3
y	3	1	3	9

The curve joining the plotted points is shown by (iii) (Fig. 181); this does not intersect the axis of x, and the roots are imaginary.

Much unnecessary labour will result if the attempt is made to obtain unity as the coefficient of x^2 in all equations. It may be found better to use another letter, such as y or z, and then to proceed to solve the equation in the ordinary manner, finally solving the equation for x. The following examples will illustrate some of the methods which may be adopted.

Ex. 5. Solve the equation $\left(\dfrac{a-x}{x-b}\right)^2 = 8\left(\dfrac{a-x}{x-b}\right) - 15.$

By transposition, we obtain

$$\left(\frac{a-x}{x-b}\right)^2 - 8\left(\frac{a-x}{x-b}\right) = -15.$$

If we write y for $\dfrac{a-x}{x-b}$ the equation becomes

$$y^2 - 8y = -15\ ;$$
$$\therefore\ y^2 - 8y + (4)^2 = -15 + 16 = 1\ ;$$
$$\therefore\ y = 4 \pm 1 = 3,\ \text{or}\ 5.$$

Hence $\qquad \dfrac{a-x}{x-b} = 3\ ;\quad \therefore\ x = \dfrac{a+3b}{4}\ ;$

or $\qquad \dfrac{a-x}{x-b} = 5\ ;\quad \therefore\ x = \dfrac{a+5b}{6}.$

Instead of using the letter y, the equation could be written as

$$\left(\frac{a-x}{x-b}\right)^2 - 8\left(\frac{a-x}{x-b}\right) + (4)^2 = -15 + 16 = 1\ ;$$

$$\therefore\ \frac{a-x}{x-b} = 4 \pm 1 = 3,\ \text{or}\ 5.$$

Two simultaneous quadratics.—Some methods which may be adopted to obtain the solution of simultaneous equations of the first degree are explained in Chap. IX., p. 91. Similar processes are applicable in equations of the second degree. That is to say we can, by multiplication, division, or substitution, obtain an equation involving only one unknown quantity. From this equation the value of the unknown quantity can be determined, and by substitution the value of the remaining unknown can be found.

Ex. 6.   Solve the equation $x^2 + y = 8$, $3x + 2y = 7$.

$$x^2 + y = 8. \quad\dots\dots\dots\dots\dots\dots(i)$$
$$3x + 2y = 7. \quad\dots\dots\dots\dots\quad\dots\dots\dots(ii)$$

Multiply (i) by 2 and subtract (ii) from it ,

$$\therefore\ 2x^2 + 2y = 16$$
$$\underline{3x\ + 2y = \ 7}$$
$$2x^2 - 3x = \ 9$$

$$\therefore\ x^2 - \frac{3}{2}x + \left(\frac{3}{4}\right)^2 = \frac{9}{2} + \frac{9}{16} = \frac{81}{16},$$

$$x = \frac{3}{4} \pm \frac{9}{4} = 3, \text{ or } -1\cdot5.$$

From (ii), when x is 3 ; $2y = 7 - 9$; $\therefore\ y = -1$;
when x is $-1\cdot5$; $2y = 7 + 4\cdot5$; $\therefore\ y = 5\cdot75$.

Ex. 7.   Solve the equation
      (i) $x^2 + xy = 84$;   (ii) $xy + y^2 = 60$.

Adding (ii) to (i) we get

$$x^2 + 2xy + y^2 = 144 ;$$
$$\therefore\ x + y = \pm 12. \quad\dots\dots\dots\dots\dots\dots(iii)$$

From (i), $x(x + y) = 84$.
From (ii), $y(x + y) = 60$.
Substituting from (iii), $\pm 12x = 84$ and $\pm 12y = 60$.
Hence $x = \pm 7,\ y = \pm 5$;
therefore the four values are $x = 7$, $x = -7$, $y = 5$, $y = -5$.

Equations reducible to quadratics.—Equations of the fourth degree can in some cases be solved as two quadratic equations.

Ex. 8.   Solve $x^4 - 17x^2 + 16 = 0$.
The equation may be written

$$(x^4 - 8x^2 + 16) - 9x^2 = 0, \text{ or } (x^2 - 4)^2 - (3x)^2 = 0 ;$$
$$\therefore\ (x^2 + 3x - 4)(x^2 - 3x - 4) = 0.$$

Hence . $x^2 + 3x - 4 = 0, \dots\dots\dots\ .\ \dots\dots\dots\dots\dots(i)$
or $x^2 - 3x - 4 = 0. \quad\dots\dots\dots\dots\dots\dots\dots(ii)$
    From (i), $x^2 + 3x - 4 = (x + 4)(x - 1) ;$
                $\therefore\ x = -4, \text{ or } 1.$

From (ii), $\qquad x^2 - 3x - 4 = (x-4)(x+1)$;

$\qquad\qquad\qquad \therefore \; x = 4, \; \text{or} \; -1.$

Hence the values of x which satisfy the given equation are

$$x = \pm 4, \; x = \pm 1.$$

Relations between the coefficients and the roots of a quadratic equation.—In the preceding examples we have been·able, from a given quadratic equation, to find the roots, or the values which satisfy the given equation. The converse of this is often required, *i.e.* to form a quadratic equation with given roots.

It has been already seen that if we can resolve the left-hand side of the given equation, when reduced to its simplest form, into factors, then the value of x which makes either of these factors zero, is a value of x which satisfies the given equation.

Thus the roots of the equation $(x-a)(x-\beta) = 0$ are a and β.

Conversely, an equation having for its roots a and β is

$$(x-a)(x-\beta) = 0.$$

Hence if a and β denote the roots of the equation

$$ax^2 + bx + c = 0.$$

We have $\qquad ax^2 + bx + c = a(x-a)(x-\beta)$;

$\qquad \therefore \; ax^2 + bx + c = a(x^2 - ax - \beta x + a\beta)$

$\qquad\qquad\qquad\qquad = a(x^2 - (a+\beta)x + a\beta).$

Comparing coefficients on both sides we have

$$a(a+\beta) = -b \; \text{ and } \; aa\beta = c \; ;$$

$$\therefore \; a + \beta = -\frac{b}{a} \; \text{ and } \; a\beta = \frac{c}{a} \; ;$$

therefore when the coefficient of x^2 is unity the sum of the roots is equal to the coefficient of x ; and the product of the roots is equal to the remaining term.

Ex. 9. Form the quadratic equations having roots 1 and 4.

Here $\qquad\qquad (x-1)(x-4) = x^2 - 5x + 4.$

Ex. 10. Form the quadratic equation having roots

$$-3 + \sqrt{2} \; \text{ and } \; -3 - \sqrt{2}.$$

Here we have $(x + 3 - \sqrt{2})(x + 3 + \sqrt{2}) = (x+3)^2 - 2$;

$\qquad \therefore \;$ the required equation is $x^2 + 6x + 7 = 0.$

Ex. 11.   Form the quadratic equation having roots a and $\dfrac{1}{a}$.

Here we have $\qquad (x-a)\left(x-\dfrac{1}{a}\right)$;

$\qquad \therefore$ required equation is $x^2 - \dfrac{a^2+1}{a}x + 1 = 0$.

EXERCISES.   L.

Solve the equations :

1. $x^2 - 60 = 80 - 4x$.

2. $x^2 + 32x = 320$.

3. $\dfrac{x^2}{12} - x + \dfrac{5}{3} = 0$.

4. $2x^2 - 4x - 6 = 0$.

5. $3x^2 - 84 = 9x$.

6. $x^2 + {\cdot}402x = {\cdot}163{\cdot}$

7. $6x^2 - 13x + 6 = 0$.

8. $x^2 - (a+b)x + ab = 0$.

9. $\dfrac{1}{\sqrt{x+2}} = \dfrac{1}{2}\sqrt{3-x}$.

10. $19x^2 - 4x - 288 = 0$.

11. $4x^2 + 4x - 3 = 0$.

12. $\sqrt{(5x+9)} - \sqrt{3x+1} = \sqrt{2(x-6)}$.

13. $x^3 - 2x^2 - 3x + 4 = 0$.

14. $(x^2 - 4x + 3)^2 - 8(x^2 - 4x + 3) = 0$.

15. $1 + 2x\sqrt{(1-x^2)} = 9x^2$.

16. $x + \dfrac{1}{x} = 2(1 + \sqrt{2})$.

17. $\dfrac{1}{1+x} + \dfrac{1}{2+x} = \dfrac{1}{1-x} + \dfrac{1}{2-x}$.

18. $40\left(x + \dfrac{1}{x}\right)^2 - 286\left(x + \dfrac{1}{x}\right) + 493 = 0$.

19. $x^2 + \dfrac{1}{x^2} + \dfrac{1}{3}\left(x + \dfrac{1}{x}\right) = 3\tfrac{5}{12}$.

20. $\left(\dfrac{a-x}{x-b}\right)^2 = 8\left(\dfrac{a-x}{x-b}\right) - 15$.

21. $x^3 + y^3 = 72$, $\quad xy(x+y) = 48$.

22. $x^2 - 4y^2 = 8$, $\quad 2(x+y) = 7$.

23. $2x^2 - 3y^2 = 5$, $\quad 3x + y = 15$.

24.   (i) Find the roots of the equation $x^2 - 2ax + (a-b)(a+b) = 0$.
      (ii) Form the equation the roots of which are the squares of the roots of the given equation.

25.   Find the roots of the equation $x^2 + 7x\sqrt{2} = 60$.   Form the quadratic equation having roots a and $\dfrac{1}{a}$.

26.   If a and β are the roots of the equation $ax^2 + bx + c = 0$, show that
$\qquad a + \beta = -\dfrac{b}{a}$ and $a\beta = \dfrac{c}{a}$.

Problems leading to quadratic equations.—As already indicated on p. 81, one of the greatest difficulties experienced by

a beginner in Algebra is to express the conditions of a given problem by means of algebraical symbols. The equations themselves may be obtained more or less readily since the conditions are generally similar to those already explained, but some difficulty may be experienced in the interpretation of the results derived from such equations. Since a quadratic equation which involves one unknown quantity has two solutions, and simultaneous quadratics involving two unknown quantities may have four values, or solutions, it is clear that ambiguity may arise. It will be found, however, that although the equations may have general solutions only one solution may be applicable to the particular problem. The fact that several solutions can be found and only one applies to the problem is due to the circumstance that algebraical language is far more general than ordinary methods of expression. Usually no difficulty will be experienced in deciding which of the solutions are applicable to the problem in hand.

Ex. 1. A boat's crew can row at the rate of 9 miles an hour. What is the speed of the river's current if it takes them $2\frac{1}{4}$ hours to row 9 miles up stream and 9 miles down?

Let s denote the speed of the current in miles per hour.

Then, $9-s$ and $9+s$ represent the crew's rate up and down stream respectively;

$$\therefore \frac{9}{9-s}+\frac{9}{9+s}=2\frac{1}{4}=\frac{9}{4}.$$
$$36+4s+36-4s=81-s^2.$$
$$s^2=9, \quad s=\pm 3.$$

Only the positive value is applicable to the problem.

Ex. 2. A certain number of articles are bought for £1, and £1. 0s. 7d. is made by selling all but one at 1d. each more than they cost. How many are bought?

Let x denote the number bought.

Then $\dfrac{240}{x}=$ price per article in pence;

$$\therefore (x-1)\left(\frac{240}{x}+1\right)=247;$$
$$\therefore (x-1)(240+x)=247x;$$
$$\therefore x^2+239x-240=247x,$$
or
$$x^2-8x-240=0;$$
$$\therefore (x-20)(x+12)=0.$$

The two values obtained are $x=20$ and $x=-12$. Obviously only the former is applicable to the problem, hence $x=20$.

Ex. 3. In the equation $s=Vt+\frac{1}{2}ft^2$. Given $s=80$, $V=64$, and $f=32$, find t.

Substituting the given values

$$80=64t+\tfrac{1}{2}\times 32t^2=64t+16t^2\ ;$$
$$\therefore\ t^2+4t+(2)^2=5+4=9\ ;$$
$$\therefore\ t=-2\pm 3=1,\ \text{or}\ -5.$$

In the case of a body projected upwards with a vertical velocity 64, then, when f is 32, the body is at a distance 80 from the starting point when $t=1$ and is moving upwards. The same conditions hold true again when $t=-5$, and the body is moving in the opposite direction.

EXERCISES. LI.

1. In the formula $t=\pi\sqrt{\dfrac{l}{g}}$.

(i) given $t=\dfrac{12}{7}$, $g=32$, $\pi=\frac{22}{7}$, find the numerical value of l.

(ii) $t=\dfrac{30}{19}$, $l=8$, find g.

2. In the formula $s=Vt+\frac{1}{2}ft^2$.

(i) given $V=12$, $s=470$, $f=7$, find t.

(ii) $V=172$, $s=90$, $f=32$, find t.

3. The area of a certain rectangle is equal to the area of a square whose side is six inches shorter than one of the sides of the rectangle. If the breadth of the rectangle be increased by one inch and its length diminished by two inches, its area is unaltered. Find lengths of its sides.

4. The perimeter of a rectangular field is to its diagonal as 34 to 13, and the length exceeds the breadth by 70 yards. What is its area?

5. A traveller starts from A towards B at 12 o'clock, and another starts at the same time from B towards A. They meet at 2 o'clock at 24 miles from A, and the one arrives at A while the other is still 20 miles from B. What is the distance between A and B?

6. From a catalogue it is found that the prices of two kinds of motors are such that seven of one kind and twelve of the other can be obtained for £250. Also that three more of the former can be purchased for £50 than can three of the latter for £30. Find the price of each.

7. A boat's crew can row at the rate of 8 miles per hour. What is the speed of the river's current if it takes them 2 hours and 20 minutes to row 8 miles up stream and 8 miles down?

8. A person lends £1500 in two separate sums at the same rate of interest. The first sum with interest is repaid at the end of eight months, and amounts to £936 ; the second sum with interest is repaid at the end of ten months, and amounts to £630. Find the separate sums lent and rate of interest.

9. Show that if the sum of two numbers be multiplied by the sum of their reciprocals the product cannot be less than 4.

10. Divide £490 among A, B, and C, so that B shall have £2 more than A, and C as many times B's share as there are shillings in A's share.

11. If in the equation $ax^2 + bx + c = 0$ the relations between a, b, and c are such that $a + b + 3 = 0$ and $2a - c + = 0$, what must be the value of a in order that one of the roots may be 5, and what is then the value of the other root?

Series.—The term **series** is applied to any expression in which each term is formed according to some law.

Thus, in the series 1, 3, 5, 7 ... each term is formed by adding 2 to the preceding term. In 1, 2, 4, 8 ... each term is formed by multiplying the preceding term by 2.

Usually only a few terms are given sufficient to indicate the law which will produce the given terms.

The first series is called an *arithmetical progression*, the constant quantity which is added to each term to produce the next is called the *common difference*. The letters A.P. are usually used to designate such a series.

The second series is called a *geometrical progression*, the constant quotient obtained by dividing any term by the preceding term is called the *common ratio* or constant factor of the series. The letters G.P. are used to denote a geometrical progression.

Arithmetical Progression.—A series is said to be an arithmetical progression when the difference between any two consecutive terms is always the same.

Thus the series 1, 2, 3, 4 ... is an arithmetical series, the constant difference obtained by subtracting from any term the preceding term is unity.

In the series 21, 18, 15, ... the constant difference is -3.

Again in a, $a+d$, $a+2d$, ... and a, $a-d$, $a-2d$, ... the first increases and the second diminishes by a common difference d.

In writing such a series it will be obvious that if a is the first term, $a+d$ the second, $a+2d$ the third, etc., any term such as the seventh is the first term a together with the addition of d repeated $(7-1)$ times or is $a+6d$.

If l denote the last term, and n the number of terms, then

$$l=a+(n-1)d \dots\dots\dots\dots\dots \text{(i)}$$

Let S denote the sum of n terms, then

$$S=a+(a+d)+(a+2d)+\dots+(l-2d)+(l-d)+l.$$

Writing the series in the reverse order we obtain

$$S=l+(l-d)+(l-2d)+\dots(a+2d)+(a+d)+a.$$

Adding we obtain

$$2S=(a+l)+(a+l)+\dots \text{ to } n \text{ terms}$$
$$=n(a+l);$$

$$\therefore\ S=\frac{n}{2}(a+l)\dots\dots\dots\dots\dots\text{(ii)}$$

From this when a and l are known the sum of n terms can be obtained.

Again, substituting in (ii) the value of l from (i) we have

$$S=\frac{n}{2}\{2a+(n-1)d\}\dots\dots\dots\dots\text{(iii)}$$

From Eq. (iii) the sum of n terms can be obtained when the first term and the common difference are known.

Arithmetical Mean.—The middle term of any three quantities in an arithmetical progression is the arithmetical mean of the remaining two.

Thus if a, A, and b form three quantities in arithmetical progression, then

$$A-a=b-A ;$$
$$\therefore\ A=\frac{a+b}{2} ;$$

or, *the arithmetical means of two quantities is one-half their sum.*

Ex. 1.   Find the 9th term of the series 2, 4, 6 ... , also the sum of nine terms.

Here, from (i), $l=a+(n-1)d.$
$$a=2,\ \ n=9,\ \text{ and }\ d=2 ;$$
$$\therefore\ l=2+(9-1)2=18.$$

From (ii), $S=\dfrac{n}{2}(a+l)=\dfrac{9}{2}(2+18)=90.$

Ex. 2. The second term of an A.P. is 24. The fifth term is 81. Find the series.

Here $a+d=24,$

also $a+4d=81$;

$\therefore 3d=57,$ or $d=19.$

As the second term is 24, the first term is $24-19=5.$ Hence the series is 5, 24, 43 ... .

Ex. 3. The twentieth term of an A.P. is 15 and the thirtieth is 20. What is the sum of the first 25 terms ?

Here. $a+19d=15$

$a+29d=20$

By subtraction, $10d= 5;\quad \therefore\ d=\tfrac{1}{2}.$

By substitution, $a=\dfrac{11}{2};$

$$\therefore\ S=\frac{n}{2}\{2a+(n-1)d\}$$

$$=\frac{25}{2}\{11+(25-1)\tfrac{1}{2}\}=287\tfrac{1}{2}.$$

EXERCISES. LII.

Sum the following series :

1. 3, $3\tfrac{1}{2}$, 4 ... to 10 terms.
2. $-2\tfrac{1}{2}$, -2, $-1\tfrac{1}{2}$ to 21 terms.
3. $7+32+57+$ ... to 20 terms.
4. $2+3\tfrac{1}{3}+4\tfrac{2}{3}+$ .. to 10 terms.
5. $\dfrac{1}{3}-\dfrac{1}{3}-1-$ ... to 20 terms.
6. $\dfrac{1}{4}-\dfrac{1}{4}-\dfrac{3}{4}$ ... to 21 terms.
7. $1-\dfrac{6}{5}-\dfrac{17}{5}-$ ... to 12 terms.

8. Find the sum of 16 terms of the series $64+96+128+.$

9. Sum the series $9+5+1-3-$ to n terms.

10. The sum of n terms of the series 2, 5, 8 ... is 950. Find n.

11. The sum of n terms of an A.P. whose first term is 5 and common difference 36 is equal to the sum of $2n$ terms of another progression whose first term is 36 and common difference is 5. Find the value of n.

12. The first term of an A.P. is 50, the fifth term 42. What is the sum of 21 terms ?

13. The fourth term is 15 and the twentieth is $23\tfrac{1}{2}$. Find the sum of the first 20 terms

14. The sum of 20 terms is 500 and the last is 45. Find the first term.

15. The sum of three numbers is 21, and their product is 315. Find the numbers.

16. If the sum of n terms be n^3 and common difference be 2, what is the first term?

17. The sum of an A.P. is 1625, the second term is 21, and the seventh 41. Find the number of terms.

18. Find the sum of the first n natural numbers.

19. Find the sum of the first n odd natural numbers.

20. Show that if unity be added to the sum of any number of terms of the series 8, 40, 72 ... the result will be the square of an odd number.

Geometrical progression.—A series of terms are said to be in geometrical progression when the quotient obtained by dividing any term by the preceding term is always the same.

The constant quotient is called the *common ratio* of the series. Let r denote the common ratio and a the first term.

The series of terms a, ar, ar^2, etc., form a geometrical progression, and any term, such as the third, is equal to a multiplied by r raised to the power $(3-1)$ or ar^2.

Thus, if l denote the last term and n the number of terms then we have
$$l = ar^{n-1}. \quad\text{.............................(i)}$$
Let S denote the sum of n terms then
$$S = a + ar + ar^2 + \dots ar^{n-2} + ar^{n-1}. \quad\text{.................(ii)}$$
Multiplying every term by r
$$Sr = ar + ar^2 + ar^3 + \dots ar^{n-1} + ar^n. \quad\text{.............(iii)}$$
(Subtract ii) from (iii).
$$\therefore\quad rS - S = ar^n - a,$$
or
$$S(r-1) = a(r^n - 1).$$
$$\therefore\quad S = \frac{a(r^n - 1)}{r-1}. \quad\text{.........................(iv)}$$

Ex. 1. The first term of a G.P. is 5; the third term is 20. Find the eighth term and the sum of eight terms.

The third term will be ar^2 where a denotes the first term and r the common ratio;
$$\therefore\quad 5r^2 = 20 \text{ or } r = 2.$$
From
$$l = ar^{m-1}$$
we get by substitution
$$l = 5r^7 = 5 \times 2^7$$
$$= 640.$$
$$S = \frac{a(r^n - 1)}{r-1} = \frac{5(2^8 - 1)}{1}$$
$$= 5 \times 255 = 1275.$$

Ex. 2. The third term of a G.P. is 20. The eighth term is 640, and the sum of all the terms is 20475. Find the number of terms.

Here $\qquad ar^2 = 20$ and $ar^7 = 640$;

$$\therefore \frac{ar^7}{ar^2} = \frac{640}{20} ;$$

or $\qquad r^5 = 32 ; \quad \therefore r = 2,$

and $\qquad a = \frac{20}{4} = 5.$

$$20475 = \frac{a(r^n - 1)}{r - 1}$$

$$= \frac{5(2^n - 1)}{2 - 1} ;$$

$$\therefore 2^n - 1 = \frac{20475}{5} = 4095 ;$$

or $\qquad 2^n = 4096 = 2^{12} ;$

$$\therefore n = 12 ;$$

or $\qquad n \log 2 = \log 4096 ;$

$$\therefore n = \frac{3 \cdot 6123}{\cdot 3010} = 12.$$

Ex. 3. The sum of a G.P. is 728, common ratio 3, and last term 486. Find the first term.

$$S = \frac{ar^n - a}{r - 1} ;$$

but, $\qquad r^{n-1} = \frac{l}{a} ; \quad \text{or } ar^n = lr ;$

$$\therefore S = \frac{rl - a}{r - 1} ;$$

or, $\qquad 728 = \frac{3 \times 486 - a}{3 - 1} ;$

$$\therefore a = 1458 - 1456 = 2.$$

By changing signs in both numerator and denominator Eq. (iv) becomes

$$S = \frac{a(1 - r^n)}{1 - r}. \quad \dots\dots\dots\dots\dots\dots\text{(v)}$$

When r is a *proper fraction* it is evident that r^n decreases as n increases. Thus when r is $\frac{1}{10}$, $r^2 = \frac{1}{100}$, $r^3 = \frac{1}{1000}$, etc., when n is indefinitely great, r^n is zero, and (v) becomes

$$S = \frac{a}{1 - r}. \quad \dots\dots\dots\dots\dots\dots\text{(vi)}$$

Sum of a G.P. containing an infinite number of terms.—Eq. (vi) is used to find the sum of an infinite number of terms, or as it is called the sum of a series of terms to infinity.

Ex. 4.   Find the sum of the series, 84, 14, $2\frac{1}{3}$ ... to infinity.

Here
$$r = \frac{14}{84} = \frac{1}{6} \; ;$$

$$\therefore \; S = \frac{a}{1-r} = \frac{84}{1 - \frac{1}{6}} = \frac{504}{5} = 100\cdot8\cdot$$

Value of a recurring decimal.—The arithmetical rules for finding the value of a recurring decimal depend on the formula for the sum of an infinite series in G.P.

Ex. 5.   Find the value of $3\cdot\dot{6}$.

$$3\cdot\dot{6} = 3\cdot666 ... = 3 + \frac{6}{10} + \frac{6}{10^2} + \frac{6}{10^3} + -$$

$$\therefore \; r = \frac{1}{10} \text{ and } a = \cdot6 \; ;$$

$$\therefore \; S = \frac{\cdot6}{1 - \frac{1}{10}} = \frac{\cdot6}{\frac{9}{10}} = \frac{6}{9} = \frac{2}{3} \; ;$$

$$\therefore \; 3\cdot\dot{6} = 3\frac{2}{3}\cdot$$

Geometrical mean.—The middle term of any three quantities in a geometrical progression is said to be a *geometric mean* between the other two.   The two initial letters G.M. may be used to denote the geometric mean.   Thus, if x and y denote two numbers, the A.M. is $\frac{x+y}{2}$ the G.M. is $\sqrt{xy}$.

In the progression 2, 4, 8 ... the middle term 4 is the G.M. of 2 and 8.   In like manner in a, ar, $ar^2 + ... ar$ is the G.M. of a and ar^2.

It will be noticed that the G.M. of two quantities is the square root of their product.

To insert a given number of geometric means between two given quantities.

From
$$l = ar^{n-1}$$
we obtain
$$r^{n-1} = \frac{l}{a}$$
from this when l and a are given r can be obtained.

Ex. 6.   Insert four geometric means between 2 and 64.

Including the two given terms the number of terms will be 6, the first term will be 2, and the last 64.

$$\therefore \quad r^{6-1}=\frac{64}{2};$$

$$\therefore \quad r^5=32, \text{ or } r=2.$$

Hence the means are 4, 8, 16, 32.

EXERCISES.   LIII.

Sum the following series :

1. $1+\dfrac{5}{6}+\dfrac{25}{36}$ to 12 terms.

2. $1-1\cdot2+1\cdot44$ to 12 terms.

3. $-\dfrac{1}{3}+\dfrac{1}{2}-\dfrac{3}{4}+..$ to 10 terms.

4. The first term of a G.P. is 3, and the third term 12.   Find the sum of the first 8 terms.

5. (i) What is the eighth term of the G.P. whose first and second terms are $2, -3$ respectively.   (ii) Find the sum of the first 12 terms of the series.

6. (i) What is the 6th term of a G.P. whose first and second terms are $3, -4$?   (ii) Find the sum of the first 10 terms.

7. Show that the arithmetical mean of two positive quantities is greater than the geometrical mean of the same quantities.

8. The arithmetical mean of two numbers is 15, and the geometrical mean 9.   Find the numbers.

Sum to infinity the series :

9. $14\cdot4, 10\cdot8, 8\cdot1\ldots$.

10. (i) $\cdot3\dot2$, (ii) $\cdot\dot7$, (iii) $2\sqrt{2}-2\sqrt{3}+3\sqrt{2}$ to 10 terms.

11. Find an A.P. first term 3, such that its second, fourth, and eighth terms may be in G.P.

12. The sum of the first 8 terms of a G.P. is 17 times the sum of the first four terms.   Find the common ratio.

13. A series whose 1st, 2nd, and 3rd terms are respectively

$$\frac{1}{\sqrt{2}}, \quad \frac{1}{1+\sqrt{2}}, \quad \frac{1}{4+3\sqrt{2}}$$

is either an A.P. or a G.P.   Determine which it is and write down the fourth term.

14. If one geometrical mean G and two arithmetical means p and q be inserted between two given quantities show that

$$G^2=(2p-q)(2q-p).$$

15. The continued product of three numbers in geometrical progression is 216, and the sum of the products of them in pairs is 156. Find the numbers.

Harmonical Progression.—A series of terms are said to be in Harmonical Progression when the reciprocals of the terms are in Arithmetical Progression.

Let the three quantities a, b, c be in H.P., then $\dfrac{1}{a}, \dfrac{1}{b}, \dfrac{1}{c}$ are in A.P.

As
$$\frac{1}{b} - \frac{1}{a} = \frac{1}{c} - \frac{1}{b} \quad\dots\dots\dots\dots\dots(i)$$

we obtain the relation $a : c = a - b : b - c$, or *three quantities are in* H.P. *when the ratio of the first to the third is equal to the ratio of the first minus the second, to the second minus the third.*

Again from (i) the harmonical mean between two quantities a and c is $b = \dfrac{2ac}{a+c}$.

In problems in harmonical progression such as to find a number of harmonical means, to continue a given series, etc.; it is only necessary to obtain the reciprocals of the given quantities and to proceed to deal with them as with quantities in arithmetical progression.

Ex. 1. Find a harmonical mean between 42 and 7.

We may use the formula H.M. $= \dfrac{2ac}{a+c} = \dfrac{2 \times 42 \times 7}{42 + 7} = 12$, or as $\dfrac{1}{42}$ and $\dfrac{1}{7}$ are in A.P.

$$\therefore \text{ mean} = \frac{\dfrac{1}{42} + \dfrac{1}{7}}{2} = \frac{1}{12}.$$

Hence the required mean is 12 and 42, 12 and 7 are three terms in H.P.

Ex. 2. Insert two harmonical means between 3 and 12.

Inverting the given terms $\dfrac{1}{3}$ and $\dfrac{1}{12}$ are the first and last terms of an A.P. of four terms; therefore from
$$l = a + (n-1)d$$
we have
$$\frac{1}{12} = \frac{1}{3} + (4-1)d;$$
$$\therefore 3d = -\frac{1}{4}, \text{ or } d = -\frac{1}{12}.$$

Hence the common difference is $-\dfrac{1}{12}$: therefore the terms are

$$\frac{1}{3} - \frac{1}{12} = \frac{1}{4} \text{ and } \frac{1}{3} - \frac{2}{12} = \frac{1}{6},$$

or the arithmetical means are $\dfrac{1}{4}$ and $\dfrac{1}{6}$.

Hence the harmonic means are 4 and 6.

EXERCISES. LIV.

1. Define harmonic progression ; insert 4 harmonic means between 2 and 12.

2. Find the arithmetic, geometric, and harmonic means between 2 and 8.

3. Find a third term to 42 and 12.

4. Find a first term to 8 and 20.

5. The sum of three terms is $11\frac{1}{2}$, if the first term is $\frac{1}{2}$, what is the series?

6. The arithmetical mean between two numbers exceeds the geometric by two, and the geometrical exceeds the harmonical by 1·6. Find the numbers.

7. A H.P. consists of six terms ; the last three terms are 2, 3, and 6, find the first three.

8. Find the fourth term to 6, 8, and 12.

9. Insert three harmonic means between 2 and 3.

10. Find the arithmetic, geometric, and harmonic means between 2 and $\dfrac{9}{2}$, and write down three terms of each series.

MATHEMATICAL TABLES.

Each candidate at the Examinations of the Board of Education (Secondary Branch) in Practical Mathematics, Applied Mechanics, and Steam is supplied with a copy of Mathematical Tables similar to those here given.

TABLE II. USEFUL CONSTANTS.

1 inch = 25·4 millimetres.

1 gallon = ·1604 cubic foot = 10 lbs. of water at 62° F.

1 knot = 6080 feet per hour.

Weight of 1 lb. in London = 445,000 dynes.

One pound avoirdupois = 7000 grains = 453·6 grammes.

1 cubic foot of water weighs 62·3 lbs.

1 cubic foot of air at 0° C. and 1 atmosphere, weighs ·0807 lb.

1 cubic foot of hydrogen at 0° C. and 1 atmosphere, weighs ·00559 lb.

1 foot-pound = 1·3562 × 10^7 ergs.

1 horse-power-hour = 33000 × 60 foot-pounds.

1 electrical unit = 1000 watt-hours.

Joule's equivalent to suit Regnault's H, is $\begin{cases} 774 \text{ ft.-lb.} = 1 \text{ Fah. unit.} \\ 1393 \text{ ft.-lb.} = 1 \text{ Cent. ,,} \end{cases}$

1 horse-power = 33000 foot-pounds per minute = 746 watts.

Volts × ampères = watts.

1 atmosphere = 14·7 lbs. per square inch = 2116 lbs. per square foot
  = 760 mm. of mercury = 10^6 dynes per square cm. nearly.

A column of water 2·3 feet high corresponds to a pressure of 1 lb.
  per square inch.

Absolute temp., $t = \theta°$ C. + 273°·7.

Regnault's $H = 606·5 + ·305 \ \Theta°$ C. $= 1082 + ·305 \ \Theta°$ F.

$pu^{1·0616} = 479$.

$$\log_{10} p = 6·1007 - \frac{B}{t} - \frac{C}{t^2},$$

  where $\log_{10} B = 3·1812$, $\log_{10} C = 5·0871$,

  p is in pounds per square inch, t is absolute temperature
  Centigrade, u is the volume in cubic feet per pound of steam.

$\pi = 3·1416$.

1 radian = 57·3 degrees.

To convert common into Napierian logarithms, multiply by 2·3026.

The base of the Napierian logarithms is $e = 2·7183$.

The value of g at London = 32·182 feet per sec. per sec.

TABLE III. LOGARITHMS.

	0	1	2	3	4	5	6	7	8	9	1 2 3	4 5 6	7 8 9
10	0000	0043	0086	0128	0170	0212	0253	0294	0334	0374	4 8 12	17 21 25	29 33 37
11	0414	0453	0492	0531	0569	0607	0645	0682	0719	0755	4 8 11	15 19 23	26 30 34
12	0792	0828	0864	0899	0934	0969	1004	1038	1072	1106	3 7 10	14 17 21	24 28 31
13	1139	1173	1206	1239	1271	1303	1335	1367	1399	1430	3 6 10	13 16 19	23 26 29
14	1461	1492	1523	1553	1584	1614	1644	1673	1703	1732	3 6 9	12 15 18	21 24 27
15	1761	1790	1818	1847	1875	1903	1931	1959	1987	2014	3 6 8	11 14 17	20 22 25
16	2041	2068	2095	2122	2148	2175	2201	2227	2253	2279	3 5 8	11 13 16	18 21 24
17	2304	2330	2355	2380	2405	2430	2455	2480	2504	2529	2 5 7	10 12 15	17 20 22
18	2553	2577	2601	2625	2648	2672	2695	2718	2742	2765	2 5 7	9 12 14	16 19 21
19	2788	2810	2833	2856	2878	2900	2923	2945	2967	2989	2 4 7	9 11 13	16 18 20
20	3010	3032	3054	3075	3096	3118	3139	3160	3181	3201	2 4 6	8 11 13	15 17 19
21	3222	3243	3263	3284	3304	3324	3345	3365	3385	3404	2 4 6	8 10 12	14 16 18
22	3424	3444	3464	3483	3502	3522	3541	3560	3579	3598	2 4 6	8 10 12	14 15 17
23	3617	3636	3655	3674	3692	3711	3729	3747	3766	3784	2 4 6	7 9 11	13 15 17
24	3802	3820	3838	3856	3874	3892	3909	3927	3945	3962	2 4 5	7 9 11	12 14 16
25	3979	3997	4014	4031	4048	4065	4082	4099	4116	4133	2 3 5	7 9 10	12 14 15
26	4150	4166	4183	4200	4216	4232	4249	4265	4281	4298	2 3 5	7 8 10	11 13 15
27	4314	4330	4346	4362	4378	4393	4409	4425	4440	4456	2 3 5	6 8 9	11 13 14
28	4472	4487	4502	4518	4533	4548	4564	4579	4594	4609	2 3 5	6 8 9	11 12 14
29	4624	4639	4654	4669	4683	4698	4713	4728	4742	4757	1 3 4	6 7 9	10 12 13
30	4771	4786	4800	4814	4829	4843	4857	4871	4886	4900	1 3 4	6 7 9	10 11 13
31	4914	4928	4942	4955	4969	4983	4997	5011	5024	5038	1 3 4	6 7 8	10 11 12
32	5051	5065	5079	5092	5105	5119	5132	5145	5159	5172	1 3 4	5 7 8	9 11 12
33	5185	5198	5211	5224	5237	5250	5263	5276	5289	5302	1 3 4	5 6 8	9 10 12
34	5315	5328	5340	5353	5366	5378	5391	5403	5416	5428	1 3 4	5 6 8	9 10 11
35	5441	5453	5465	5478	5490	5502	5514	5527	5539	5551	1 2 4	5 6 7	9 10 11
36	5563	5575	5587	5599	5611	5623	5635	5647	5658	5670	1 2 4	5 6 7	8 10 11
37	5682	5694	5705	5717	5729	5740	5752	5763	5775	5786	1 2 3	5 6 7	8 9 10
38	5798	5809	5821	5832	5843	5855	5866	5877	5888	5899	1 2 3	5 6 7	8 9 10
39	5911	5922	5933	5944	5955	5966	5977	5988	5999	6010	1 2 3	4 5 7	8 9 10
40	6021	6031	6042	6053	6064	6075	6085	6096	6107	6117	1 2 3	4 5 6	8 9 10
41	6128	6138	6149	6160	6170	6180	6191	6201	6212	6222	1 .2 3	4 5 6	7 8 9
42	6232	6243	6253	6263	6274	6284	6294	6304	6314	6325	1 2 3	4 5 6	7 8 9
43	6335	6345	6355	6365	6375	6385	6395	6405	6415	6425	1 2 3	4 5 6	7 8 9
44	6435	6444	6454	6464	6474	6484	6493	6503	6513	6522	1 2 3	4 5 6	7 8 9
45	6532	6542	6551	6561	6571	6580	6590	6599	6609	6618	1 2 3	4 5 6	7 8 9
46	6628	6637	6646	6656	6665	6675	6684	6693	6702	6712	1 2 3	4 5 6	7 7 8
47	6721	6730	6739	6749	6758	6767	6776	6785	6794	6803	1 2 3	4 5 5	6 7 8
48	6812	6821	6830	6839	6848	6857	6866	6875	6884	6893	1 2 3	4 4 5	6 7 8
49	6902	6911	6920	6928	6937	6946	6955	6964	6972	6981	1 2 3	4 4 5	6 7 8
50	6990	6998	7007	7016	7024	7033	7042	7050	7059	7067	1 2 3	3 4 5	6 7 8
51	7076	7084	7093	7101	7110	7118	7126	7135	7143	7152	1 2 3	3 4 5	6 7 8
52	7160	7168	7177	7185	7193	7202	7210	7218	7226	7235	1 2 2	3 4 5	6 7 7
53	7243	7251	7259	7267	7275	7284	7292	7300	7308	7316	1 2 2	3 4 5	6 6 7
54	7324	7332	7340	7348	7356	7364	7372	7380	7388	7396	1 2 2	3 4 5	6 6 7

TABLE III. LOGARITHMS.

	0	1	2	3	4	5	6	7	8	9	1	2	3	4	5	6	7	8	9
55	7404	7412	7419	7427	7435	7443	7451	7459	7466	7474	1	2	2	3	4	5	5	6	7
56	7482	7490	7497	7505	7513	7520	7528	7536	7548	7551	1	2	2	3	4	5	5	6	7
57	7559	7566	7574	7582	7589	7597	7604	7612	7619	7627	1	2	2	3	4	5	5	6	7
58	7634	7642	7649	7657	7664	7672	7679	7686	7694	7701	1	1	2	3	4	4	5	6	7
59	7709	7716	7723	7731	7738	7745	7752	7760	7767	7774	1	1	2	3	4	4	5	6	7
60	7782	7789	7796	7803	7810	7818	7825	7832	7839	7846	1	1	2	3	4	4	5	6	6
61	7853	7860	7868	7875	7882	7889	7896	7903	7910	7917	1	1	2	3	4	4	5	6	6
62	7924	7931	7938	7945	7952	7959	7966	7973	7980	7987	1	1	2	3	3	4	5	6	6
63	7993	8000	8007	8014	8021	8028	8035	8041	8048	8055	1	1	2	3	3	4	5	5	6
64	8062	8069	8075	8082	8089	8096	8102	8109	8116	8122	1	1	2	3	3	4	5	5	6
65	8129	8136	8142	8149	8156	8162	8169	8176	8182	8189	1	1	2	3	3	4	5	5	6
66	8195	8202	8209	8215	8222	8228	8235	8241	8248	8254	1	1	2	3	3	4	5	5	6
67	8261	8267	8274	8280	8287	8293	8299	8306	8312	8319	1	1	2	3	3	4	5	5	6
68	8325	8331	8338	8344	8351	8357	8363	8370	8376	8382	1	1	2	3	3	4	4	5	6
69	8388	8395	8401	8407	8414	8420	8426	8432	8439	8445	1	1	2	2	3	4	4	5	6
70	8451	8457	8463	8470	8476	8482	8488	8494	8500	8506	1	1	2	2	3	4	4	5	6
71	8513	8519	8525	8531	8537	8543	8549	8555	8561	8567	1	1	2	2	3	4	4	5	5
72	8573	8579	8585	8591	8597	8603	8609	8615	8621	8627	1	1	2	2	3	4	4	5	5
73	8633	8639	8645	8651	8657	8663	8669	8675	8681	8686	1	1	2	2	3	4	4	5	5
74	8692	8698	8704	8710	8716	8722	8727	8733	8739	8745	1	1	2	2	3	4	4	5	5
75	8751	8756	8762	8768	8774	8779	8785	8791	8797	8802	1	1	2	2	3	3	4	5	5
76	8808	8814	8820	8825	8831	8837	8842	8848	8854	8859	1	1	2	2	3	3	4	5	5
77	8865	8871	8876	8882	8887	8893	8899	8904	8910	8915	1	1	2	2	3	3	4	4	5
78	8921	8927	8932	8938	8943	8949	8954	8960	8965	8971	1	1	2	2	3	3	4	4	5
79	8976	8982	8987	8993	8998	9004	9009	9015	9020	9025	1	1	2	2	3	3	4	4	5
80	9031	9036	9042	9047	9053	9058	9063	9069	9074	9079	1	1	2	2	3	3	4	4	5
81	9085	9090	9096	9101	9106	9112	9117	9122	9128	9133	1	1	2	2	3	3	4	4	5
82	9138	9143	9149	9154	9159	9165	9170	9175	9180	9186	1	1	2	2	3	3	4	4	5
83	9191	9196	9201	9206	9212	9217	9222	9227	9232	9238	1	1	2	2	3	3	4	4	5
84	9243	9248	9253	9258	9263	9269	9274	9279	9284	9289	1	1	2	2	3	3	4	4	5
85	9294	9299	9304	9309	9315	9320	9325	9330	9335	9340	1	1	2	2	3	3	4	4	5
86	9345	9350	9355	9360	9365	9370	9375	9380	9385	9390	1	1	2	2	3	3	4	4	5
87	9395	9400	9405	9410	9415	9420	9425	9430	9435	9440	0	1	1	2	2	3	3	4	4
88	9445	9450	9455	9460	9465	9469	9474	9479	9484	9489	0	1	1	2	2	3	3	4	4
89	9494	9499	9504	9509	9513	9518	9523	9528	9533	9538	0	1	1	2	2	3	3	4	4
90	9542	9547	9552	9557	9562	9566	9571	9576	9581	9586	0	1	1	2	2	3	3	4	4
91	9590	9595	9600	9605	9609	9614	9619	9624	9628	9633	0	1	1	2	2	3	3	4	4
92	9638	9643	9647	9652	9657	9661	9666	9671	9675	9680	0	1	1	2	2	3	3	4	4
93	9685	9689	9694	9699	9703	9708	9713	9717	9722	9727	0	1	1	2	2	3	3	4	4
94	9731	9736	9741	9745	9750	9754	9759	9763	9768	9773	0	1	1	2	2	3	3	4	4
95	9777	9782	9786	9791	9795	9800	9805	9809	9814	9818	0	1	1	2	2	3	3	4	4
96	9823	9827	9832	9836	9841	9845	9850	9854	9859	9863	0	1	1	2	2	3	3	4	4
97	9868	9872	9877	9881	9886	9890	9894	9899	9903	9908	0	1	1	2	2	3	3	4	4
98	9912	9917	9921	9926	9930	9934	9939	9943	9948	9952	0	1	1	2	2	3	3	4	4
99	9956	9961	9965	9969	9974	9978	9983	9987	9991	9996	0	1	1	2	2	3	3	3	4

TABLE IV.   ANTILOGARITHMS.

	0	1	2	3	4	5	6	7	8	9	1	2	3	4	5	6	7	8	9
·00	1000	1002	1005	1007	1009	1012	1014	1016	1019	1021	0	0	1	1	1	1	2	2	2
·01	1023	1026	1028	1030	1033	1035	1038	1040	1042	1045	0	0	1	1	1	1	2	2	2
·02	1047	1050	1052	1054	1057	1059	1062	1064	1067	1069	0	0	1	1	1	1	2	2	2
·03	1072	1074	1076	1079	1081	1084	1086	1089	1091	1094	0	0	1	1	1	1	2	2	2
·04	1096	1099	1102	1104	1107	1109	1112	1114	1117	1119	0	1	1	1	1	2	2	2	2
·05	1122	1125	1127	1130	1132	1135	1138	1140	1143	1146	0	1	1	1	1	2	2	2	2
·06	1148	1151	1153	1156	1159	1161	1164	1167	1169	1172	0	1	1	1	1	2	2	2	2
·07	1175	1178	1180	1183	1186	1189	1191	1194	1197	1199	0	1	1	1	1	2	2	2	2
·08	1202	1205	1208	1211	1213	1216	1219	1222	1225	1227	0	1	1	1	1	2	2	2	3
·09	1230	1233	1236	1239	1242	1245	1247	1250	1253	1256	0	1	1	1	1	2	2	2	3
·10	1259	1262	1265	1268	1271	1274	1276	1279	1282	1285	0	1	1	1	1	2	2	2	3
·11	1288	1291	1294	1297	1300	1303	1306	1309	1312	1315	0	1	1	1	2	2	2	2	3
·12	1318	1321	1324	1327	1330	1334	1337	1340	1343	1346	0	1	1	1	2	2	2	2	3
·13	1349	1352	1355	1358	1361	1365	1368	1371	1374	1377	0	1	1	1	2	2	2	3	3
·14	1380	1384	1387	1390	1393	1396	1400	1403	1406	1409	0	1	1	1	2	2	2	3	3
·15	1413	1416	1419	1422	1426	1429	1432	1435	1439	1442	0	1	1	1	2	2	2	3	3
·16	1445	1449	1452	1455	1459	1462	1466	1469	1472	1476	0	1	1	1	2	2	2	3	3
·17	1479	1483	1486	1489	1493	1496	1500	1503	1507	1510	0	1	1	1	2	2	2	3	3
·18	1514	1517	1521	1524	1528	1531	1535	1538	1542	1545	0	1	1	1	2	2	2	3	3
·19	1549	1552	1556	1560	1563	1567	1570	1574	1578	1581	0	1	1	1	2	2	3	3	3
·20	1585	1589	1592	1596	1600	1603	1607	1611	1614	1618	0	1	1	1	2	2	3	3	3
·21	1622	1626	1629	1633	1637	1641	1644	1648	1652	1656	0	1	1	2	2	2	3	3	3
·22	1660	1663	1667	1671	1675	1679	1683	1687	1690	1694	0	1	1	2	2	2	3	3	3
·23	1698	1702	1706	1710	1714	1718	1722	1726	1730	1734	0	1	1	2	2	2	3	3	4
·24	1738	1742	1746	1750	1754	1758	1762	1766	1770	1774	0	1	1	2	2	2	3	3	4
·25	1778	1782	1786	1791	1795	1799	1803	1807	1811	1816	0	1	1	2	2	2	3	3	4
·26	1820	1824	1828	1832	1837	1841	1845	1849	1854	1858	0	1	1	2	2	3	3	3	4
·27	1862	1866	1871	1875	1879	1884	1888	1892	1897	1901	0	1	1	2	2	3	3	3	4
·28	1905	1910	1914	1919	1923	1928	1932	1936	1941	1945	0	1	1	2	2	3	3	4	4
·29	1950	1954	1959	1963	1968	1972	1977	1982	1986	1991	0	1	1	2	2	3	3	4	4
·30	1995	2000	2004	2009	2014	2018	2023	2028	2032	2037	0	1	1	2	2	3	3	4	4
·31	2042	2046	2051	2056	2061	2065	2070	2075	2080	2084	0	1	1	2	2	3	3	4	4
·32	2089	2094	2099	2104	2109	2113	2118	2123	2128	2133	0	1	1	2	2	3	3	4	4
·33	2138	2143	2148	2153	2158	2163	2168	2173	2178	2183	0	1	1	2	2	3	3	4	4
·34	2188	2193	2198	2203	2208	2213	2218	2223	2228	2234	1	1	2	2	3	3	4	4	5
·35	2239	2244	2249	2254	2259	2265	2270	2275	2280	2286	1	1	2	2	3	3	4	4	5
·36	2291	2296	2301	2307	2312	2317	2323	2328	2333	2339	1	1	2	2	3	3	4	4	5
·37	2344	2350	2355	2360	2366	2371	2377	2382	2388	2393	1	1	2	2	3	3	4	4	5
·38	2399	2404	2410	2415	2421	2427	2432	2438	2443	2449	1	1	2	2	3	3	4	4	5
·39	2455	2460	2466	2472	2477	2483	2489	2495	2500	2506	1	1	2	2	3	3	4	5	5
·40	2512	2518	2523	2529	2535	2541	2547	2553	2559	2564	1	1	2	2	3	4	4	5	5
·41	2570	2576	2582	2588	2594	2600	2606	2612	2618	2624	1	1	2	2	3	4	4	5	5
·42	2630	2636	2642	2649	2655	2661	2667	2673	2679	2685	1	1	2	2	3	4	4	5	6
·43	2692	2698	2704	2710	2716	2723	2729	2735	2742	2748	1	1	2	3	3	4	4	5	6
·44	2754	2761	2767	2773	2780	2786	2793	2799	2805	2812	1	1	2	3	3	4	4	5	6
·45	2818	2825	2831	2838	2844	2851	2858	2864	2871	2877	1	1	2	3	3	4	5	5	6
·46	2884	2891	2897	2904	2911	2917	2924	2931	2938	2944	1	1	2	3	3	4	5	5	6
·47	2951	2958	2965	2972	2979	2985	2992	2999	3006	3013	1	1	2	3	3	4	5	5	6
·48	3020	3027	3034	3041	3048	3055	3062	3069	3076	3083	1	1	2	3	4	4	5	6	6
·49	3090	3097	3105	3112	3119	3126	3133	3141	3148	3155	1	1	2	3	4	4	5	6	6

TABLE IV. ANTILOGARITHMS.

	0	1	2	3	4	5	6	7	8	9	1	2	3	4	5	6	7	8	9
·50	3162	3170	3177	3184	3192	3199	3206	3214	3221	3228	1	1	2	3	4	4	5	6	7
·51	3236	3243	3251	3258	3266	3273	3281	3289	3296	3304	1	2	2	3	4	5	5	6	7
·52	3311	3319	3327	3334	3342	3350	3357	3365	3373	3381	1	2	2	3	4	5	5	6	7
·53	3388	3396	3404	3412	3420	3428	3436	3443	3451	3459	1	2	2	3	4	5	6	6	7
·54	3467	3475	3483	3491	3499	3508	3516	3524	3532	3540	1	2	2	3	4	5	6	6	7
·55	3548	3556	3565	3573	3581	3589	3597	3606	3614	3622	1	2	2	3	4	5	6	7	7
·56	3631	3639	3648	3656	3664	3673	3681	3690	3698	3707	1	2	3	3	4	5	6	7	8
·57	3715	3724	3733	3741	3750	3758	3767	3776	3784	3793	1	2	3	3	4	5	6	7	8
·58	3802	3811	3819	3828	3837	3846	3855	3864	3873	3882	1	2	3	4	4	5	6	7	8
·59	3890	3899	3908	3917	3926	3936	3945	3954	3963	3972	1	2	3	4	5	5	6	7	8
·60	3981	3990	3999	4009	4018	4027	4036	4046	4055	4064	1	2	3	4	5	6	6	7	8
·61	4074	4083	4093	4102	4111	4121	4130	4140	4150	4159	1	2	3	4	5	6	7	8	9
·62	4169	4178	4188	4198	4207	4217	4227	4236	4246	4256	1	2	3	4	5	6	7	8	9
·63	4266	4276	4285	4295	4305	4315	4325	4335	4345	4355	1	2	3	4	5	6	7	8	9
·64	4365	4375	4385	4395	4406	4416	4426	4436	4446	4457	1	2	3	4	5	6	7	8	9
·65	4467	4477	4487	4498	4508	4519	4529	4539	4550	4560	1	2	3	4	5	6	7	8	9
·66	4571	4581	4592	4603	4613	4624	4634	4645	4656	4667	1	2	3	4	5	6	7	9	10
·67	4677	4688	4699	4710	4721	4732	4742	4753	4764	4775	1	2	3	4	5	7	8	9	10
·68	4786	4797	4808	4819	4831	4842	4853	4864	4875	4887	1	2	3	4	6	7	8	9	10
·69	4898	4909	4920	4932	4943	4955	4966	4977	4989	5000	1	2	3	5	6	7	8	9	10
·70	5012	5023	5035	5047	5058	5070	5082	5093	5105	5117	1	2	4	5	6	7	8	9	11
·71	5129	5140	5152	5164	5176	5188	5200	5212	5224	5236	1	2	4	5	6	7	8	10	11
·72	5248	5260	5272	5284	5297	5309	5321	5333	5346	5358	1	2	4	5	6	7	9	10	11
·73	5370	5383	5395	5408	5420	5433	5445	5458	5470	5483	1	3	4	5	6	8	9	10	11
·74	5495	5508	5521	5534	5546	5559	5572	5585	5598	5610	1	3	4	5	6	8	9	10	12
·75	5623	5636	5649	5662	5675	5689	5702	5715	5728	5741	1	3	4	5	7	8	9	10	12
·76	5754	5768	5781	5794	5808	5821	5834	5848	5861	5875	1	3	4	5	7	8	9	11	12
·77	5888	5902	5916	5929	5943	5957	5970	5984	5998	6012	1	3	4	5	7	8	10	11	12
·78	6026	6039	6053	6067	6081	6095	6109	6124	6138	6152	1	3	4	6	7	8	10	11	13
·79	6166	6180	6194	6209	6223	6237	6252	6266	6281	6295	1	3	4	6	7	9	10	11	13
·80	6310	6324	6339	6353	6368	6383	6397	6412	6427	6442	1	3	4	6	7	9	10	12	13
·81	6457	6471	6486	6501	6516	6531	6546	6561	6577	6592	2	3	5	6	8	9	11	12	14
·82	6607	6622	6637	6653	6668	6683	6699	6714	6730	6745	2	3	5	6	8	9	11	12	14
·83	6761	6776	6792	6808	6823	6839	6855	6871	6887	6902	2	3	5	6	8	9	11	13	14
·84	6918	6934	6950	6966	6982	6998	7015	7031	7047	7063	2	3	5	6	8	10	11	13	15
·85	7079	7096	7112	7129	7145	7161	7178	7194	7211	7228	2	3	5	7	8	10	12	13	15
·86	7244	7261	7278	7295	7311	7328	7345	7362	7379	7396	2	3	5	7	8	10	12	13	15
·87	7413	7430	7447	7464	7482	7499	7516	7534	7551	7568	2	3	5	7	9	10	12	14	16
·88	7586	7603	7621	7638	7656	7674	7691	7709	7727	7745	2	4	5	7	9	11	12	14	16
·89	7762	7780	7798	7816	7834	7852	7870	7889	7907	7925	2	4	5	7	9	11	13	14	16
·90	7943	7962	7980	7998	8017	8035	8054	8072	8091	8110	2	4	6	7	9	11	13	15	17
·91	8128	8147	8166	8185	8204	8222	8241	8260	8279	8299	2	4	6	8	9	11	13	15	17
·92	8318	8337	8356	8375	8395	8414	8433	8453	8472	8492	2	4	6	8	10	12	14	15	17
·93	8511	8531	8551	8570	8590	8610	8630	8650	8670	8690	2	4	6	8	10	12	14	16	18
·94	8710	8730	8750	8770	8790	8810	8831	8851	8872	8892	2	4	6	8	10	12	14	16	18
·95	8913	8933	8954	8974	8995	9016	9036	9057	9078	9099	2	4	6	8	10	12	15	17	19
·96	9120	9141	9162	9183	9204	9226	9247	9268	9290	9311	2	4	6	8	11	13	15	17	19
·97	9333	9354	9376	9397	9419	9441	9462	9484	9506	9528	2	4	7	9	11	13	15	17	20
·98	9550	9572	9594	9616	9638	9661	9683	9705	9727	9750	2	4	7	9	11	13	16	18	20
·99	9772	9795	9817	9840	9863	9886	9908	9931	9954	9977	2	5	7	9	11	14	16	18	20

TABLE V.

ngle. Radians.	Chords.	Sine.	Tangent.	Cotangent.			
0	0	0	0	∞	1	1·414	1·5708
·0175	·017	·0175	·0175	57·2900	·9998	1·402	1·5533
·0349	·035	·0349	·0349	28·6363	·9994	1·389	1·5359
·0524	·052	·0523	·0524	19·0811	·9986	1·377	1·5184
·0698	·070	·0698	·0699	14·3006	·9976	1·364	1·5010
·0873	·087	·0872	·0875	11·4301	·9962	1·351	1·4835
·1047	·105	·1045	·1051	9·5144	·9945	1·338	1·4661
·1222	·122	·1219	·1228	8·1443	·9925	1·325	1·4486
·1396	·139	·1392	·1405	7·1154	·9903	1·312	1·4312
·1571	·157	·1564	·1584	6·3138	·9877	1·299	1·4137
·1745	·174	·1736	·1763	5·6713	·9848	1·286	1·3963
·1920	·192	·1908	·1944	5·1446	·9816	1·272	1·3788
·2094	·209	·2079	·2126	4·7046	·9781	1·259	1·3614
·2269	·226	·2250	·2309	4·3315	·9744	1·245	1·3439
·2443	·244	·2419	·2493	4·0108	·9703	1·231	1·3265
·2618	·261	·2588	·2679	3·7321	·9659	1·217	1·3090
·2793	·278	·2756	·2867	3·4874	·9613	1·204	1·2915
·2967	·296	·2924	·3057	3·2709	·9563	1·190	1·2741
·3142	·313		·3249	3·0777	·9511	1·176	1·2566
·3316	·330	·3256	·3443	2·9042	·9455	1·161	1·2392
·3491	·347	·3420	·3640	2·7475	·9397	1·147	1·2217
·3665	·364	·3584	·3839	2·6051	·9336	1·133	1·2043
·3840	·382	·3746	·4040	2·4751	·9272	1·118	1·1868
·4014	·399		·4245	2·3559		1·104	1·1694
·4189	·416		·4452	2·2460		1·089	1·1519
·4363	·433		·4663	2·1445		1·075	1·1345
·4538	·450		·4877	2·0503		1·060	1·1170
·4712	·467		·5095	1·9626		1·045	1·0996
·4887	·484		·5317	1·8807		1·030	1·0821
·5061	·501		·5543	1·8040		1·015	1·0647
·5236	·518		·5774	1·7321		1·000	1·0472
·5411	·534		·6009	1·6643		·985	1·0297
·5585	·551		·6249	1·6003		·970	1·0123
·5760	·568		·6494	1·5399			
·5934	·585		·6745	1·4826			
·6109	·601		·7002	1·4281			
·6283	·618		·7265	1·3764			
·6458	·635		·7536	1·3270			
·6632	·651		·7813	1·2799			
·6807	·668		·8098	1·2349			
·6981	·684		·8391	1·1918			
·7156	·700		·8693	1·1504			
·7330	·717		·9004	1·1106			
·7505	·733		·9325	1·0724			
·7679	·749		·9657	1·0355			
·7854	·765		1·0000	1·0000			

| | Cotangent. | Tangent. | |

ELEMENTARY PRACTICAL MATHEMATICS. 1901.

Only EIGHT *questions are to be answered.   Two of these should be Nos.* 1 *and* 2.

1. Compute $30{\cdot}56 \div 4{\cdot}105$, $0{\cdot}03056 \times 0{\cdot}4105$, $4{\cdot}105^{1{\cdot}23}$, $\cdot04105^{-23}$.
The answers must be right to three significant figures.
Why do we multiply log a by b to obtain the logarithm of a^b?

2. Answer only *one* of the following, (*a*) or (*b*):

(*a*) Find the value of
$$ae^{-bt} \sin (ct + g)$$
if $a=5$, $b=200$, $c=600$, $g=-0{\cdot}1745$ radian, $t=\cdot001$. (Of course the angle is in radians.)

(*b*) Find the value of $\sin A \cos B - \cos A \sin B$ if A is 65° and B is 34°.

3. A tube of copper (0·32 lb. per cubic inch) is 12 feet long and 3 inches inside diameter; it weighs 100 lb. Find its outer diameter, and the area of its curved outer surface.

4. ABC is a triangle. The angle A is 37°, the angle C is 90°, and the side AC is 5·32 inches. Find the other sides, the angle B, and the area of the triangle.

5. An army of 5000 men costs a country £800,000 per annum to maintain it, an army of 10,000 men costs £1,300,000 per annum to maintain it, what is the annual cost of an army of 8000? Take the simplest law which is consistent with the figures given. Use squared paper or not, as you please.

6. In any class of turbine if P is power of the waterfall and H the height of the fall, and n the rate of revolution, then it is known that for any particular class of turbines of all sizes
$$n \propto H^{1{\cdot}25} P^{-0{\cdot}5}.$$
In the list of a particular maker I take a turbine at random for a fall of 6 feet, 100 horse-power, 50 revolutions per minute. By means of this I find I can calculate n for all the other turbines of the list. Find n for a fall of 20 feet and 75 horse-power.

7. At the following draughts in sea water a particular vessel has the following displacements:

Draught h feet - -	15	12	9	6.3
Displacement T tons -	2098	1512	1018	586

What are the probable displacements when the draughts are 11 and 13 feet respectively?

8. The three parts (a), (b), (c) must all be answered to get full marks.

(a) There are two quantities, a and b. The square of a is to be multiplied by the sum of the squares of a and b ; add 3 ; extract the cube root ; divide by the product of a and the square root of b. Write down this algebraically.

(b) Express $\dfrac{1}{x^2 - 7x + 12}$ as the sum of two simpler fractions.

(c) A crew which can pull at the rate of six miles an hour finds that it takes twice as long to come up a river as to go down ; at what rate does the river flow ?

9. A number is added to $2 \cdot 25$ times its reciprocal ; for what number is this a minimum ? Use squared paper or the calculus as you please.

10. If $y = \frac{1}{2}x^2 - 3x + 3$, show, by taking some values of x and calculating y and plotting on squared paper, the nature of the relationship between x and y. For what values of x is $y = 0$?

11. The keeper of a restaurant finds that when he has G guests a day his total daily profit (the difference between his actual receipts and expenditure including rent, taxes, wages, wear and tear, food and drink) is P pounds, the following numbers being averages obtained by comparison of many days' accounts, what simple law seems to connect P and G ?

G	P
210	$- 0 \cdot 9$
270	$+ 1 \cdot 8$
320	$+ 4 \cdot 8$
360	$+ 6 \cdot 4$

For what number of guests would he just have no profit ?

12. At the end of a time t seconds it is observed that a body has passed over a distance s feet reckoned from some starting point. If it is known that $s = 25 + 150t - 5t^2$ what is the velocity at the time t ?

Prove the rule that you adopt to be correct. If corresponding values of s and t are plotted on squared paper what indicates the velocity and why ?

13. The three rectangular co-ordinates of a point P are $2 \cdot 5$, $3 \cdot 1$ and 4. Find (1) the length of the line joining P with O the origin, (2) the cosines of the angles which OP makes with the three axes, and (3) the sum of the squares of the three cosmes.

ELEMENTARY PRACTICAL MATHEMATICS.
1902.

1. Compute by contracted methods without using logarithms $23 \cdot 07 \times 0 \cdot 1354$, $2307 \div 1 \cdot 354$.

Compute $2 \cdot 307^{0 \cdot 65}$ and $23 \cdot 07^{-1 \cdot 25}$ using logarithms.

The answers to consist of four significant figures.

Why do we add logarithms to obtain the logarithm of a product?

2. Answer only *one* of the following (a) or (b) :

(a) If $w = 144 \{ p_1 (1 + \log r) - r (p_3 + 10) \}$ and if $p_1 = 100$, $p_3 = 17$; find w for the four values of r, $1\frac{1}{2}$, 2, 3, 4.

Tabulate your answers.

(b) If c is 20 feet, $D = 6$ feet, $d = 3$ feet, find θ in radians if

$$\sin \theta = \frac{D + d}{2c}.$$

Now calculate L the length of a belt, if

$$L = (D + d) \left\{ \frac{\pi}{2} + \theta + \frac{1}{\tan \theta} \right\}.$$

3. The three parts (a), (b), and (c) must all be answered to get full marks.

(a) Let x be multiplied by the square of y, and subtracted from the cube of z, the cube root of the whole is taken and is then squared. This is divided by the sum of x, y, and z. Write all this down algebraically.

(b) Express $\dfrac{x - 13}{x^2 - 2x - 15}$ as the sum of two simpler fractions.

(c) The sum of two numbers is 76, and their difference is equal to one-third of the greater, find them.

4. What is the idea on which compound interest is calculated? Explain, as if to a beginner, how it is that

$$A = P \left(1 + \frac{r}{100} \right)^n$$

where P is the money lent, and A is what it amounts to in n years at r per cent. per annum.

If A is 130, and P is 100, and n is $7 \cdot 5$, find r.

5. Suppose s the distance in feet passed through by a body in the time of t seconds is $s = 10t^2$. Find s when t is 2, find s when t is 2·01, and also when t is 2·001. What is the average speed in each of the two short intervals of time after $t = 2$? When the interval of time is made shorter and shorter, what does the average speed approximate to?

6. If $z = ax - by^3 x^{\frac{1}{2}}$.
If $z = 1·32$ when $x = 1$ and $y = 2$; and if $z = 8·58$ when $x = 4$ and $y = 1$; find a and b.
Then find z when $x = 2$ and $\dot{y} = 0$.

7. A prism has a cross-section of 50·32 square inches. There is a section making an angle of 20° with the cross-section: what is its area? Prove the rule that you use.

8. In a triangle ABC, AD is the perpendicular on BC; AB is 3·25 feet; the angle B is 55°. Find the length of AD. If BC is 4·67 feet, what is the area of the triangle?
Find also BD and DC and AC. Your answers must be right to three significant figures.

9. It is known that the weight of coal in tons consumed per hour in a certain vessel is $0·3 + 0·001v^3$ where v is the speed in knots (or nautical miles per hour). For a voyage of 1000 nautical miles tabulate the time in hours and the total coal consumption for various values of v. If the wages, interest on cost of vessel, etc., are represented by the value of 1 ton of coal per hour, tabulate for each value of v the total cost, stating it in the value of tons of coal, and plot on squared paper. About what value of v gives greatest economy?

10. An examiner has given marks to papers; the highest number of marks is 185, the lowest 42. He desires to change all his marks according to a linear law converting the highest number of marks into 250 and the lowest into 100; show how he may do this, and state the converted marks for papers already marked 60, 100, 150.
Use squared paper, or mere algebra, as you please.

11. A is the horizontal sectional area of a vessel in square feet at the water level, h being the vertical draught in feet.

A	14,850	14,400	13,780	13,150
h	23·6	20·35	17·1	14·6

Plot on squared paper and read off and tabulate A for values of h, 23, 20, 16.
If the vessel changes in draught from 20·5 to 19·5, what is the diminution of its displacement in cubic feet?

12. Find a value of x which satisfies the equation
$$x^2 - 5\log_{10}x - 2·531 = 0.$$

13. If $x = a(\phi - \sin\phi)$ and $y = a(1 - \cos\phi)$, and if $a = 5$; taking various values of ϕ between 0 and, say 1·5, calculate x and y and plot this part of the curve.

PRACTICAL MATHEMATICS.   1903.

STAGE I.

Only EIGHT *questions to be answered.   Three of these must be Nos. 1, 2 and 3.*

1. Compute by contracted methods to four significant figures only, and without using logarithms,

$$8{\cdot}102 \times 35{\cdot}14, \quad 254{\cdot}3 \div 0{\cdot}09027.$$

Compute, using logarithms,

$$\sqrt[3]{37{\cdot}24}, \quad \sqrt[2]{3{\cdot}724}, \quad 372{\cdot}4^{2{\cdot}43}, \quad 0{\cdot}3724^{-2{\cdot}43}.$$

What is the theory underlying the use of logarithms in helping us to multiply, divide, and raise a number to any power?

2. Answer only *one* of the following (*a*), (*b*), or (*c*):

(*a*) If $x = \tan\theta \div \tan(\theta + \phi)$ where ϕ is always 10°, find x when θ has the values 30°, 40°, 50°, 60°, and plot the values of x and of θ on squared paper.   About what value of θ seems to give the largest value of x?

(*b*) At speeds greater than the velocity of sound, the air resistance to the motion of a projectile of the usual shape of weight w lb., diameter d inches, is such that when the speed diminishes from v_1 feet per second to v, if t is the time in seconds and s is the space passed over in feet,

$$t = 7{,}000\,\frac{w}{d^2}\left(\frac{1}{v} - \frac{1}{v^1}\right),$$

$$s = 7{,}000\,\frac{w}{d^2}\log_e \frac{v_1}{v}.$$

If v_1 is 2,000, find s and t when $v = 1{,}500$ for a projectile of 12 lb. whose diameter is 3 inches.

(*c*) Find the value of

$$t_1 - t_3 - t_3 \log_e \frac{t_1}{t_3} + l_1\left(1 - \frac{t_3}{t_1}\right)$$

if $t_1 = 458$, $t_3 = 373$, $l_1 = 796 - 0{\cdot}695\,t_1.$

P.M B   X

3. The four parts (a), (b), (c), and (d) must all be answered to get full marks.

(a) Write down algebraically : Add twice the square root of the cube of x to the product of y squared and the cube root of z. Divide by the sum of x and the square root of y. Add four and extract the square root of the whole.

(b) Express
$$\frac{3x - 2}{x^2 - 3x - 4}$$
as the sum of two simpler fractions.

(c) Find two numbers such that if four times the first be added to two and a half times the second the sum is $17\cdot3$, and if three times the second be subtracted from twice the first the difference is $1\cdot2$.

(d) In a triangle ABC, C being a right angle, AB is $14\cdot85$ inches, AC is $8\cdot32$ inches. Compute the angle A in degrees, using your tables.

4. The following are the areas of cross section of a body at right angles to its straight axis :

A in square inches -	250	292	310	273	215	180	135	120
x inches from one end	0	22	41	70	84	102	130	145

Plot A and x on squared paper. What is the probable cross section at $x=50$? What is the average cross section and the whole volume?

5. The following table records the heights in inches of a girl A (born January, 1890) and a boy B (born May, 1894). Plot these records. The intervals of time may be taken as exactly four months.

Year	1900.	1901.			1902.			1903.
Month	Sept.	Jan.	May.	Sept.	Jan.	May.	Sept.	Jan.
A	54·8	55·6	56·6	58·0	59·2	60·2	60·9	61·3
B	48·3	49·0	49·8	50·6	51·5	52·3	53·1	53·9

Find in inches per year the *average* rates of growth of A and B during the given period. At about what age was the growth of A most rapid? State this rate; divide it by her average rate.

6. In any such question as Question 5, where points on a curve have coordinates like h (height) and t (time), show exactly how it is that the slope of a curve at a point represents there the rate of growth of h as t increases.

7. Find accurately to three significant figures a value of x which satisfies the equation
$$2x^2 - 10\log_{10} x - 3\cdot 25 = 0.$$

8. Answer only *one* of the following (*a*) or (*b*):

(*a*) A cast-iron flywheel rim (0·26 lb. per cubic inch) weighs 13,700 lbs. The rim is of rectangular section, thickness radially x, size the other way $1\cdot 6x$. The inside radius of the rim is $14x$. Find the actual sizes.

(*b*) The electrical resistance of copper wire is proportional to its length divided by its cross section. Show that the resistance of a pound of wire of circular section all in one length is inversely proportional to the fourth power of the diameter of the wire.

9. It is thought that the following observed quantities, in which there are probably errors of observation, follow a law like
$$y = ae^{bx}.$$
Test if this is so, and find the most probable values of a and b.

x	2·30	3·10	4·00	4·92	5·91	7·20
y	33·0	39·1	50·3	67·2	85·6	125·0

10.   Plot $3y = 4\cdot 8x + 0\cdot 9$
     Plot $y = 2\cdot 24 - 0\cdot 7x.$

Find the point where they cross. What angle does each of them make with the axis of x? At what angle do they meet?

11. A firm is satisfied from its past experience and study that its expenditure per week in pounds is
$$120 + 3\cdot 2x + \frac{C}{x+5} + 0\cdot 01 C,$$
where x is the number of horses employed by the firm, and C is the usual turnover.

If C is 2,150 pounds, find for various values of x what is the weekly expenditure, and plot on squared paper to find the number of horses which will cause the expenditure to be a minimum.

12. Assuming the earth to be a sphere, if its circumference is 360×60 nautical miles, what is the circumference of the parallel of latitude 56°? What is the length there of a degree of longitude? If a small map is to be drawn in this latitude, with north and south and east and west distances to the same scale, and if a degree of latitude (which is of course 60 miles) is shown as 10 inches, what distance will represent a degree of longitude?

13. At a certain place where all the months of the year are assumed to be of the same length (30·44 days each), at the same time in each month the length of the day (interval from sunrise to sunset in hours) was measured, as in this table.

Nov.	Dec.	Jan.	Feb.	Mar.	April.	May.	June.	July.
8·35	7·78	8·35	9·87	12	14·11	15·65	16·22	15·65

What is the average increase of the length of the day (state in decimals of an hour per day) from the shortest day which is 7·78 hours to the longest which is 16·22 hours?   When is the increase of the day most rapid, and what is it?

14. At an electricity works, where new plant has been judiciously added, if W is the annual works cost in millions of pence, and T is the annual total cost, and U the number of millions of electrical units sold. the following results have been found :

U	W	T
0·3	0·47	0·78
1·2	1·03	1·64
2·3	1·70	2·73
3·4	2·32	3·77

Find approximately the rule connecting T and W with U.   Also find the probable values of W and T when U becomes 5, if there is the same judicious management.

PRACTICAL MATHEMATICS. 1904.

STAGE I.

Answer questions Nos. 1, 2 and 3 and FIVE *others.*

1. The three parts (*a*), (*b*) and (*c*) must be answered to get full marks.

(*a*) Compute by contracted methods to four significant figures only, and without using logarithms, $3 \cdot 405 \times 9 \cdot 123$ and $3 \cdot 405 \div 9 \cdot 123$.

(*b*) Compute, using logarithms, $\sqrt[3]{2 \cdot 354 \times 1 \cdot 607}$ and $(32 \cdot 15)^{1 \cdot 52}$.

(*c*) Write down the values of $\sin 23°$, $\tan 53°$, $\log_{10} 153 \cdot 4$, $\log_e 153 \cdot 4$.

2. Both (*a*) and (*b*) must be answered to get full marks.

(*a*) If $\qquad F = EI\pi^2 \div 4l^2$,

If $\qquad I = bt^3 \div 12$,

If $\quad E = 3 \times 10^7$, $\pi = 3 \cdot 142$, $l = 62$, $b = 2$, $t = 0 \cdot 5$, find F.

(*b*) Two men measure a rectangular box; one finds its length, breadth, and depth in inches to be $5 \cdot 32$, $4 \cdot 15$, $3 \cdot 29$. The other finds them to be $5 \cdot 35$, $4 \cdot 17$, $3 \cdot 33$. Calculate the volume in each case; what is the mean of the two, what is the percentage difference of either from the mean?

3. All of these (*a*), (*b*) and (*c*), must be answered to get full marks.

(*a*) Write down algebraically: Square *a*, divide by the square of *b*, add 1, extract the square root, multiply by *w*, divide by the square of *n*.

(*b*) The ages of a man and his wife added together amount to $72 \cdot 36$ years; fifteen years ago the man's age was $2 \cdot 3$ times that of his wife; what are their ages now?

(*c*) *ABC* is a triangle, *C* being a right angle. The side *AB* is $15 \cdot 34$ inches, the side *BC* is $10 \cdot 15$ inches. What is the length of *AC*? Express the angles *A* and *B* in degrees. What is the area of the triangle in square inches? If this is the shape of a piece of sheet brass $0 \cdot 13$ inch thick, and if brass weighs $0 \cdot 3$ lb. per cubic inch, what is its weight?

4. If $\qquad y = 3x^2 - 20 \log_{10} x - 7 \cdot 077$,
find the values of y when x is $1 \cdot 5$, 2, $2 \cdot 3$. Plot the values of y and x on squared paper, and draw the probable curve in which these points lie. State approximately what value of x would cause y to be 0.

5. It has been found that if P is the horse power wasted in air friction when a disc d feet diameter is revolving at n revolutions per minute,

$$P = cd^{5\cdot5}n^{3\cdot5}.$$

If P is 0·1 when $d=4$ and $n=500$, find the constant c.   Now find P when d is 9 and n is 400.

6. There is a district in which the surface of the ground may be regarded as a sloping plane ; its actual area is 3·246 square miles ; it is shown on the map as an area of 2·875 square miles ; at what angle is it inclined to the horizontal ?

There is a straight line 20·17 feet long which makes an angle of 52° with the horizontal plane ; what is the length of its projection on the horizontal plane ?

7. A British man or woman of age x years may on the average expect to live for an additional y years.

Age x.	Expected further Life y.	
	Man.	Woman.
70	8·27	8·95
60	13·14	14·24
50	18·93	20·68
40	25·30	27·46
30	32·10	34·41

Plot a curve for men and one for women, and find the expectations of life for a man and for a woman aged 54 years.

8. The following tests were made upon a condensing-steam-turbine-electric-generator.   There are probably some errors of observation, as the measurement of the steam is troublesome. The figures are given just as they were published in a newspaper.

Output in Kilowatts K, - - -	1,190	995	745	498	247	0
Weight W lb. of steam consumed per hour,	23,120	20,040	16,630	12,560	8,320	4,065

Plot on squared paper. Find if there is a simple approximate law connecting K and W, but do not state it algebraically. What are the probablo values of K when W is 22,000 and when W is 6,000?

9. If
$$y = 2x + \frac{1\cdot5}{x},$$

for various values of x, calculate y; plot on squared paper; state approximately the value of x which causes y to be of its smallest value.

10. A series of soundings taken across a river channel is given by the following table, x feet being distance from one shore and y feet the corresponding depth. Draw the section. Find its area.

x	0	10	16	23	30	38	43	50	55	60	70	75	80
y	5	10	13	14	15	16	14	12	8	6	4	3	0

11. The value of a ruby is said to be proportional to the $1\frac{1}{2}$ power of its weight. If one ruby is exactly of the same shape as another, but of 2·20 times its linear dimensions, of how many times the value is it?

[Note that the weights of similar things made of the same stuff are as the cubes of their linear dimensions.]

12. x and t are the distance in miles and the time in hours of a train from a railway station. Plot on squared paper. State how the curve shows where the speed is greatest and where it is least. What is the average speed in miles per hour during the whole time tabulated?

t	0	·05	·10	·15	·2	·25	·3	·35	·40	·45	·5
x	0	·25	1·00	3·05	5·00	5·85	6·10	6·10	6·35	7·00	7·65

PRACTICAL MATHEMATICS. 1905.

STAGE I.

Answer questions Nos. 1, 2 and 3, and FIVE *others.*

1. The three parts (*a*), (*b*) and (*c*) must all be answered to get full marks.

(*a*) Compute by contracted methods to four significant figures only, and without using logarithms, $12 \cdot 39 \times 5 \cdot 024$ and $5 \cdot 024 \div 12 \cdot 39$.

(*b*) Compute, using logarithms, $\sqrt[2]{2 \cdot 607}$ and $26 \cdot 07^{1 \cdot 13}$.

(*c*) Write down the values of $\cos 35°$, $\tan 52°$, $\sin^{-1} 0 \cdot 4226$, $\log_{10} 14 \cdot 36$, $\log_e 14 \cdot 36$.

[NOTE. $\sin^{-1} n$ means the angle whose sine is n.]

2. The three parts (*a*), (*b*) and (*c*) must all be answered to get full marks.

(*a*) If $x = a(\phi - \sin \phi)$ and $y = a(1 - \cos \phi)$, find
x and y when a is 10 and $\phi = 0 \cdot 5061$ radian.

(*b*) In a piece of coal there was found to be $11 \cdot 30$ lb. of carbon, $0 \cdot 92$ lb. of hydrogen, $0 \cdot 84$ lb. of oxygen, $0 \cdot 56$ lh. of nitrogen, $0 \cdot 71$ lb. of ash. There being nothing else, state the percentage composition of the coal.

(*c*) A brass tube, 8 feet long, has an outside diameter 3 inches, inside $2 \cdot 8$ inches. What is the volume of the brass in cubic inches? If a cubic inch of brass weighs $0 \cdot 3$ lb., what is the weight of the tube?

3. The four parts (*a*), (*b*), (*c*) and (*d*) must all be answered to get full marks.

(*a*) Write down algebraically: Three times the square of x, multiplied by the square root of y; from this subtract a times the Napierean logarithm of x; again, subtract b times the sine of cx; divide the result by the sum of the cube of x and the square of y.

(*b*) Express
$$\frac{3x + 5}{x^2 + x - 12}$$
as the sum of two simpler fractions.

(*c*) Some men agree to pay equally for the use of a boat, and each pays 15 pence. If there had been two more men in the party, each would have paid 10 pence. How many men were there, and how much was the hire of the boat?

(d) The altitude of a tower observed from a point distant 150 feet horizontally from its foot is 26°; find its height.

4. If $p_1v_1^{1\cdot13} = p_2v_2^{1\cdot13}$ and if v_2/v_1 be called r.

If $p_2 = 6$, find r if $p_1 = 150$.

5. If $y = \dfrac{2}{x} + 5\log_{10}x - 2\cdot70$, find the values of y when x has the values 2, 2·5, 3.

Plot the values of y and x on squared paper, and draw the probable curve in which these points lie. State approximately what value of x would cause y to be 0.

6. x and t are the distance in miles and the time in hours of a train from a railway station. Plot on squared paper. Describe why it is that the *slope* of the curve shows the speed; where is the speed greatest and where is it least?

x	0	0·12	0·50	1·52	2·50	2·92	3·05	3·17	3·50	3·82	4·15
t	0·00	0·05	0·10	0·15	0·20	0·25	0·30	0·35	0·40	0·45	0·50

7. A vessel is shaped like the frustum of a cone, the circular base is 10 inches diameter, the top is 5 inches diameter, the vertical axial height is 8 inches. By drawing, find the axial height to the imaginary vertex of the cone. If x is the height of the surface of a liquid from the bottom, plot a curve, to any scales you please, showing for any value of x the area of the horizonal section there. Three points of the curve will be enough to find.

8. A circle is 3 inches diameter, its centre is 4 inches from a line in its plane. The circle revolves about the line as an axis and so generates a ring. Find the volume of the ring, also its surface area.

9. If u is usefulness of flywheels, $u \propto d^5n^2$, if d is the linear size (say diameter) and n the speed. We assume all flywheels to be similar in shape. I wish to have the usefulness one hundred times as great, the speed being trebled, what is the ratio of the new diameter to the old one?

10. The total cost C of a ship per hour (including interest and depreciation on capital, wages, coal, etc.) is $C = a + bs^3$, where s is the speed in knots (or nautical miles per hour).

When s is 10, C is found to be 5·20 pounds.

When s is 15, C is found to be 7·375 pounds.

Calculate a and b. What is C when s is 12?

How many hours are spent in a passage of 3,000 nautical miles at a speed of 12 knots, and what is the total cost of the passage?

11. A feed pump of variable stroke driven by an electromotor at constant speed; the following experimental results were obtained:

Electrical Horse Power.	Power given to Water.
3·12	1·19
4·5	2·21
7·5	4·26
10·74	6·44

Plot on squared paper, and state the probable electrical power when the power given to the water was 5.

12. Mr. Scott Russell found that at the following speeds of a canal-boat the tow-rope pull was as follows:

Speed in miles per hour, -	6·19	7·57	8·52	9·04
Tow-rope pull in pounds, -	250	500	400	280

What was the probable pull when the speed was 8 miles per hour? There was reason to believe that the pull was at its maximum at 8 miles per hour, because this was the natural speed of a long wave in that canal.

UNIVERSITY OF LONDON.

MATRICULATION EXAMINATION.

SEPTEMBER, 1902.

ARITHMETIC AND ALGEBRA.

1. An iron bar is 117 centimetres long and its cross-section is a square of which the side measures 9 millimetres. Find its weight to the nearest gram, supposing the iron to weigh 7·6 grams per cubic centimetre.

2. The average of a certain set of p numbers is a, and that of another set of q numbers is b; find an expression for the average of all the numbers taken together.

The population of two towns are 107,509 and 189,160; their birth-rates per thousand are 27·9 and 25·7. Find to the same degree of exactness the birth-rate for the two towns taken together.

3. From the equation

$$t = 2\pi \sqrt{(l \div g)}$$

find l in terms of the other quantities, and calculate its value to three significant figures when

$$t = 1, \quad g = 32\text{·}18, \quad \pi = 3\text{·}1416\text{·}$$

4. A quantity m is altered in the ratio of a to b and the result is then changed in the ratio of c to d; write down an expression for the final result.

A manufacturer reduces the price of his goods by $2\frac{1}{2}$ per cent.; what percentage increase in sales after the reduction will produce an increase of 1 per cent. in gross receipts?

5. State in words the meaning of the formula

$$m(a + b) = ma + mb$$

and prove it when m, a, b denote positive whole numbers.

6. Bring the expression

$$(1 + x) - (1 + \tfrac{1}{2}x - \tfrac{1}{8}x^2)^2$$

to its simplest form; and show that when x is a positive proper fraction the value of the expression is between 0 and $x^3 \div 8$.

7. Factorise $2x^2 - x - 1$, and find the values of x which make it equal to 0.

8. Draw the graphs of x^2 and of $3x + 1$. By means of them find approximate values for the roots of $x^2 - 3x - 1 = 0$.

Calculate the roots of this equation to three significant figures.

9. The nth term of a series is $3n - 1$, whatever whole number n may be; prove that it is an arithmetic progression, and that the sum of the first $2n$ terms is $n(6n + 1)$. Check this result by giving a particular value to n.

10. The area of a rectangular plot of land is 6,000 square feet and the diagonal of it measures 130 feet; find the length and breadth of the plot.

MATHEMATICS (MORE ADVANCED).

1. What is the meaning of a^n when n is a positive whole number?

Find meanings for $a^{\frac{1}{3}}$ and a^{-3}, stating clearly the assumption which you make.

Find the values of $128^{-\frac{3}{7}}$ and $\log_{\frac{1}{2}} 2$.

2. Why is the logarithm of ·5 written as $\overline{1}\cdot69897$ and not as $- \cdot30103$?

The logarithmic sine of an angle is 9·87314. Make use of the given tables to find the angle. May your result be regarded as correct to the nearest minute?

3. Find the 10th term of the expansion of $(2a - 3b)^{15}$.

Employ the binomial theorem to find the value of $(1\cdot012)^5$ to three places of decimals.

4. Find M and H from the following data:

$$\frac{M}{H} = \frac{d^3 \tan a}{2}, \quad MH = \frac{4\pi^2 I}{t^2},$$

where $\quad d = 20, \quad a = 18°, \quad I = 169, \quad t = 13\cdot3, \quad \pi = 3\cdot14.$

Use the tables provided.

5. Construct an equilateral triangle whose area shall be 3 times that of a given equilateral triangle, explaining every step in your work.

6. Give some method of finding the formula for the area of a circle whose radius is r.

What is the circumference of a circle whose area is 1 acre? ($\pi = 3\cdot1416$).

7. The tangent of one acute angle is 7, and the sine of another is 0·7; find graphically the cosine of the difference between the angles, explaining the constructions and measurements which you make.

Check your result by measuring the difference of the angles with your protractor and finding its cosine from the tables.

8. Solve completely a right-angled triangle in which
$$a = 68 \cdot 07, \quad A = 39°.$$
Show that Δ, the area of the triangle, may be found by the formula
$$\log 2\Delta = 2 \log a + \log \cot A.$$

9. Prove the formula
$$\cos A = \frac{b^2 + c^2 - a^2}{2bc},$$
and use it to find to the nearest degree the largest angle of a triangle of which the sides measure 3, 4, and 6 inches.

Construct a triangle of this shape with its longest side equal to 2·4 inches; measure its angles with your protractor, and check by adding the results.

10. Taking rectangular axes, plot off the points $(-1, 2)$ and $(3, 4)$, and draw the line represented by
$$2x - y - 3 = 0.$$
Find the co-ordinates of the point on the given line which is equidistant from the given points.

11. Find the equation of the circle which passes through the points $(-1, 2)$, $(3, 4)$, and has its centre on the line
$$2x - y - 3 = 0.$$
Give a diagram.

Prove that the tangents to this circle which cut the axis of x at 45° are represented by
$$x - y - 1 \pm 2\sqrt{5} = 0.$$

ANSWERS.

Exercises I., p. 3.

1. 124·971046.	**2.** 26·010801.	**3.** 706·42724.	**4.** 38·732229.
5. 280·68054.	**6.** 1290·657788.	**7.** 332·72973.	**8.** 32·04147.
9. 107·060597.	**10.** 472·979307.	**11.** 100·610704.	**12.** 98·0246457.
13. 1·15855.	**14.** 11·200568.	**15.** ·05444.	**16.** 101·68787.

Exercises II., p. 11.

1. 0·34118.	**2.** ·014955.	**3.** 501·8551.	**4.** ·0312034.
5. ·6248501.	**6.** ·2074272.	**7.** 756·872.	**8.** 5·329956.
9. 5·20163.	**10.** 2·824575.	**11.** ·1481883·	**12.** 4·41063.
13. 3·349313.	**14.** 183·6587.	**15.** ·049265.	**16.** 10·84589.
17. 1·5581.	**18.** 1·15421.	**19.** 73·93787.	**20.** 6·955714.
21. 2·114.	**22.** ·0560682.	**23.** 2·332714.	**24.** ·014056093.
25. 189.	**26.** ·3472.	**27.** 8·304 pence.	**28.** 33·7708 hrs.
29. ·37·072 lbs.	**30.** 4621·32 ft.	**31.** 8s. 7¾d.	**32.** ·75·
33. ·325.	**34.** 759·725.		

Exercises III., p. 14.

1. ·00198.	**2.** ·02665.	**3.** 575.	**4.** 30·16.
5. 470.	**6.** ·012.	**7.** ·0645.	**8.** 296000.
9. ·00545.	**10.** ·0125.	**11.** ·00892.	**12.** ·01733.

13. 846 ; remʳ, ·0047 ft.   **14.** 6·453.   **15.** ·34118, ·01733.

16. 29·7.   **17.** 17404.   **18.** 1217·6.   **19.** 53·05.

20. 563·54.   **21.** (i) 3·123, (ii) 1704.   **22.** (i) ·01495, (ii) ·007529.

Exercises IV., p. 20.

1. 485 miles.   **2.** 7⅕.   **3.** £224 ; £240, £350.   **4.** 22 cwt. 2 qrs.

5. ·6525.   **6.** £7, £11. 13s. 4d., £16. 6s. 8d., £21.   **7.** 59½, 68, 76½.

8. 5.   **9.** £7173. 6s. 8d., £8070, £8608, £8966. 13s. 4d.

11. £126. 2s. 0d.   **12.** 5 5/11 miles.   **13.** £8. 6s. 8d.

Exercises V., p. 23.

1. 543·9 lbs., 923·5 lbs. copper, 76·5 lbs. tin.

2. 3·7 %, 7·4 %, 88·9 % ; 462 lbs., 8·8 lbs., 13·4 lbs., 177·8 lbs.

3. 5s. 3d.   **4.** £80.   **5.** 70.   **6.** 2,825,761, 2,560,000.

7. Gained 11·6 per cent.   **8.** £37. 10*s*.   **9.** 72 percentage of beer.
10. 2*d*.   **11.** 1080 candidates, 432 failures.   **12.** 35 per cent.
13. £10. 18*s*. 9*d*.; $32\frac{1}{7}$ per cent.   **14.** £2000.

Exercises VI., p. 30.

1. 193.   **2.** 2·22.   **3.** 1003.   **4.** 4321.   **5.** 11·05.
6. 8·0623, 7·0711, 2·828428.   **7.** 57·13·   **8.** 671·3.   **9.** 6·25 ; 200·02.
10. 300·03.   **11.** 82929.   **12.** 9·99·   **13.** 206.   **14.** ·0708.
15. 4321.   **16.** 32·94.   **17.** 237·96·   **18.** $\frac{3}{2\,5}$·
19. (i) ·73, (ii) ·85, (iii) ·9, (iv) 1·12.

Exercises VII., p. 55.

1. 8·66″.   **2.** 28 ft.   **3.** 33·11 ft.   **4.** 4·9 ft., 9·8 sq. ft.
5. 5·3 miles.   **9.** 6·91.   **11.** 104°·5, 46°·5, 29°, 1·38″.   **13.** 1·51, 1·6.
14. 4·16.   **15.** 48° 8′.   **16.** 1·115, 109°, 34°.   **17.** 10, 6.   **18.** 29° 56′.
19. 34° 8′, 4114 sq. ft.   **20.** (i) 23·69; (ii) 1147 sq. ft. ; (iii) 22°, 126°.
21. 2·6624 ft.; 6·217 sq. ft.; 1·864 ; 2·806 ; 3·868.
22. 36° 2′, 53° 56′, 90° 2′.   **23.** 9·196.   **24.** 2·52, 1·92.

Exercises VIII., p. 61.

1. 20.   **2.** 0.   **3.** 0.   **4.** 3.   **5.** 27.   **6.** 1·058.   **7.** 172800.
8. 4·022.   **9.** 0.   **10.** 2.   **11.** $1\frac{1}{2}$.   **12.** -2.   **13.** 1.   **14.** 3.

Exercises IX., p. 63.

1. $\frac{7}{6}a+\frac{7}{6}b+\frac{7}{6}c$.   **2.** $3ax^2-3bx^2$.   **3.** $16m-11n$.   **4.** $5a+7b-6c$; 1.
5. $2x+3y$.   **6.** $3b$.   **7.** $3ax^2-x^2-dx^2-2x+bx-f$.
8. $8a+2b+4c$; $2\frac{2}{3}$.   **9.** $x^4+y^4-x^2y^2$.   **10.** $26xy-5x^2-5y^2$.
11. x^6-21x^2+20.   **12.** $2xy+x-x^2+y^2+66$; 2666.   **13.** $8x-4y$; 8.

Exercises X., p. 65.

1. $3a+b-c$.   **2.** $3a-10b$.   **3.** $3x^2-8x+8$.   **4.** $10b$.
5. $ax+cy$.   **6.** $ax-cx-ay-cy$.   **7.** $3m-n-2p$.   **8.** $xy-xz$.
9. $a+3b-c+3d+4e$.   **10.** $3y^2+7xy-11xz+z^2$.
11. $a+2b+3c+4d$; $2a+2b+2c+2d$.   **12.** $2c-a-b+d$; x^2-y^2.
13. $2a^3-3a^2b$.   **14.** $x^3y+12x^2y^2+10xy^3+21y^4$.
15. $2a^4+3a^3b+3a^2b^2+2ab^3+b^4$.   **16.** $2x^3+31ax^2-31a^2x+7a^3$.
17. $44ab+33ax+24cy+43ez$.

Exercises XI., p. 68.

1. $x^4+a^2x^2+a^4$.   **2.** $4a^6+11a^4b^2+7a^2b^4-b^6$.   **3.** $x^8+x^4y^4+y^8$.
4. x^6-21x^2+20.   **5.** $x^4-8x^2y^2+16y^4$.   **6.** $x^6-24x^4+144x^2-256$.

7. $a^{12} - 3a^{10}b^2 - 2a^8b^4 + 13a^6b^6 - 3a^4b^8 - 12a^2b^{10} + 6b^{12}$.

8. $x^6 - y^6$. 9. $8a^5b - 26a^3b^3 + 2ab^5$.

10. $a^6 - a^2b^4 - a^4b^2 + a^2c^4 + b^6 - a^4c^2 - b^2c^4 + b^4c^2 - c^6$.

11. $1 - y^2 - y^3 - y^4 - y^5 + y^6 + 2y^7 + y^8$. 12. $6x^8 - 11x^6 + 22x^4 - 4x^2 - 7$.

13. $x^5 + 4x^4 + 48x - 32$. 14. $16a^4 - 72a^2b^2 + 81b^4$.

15. $-13a^3 - 22a^2 + 96a + 135$.

Exercises XII., p. 70.

1. $2a^3 - 3a^2 + 2a$. 2. $x + 2y - z$. 3. $a + b + c$.

4. $2x^3 - 6x^2y + 18xy^2 - 27y^3$. 5. $\dfrac{x - y}{x + y}$. 6. $3a + 2b - c$.

7. $5a + b - 2c$. 8. $-9ab^4c^2$. 9. $4xy + 2y + 3x + 1$.

10. $3x^2 - 2x - 9$. 11. $2a^2 - 2b^2 + 3c^2$. 12. $a^2 - a + 1$.

13. $a^4 - 4a^2bc + 7b^2c^2$. 14. (i) $\dfrac{a^2 - b^2}{a + b}$, (ii) $\dfrac{a^3 + b^3 + c^3 - 3abc}{a + b + c}$.

Exercises XIII., p. 72.

1. $3\frac{2}{3}$. 2. 5. 3. c. 4. $10x - 7y + 16z$; $-20x + 14y - 32z$.

5. $17a$. 6. $1\frac{83}{84}$. 7. $-4x - 3y + 2z$; $-6\frac{1}{2}$.

8. $-3xy - y^2$; $1\frac{1}{4}$. 9. $-x + 3z$; -7. 10. $2(5c + a)$.

11. $1 + a^{\frac{1}{2}}$. 12. $8ab$. 13. $12b(a - b)$. 14. $-6ab - b^2 - a^2$.

Exercises XIV., p. 77.

1. $(x - 2)(x - 5)$. 2. $(x - 10)(x + 9)$. 3. $(x - 4)(x + 1)$.

4. $(x + 5)(x - 3)$. 5. $(3a + 2b)(9a^2 - 6ab + 4b^2)$.

6. $(2x - 3)(4x^2 + 6x + 9)$. 7. $(x - 6)(x + 5)$. 8. $(x + 17)(x - 5)$.

9. $(x - 2y)(x - z)$. 10. $3(x - 3y)(x + 3y)$. 11. $(x + 25)(x - 7)$.

12. $(x - 2y)(x - 3z)$. 13. $(25x^2 + y^2)(5x - y)(5x + y)$.

14. $(10x - 1)(x + 8)$. 15. $x(x - 6y)(x - 7y)$.

16. $(a + b + c)(a + b - c)(a - b + c)(a - b - c)$. 17. $(x - y)(x^2 - 5xy + 7y^2)$.

18. (i) $(a + b - c - d)(a - b - c + d)$;

   (ii) $(p + q + r)(p + q - r)(p - q + r)(p - q - r)$; (iii) $(1 - m^{\frac{1}{2}})(1 - m)$.

Exercises XV., p. 80.

1. $\dfrac{4bx}{3a}$; 2. $a - x$. 3. $\dfrac{x^2 - ax + a^2}{a + x}$. 4. $\dfrac{2 + 3x}{1 + 5x}$.

5. $\dfrac{4x^2 - 4x + 1}{4x^2 - 3x - 1}$. 6. $\dfrac{x - 1}{x + 1}$. 7. $\dfrac{4xy}{x^2 - y^2}$. 8. $\dfrac{x^2 + a^2}{x}$.

9. 1. 10. $\dfrac{2}{x - 3}$. 11. $(x^2 + 1)^2 - x^2$, $(x^2 + 1 + x)(x^2 + 1 - x)$.

12. $\dfrac{7ab}{2c^2}$.   **13.** $\dfrac{a+x}{a^2-bx}$.   **14.** $x+\dfrac{1}{x-1}$.   **15.** 1.

16. $\dfrac{2x+5}{5x+2}$.   **17.** $\dfrac{1}{3x-2y}$.   **18.** $\dfrac{2}{x-3y}$.   **19.** $\dfrac{3x^2+1}{x^2+2x-3}$.

20. $\dfrac{4xy}{x^2-y^2}$.   **21.** $\dfrac{1}{(x+1)(x^2+1)}$.   **22.** $\dfrac{2}{a-b}$.

Exercises XVI., p. 85.

1. 2.   **2.** $\frac{7}{2}$.   **3.** 2.   **4.** 8.   **5.** 9.   **6.** 24.   **7.** 8.

8. 4.   **9.** 6.   **10.** 2.   **11** 3.   **12.** 2.   **13.** 23.   **14.** 43.

15. -5.   **16.** 6.   **17.** $a(b-a)$.   **18.** ab.   **19.** b.   **20.** $\dfrac{a+b}{2}$.

Exercises XVII., p. 88.

1. 54; 21.   **2.** 16; 9.   **3.** 420.

4. $27\frac{3}{11}$ past 2; at 3, and at $32\frac{8}{11}$ min. past 3.   **5.** 9 oz., 12 oz., 16 oz.

6. A is 54, B 12.   **7.** A is £37, B is £27, C is £47.

8. 75.   **9.** 32, 48, 480.   **10.** $11\frac{1}{9}$ yards.   **11.** £6000, £5000, £3000.

12. A £400, B £160, C £140.   **13.** 30 hours.   **14.** 25, 24.

15. 10, 15.   **16.** 15.   **17.** 120.   **18.** $3\frac{1}{2}$ miles.   **19.** 19 : 16.

Exercises XVIII., p. 95.

1. 5, 6.   **2.** 5, 4.   **3.** 30, 20.   **4.** $\frac{3}{2}$, $\frac{2}{3}$.   **5.** $\frac{2}{7}$, $2\frac{1}{3}$.   **6.** 3, $\frac{1}{3}$.

7. $\frac{3}{2}$, $\frac{2}{3}$.   **8.** $\frac{1}{2}$, 2.   **9.** 7, 2.   **10.** 3, -4.   **11.** 4, 5.   **12.** 16, 35.

13. $x=\dfrac{2}{a-b+c}$, $y=\dfrac{2}{a+b-c}$, $z=\dfrac{2}{b-a+c}$.

14. $\dfrac{pq(qm+pn)}{p^2+q^2}$, $\dfrac{pq(pm-qn)}{p^2+q^2}$.

15. $\frac{5}{2}$.   **16.** $7\frac{323}{596}$, $-\frac{253}{149}$.   **17.** a, b.   **18.** $x=y=\dfrac{n}{a+b+p-q}$.

Exercises XIX., p. 98.

1. $\frac{4}{5}$.   **2.** 16, 2.   **3.** $\frac{7}{17}$.   **4.** 72.   **5.** $\frac{8}{15}$.   **6.** 13, 10.

7. $\frac{19}{51}$.   **8.** 9, 15.   **9.** $1\frac{1}{3}$ hrs.   **10** Horse costs £25 ; cow £18.

11. $\frac{6}{7}$.   **12.** A 27, B 22.   **13.** 2 : 5.   **14.** 10 gallons.   **15.** $v^2-u^2=2fs$.

16. 90 : 89.   **17.** £1250000 ; £128048. 15s. 7d.

Exercises XX., p. 105.

1. A by £1. 5s. 0d.   **2.** $\frac{5}{6}$.   **3.** $5\frac{1}{4}$ hours.   **4.** 16 days.

5. 25 lbs. per sq. in.   **6.** 15.   **7.** 4·5.   **8.** 15.   **9.** 18.

10. 35 days.   **11.** 1000.   **12.** 9.   **14.** £15012.

Exercises XXI., p. 112.

1. $a + b^2 + c^3 - 3a^{\frac{1}{3}}b^{\frac{2}{3}}c$.   **2.** (i) 21·656 ; (ii) 17·656.   **3.** $c^{\frac{1}{6}}$.

4. $a^{-\frac{3}{2}}b^{-\frac{7}{12}}$.   **5.** $2a$.   **6.** $a^{q(p+x)}$.   **7.** $a^{12(m-n)}$.

8. $x^{4m} - 4x^{3m+n} + 6x^{2m+2n} - 4x^{m+3n} + x^{4n}$.   **9.** ·000024.

10. ·0015 per cent.   **11.** 1·001, ·992, ·00162 %.   **12.** ·0008 %.

13. 4·2172, 2·3713.   **14.** 1.006, ·999.

15. 1·2432, 1·6548, 2·3758.   **16.** $x^{\frac{3}{m}-\frac{2}{n}}y^{n-m}$.

17. $a^{2\frac{1}{6}}b^{\frac{7}{6}}c^{\frac{7}{36}}$.   **18.** $2^{\frac{1}{m^3}}ab^2c^{\frac{2}{m}}$.   **19.** $b^{\frac{3}{2}}$.   **20.** $\left(\dfrac{a}{b}\right)^{mn}$.

21. (i) $a^{-1}b^{\frac{1}{6}}$, (ii) $a^{\frac{2}{3}}x^{-\frac{2}{3}} + a^{\frac{1}{3}}x^{-\frac{1}{3}} + a^{-\frac{1}{3}}x^{\frac{1}{3}} + a^{-\frac{2}{3}}x^{\frac{2}{3}}$.

22. 5·44·   **23.** $x^{-1}y^{\frac{1}{6}}$.   **24.** 12·127·

Exercises XXII., p. 119.

1. ·929, 8·361.   **2.** 1011·68.   **3.** 836·113.   **4.** ·2019.

5. ·645137.   **6.** 10.   **7.** latter, ·033.

Exercises XXIII., p. 124.

1. 4·167.   **2.** 48 to 1.   **3.** 97·25 lbs., 145·9 lbs.

4. 36·4 c.c. 7·5.   **5.** ·729 inches.

6. 29·92 in., 33·9 ft., 14·7 lbs., 2116, 1·034.

7. 4·15 kilog.   **8.** 719·6.   **9.** 92·9 tons.   **10.** 1·4.   **11.** ·72·

Exercises XXIV., p. 130.

1. ·0007736.   **2.** ·00001573.   **3.** ·07502.

4. 34·67.   **5.** ·2025·   **6.** 2·583, ·000744.

7. ·01374.   **8.** 4·08.   **9.** 78·77.

10. (i) ·1097 ; (ii) 973·6 ; (iii) ·09761 ; (iv) ·00007381.

11. (i) 157·8 ; (ii) 416·8.

Exercises XXV., p. 132.

1. 4·799·   **2.** ·00025, 250000.   **3.** ·000009687.

5. 165000.   **6.** (i) 1262, (ii) ·8042.   **7.** ·0006398.

8. ·02665.   **9.** ·06039.   **10.** (i) 50·67 ; (ii) ·0004511.

11. (i) 1·285, (ii) 33·29, (iii) 53·32·   **12.** (i) 3·468 ; (ii) 346·8.

13. (i) ·6797, (ii) 67·97.   **14.** ·624.

15. ·1394, 2·283.   **16.** ·01496, ·00753.   **17.** 7·446, ·01254.

Exercises XXVI., p. 140.

1. ·00917.   **2.** 1·078.   **3.** ·0001404.   **4.** $\bar{1}$·4779, $\bar{1}$·6797.

5. $\bar{1}$·6796.   **6.** 0·5611.   **7.** 1·1999.   **8.** 1·3865.

9. ·003176.   10. ·6869.   11. $\bar{1}$·4577, $\bar{3}$·0701.   12. 20·95.

13. 3·546.   14. 1·565 × 10⁷.   15. ·2311.   16. 9·2 × 10⁶.

17. ·4055, ·6931, ·9163, 1·0986, 1·2528, 1·3863, 1·5041, 1·6094.

18. 5·435.   19. 6·575.   20. 1·948.   21. ·4409·

22. ·2928.   24. 263·3, ·2353.   25. − 17·75.   26. 2·682·

27. 5·67.   28. a = − 57·1, b = 26·63, 671·2.

29. 5·228, 1·222, ·4956, ·2563, ·1665·   30. ·04801; (ii) ·2869; (iii) ·2291.

31. 20·78.   32. ·02147.   33. 2·885.

Miscellaneous Exercises XXVII., p. 142.

1. 2·865.   2. 88·1.   3. 1·658.   4. (i) ·894 ; (ii) ·891.

5. 33·64, 6·995.   6. 6·686.   7. 2·892.   8. 1588.

9. 1·443.   10. 1·015.   11. ·9035.   12. ·7578.

13. 98·51.   14. (i) 1503, a = 243·9, b = − 26·6, y = 671·86.

Exercises XXVIII., p. 153.

1. 13·75 ft.   2. The former.   3. 5°·73.   4. $\frac{1}{9}$, 6° 22′.   5. 108°.

6. 17°·19.   7. ·3708.   8. 2·36, 135°.   9. 1 foot.

Exercises XXIX., p. 162.

2. $\frac{4}{5}$, $\frac{3}{4}$·   3. $\frac{99}{101}$, $\frac{99}{20}$·   4. $\frac{2}{3}$, $\frac{\sqrt{5}}{3}$·

5. $\frac{3}{5}$, $\frac{15}{4}$, $\frac{21}{5}$·   7. 2·38 in.   8. 2·64 in., 5·014 sq. in.

9. 2·843, 1·991.   10. 36°·52′, ·96 sq. ft.

Exercises XXX., p. 167.

1. $\bar{1}$·4603, ·5371.   2. ·8102.   3. ·7903.   4. $\bar{1}$·8492, ·6141.

5. ·419·   6. ·6362.   7. ·2057, ·4429·   8. ·0887.   9. ·248·

10. ·8461.   11. 1·15, 1·9918, 2·2216, 2·2216, 1·9918, 1·6261, 2·2.

12. − ·4317·   13. 30140.   14. 336.   15. ·1526, 1088.

16. 2·007·   17. − ·1387.   18. 7·718.   19. ·02076.

Exercises XXXI., p. 169.

1. 117·7.   2. 488·5.   3. 43·3.   4. 120.

6. 1·225 miles.   7. 10·62 miles.   8. 173·2 ft.   9. 732·1.

10. 12·13 ft.   11. 8·869 miles.   12. 151·5 ft.   13. 8768 yds.

14. 3960.   15. 1034 ft.   16. 3·18 miles per hour.

Exercises XXXII., p. 185.

1. E = ·0427R + 4·4, 100 lbs.

2. (i) E = ·118R + 1·84, F = ·0736R + 1·83 ; (ii) E = ·042R + ·35, F = 2R + 25 ; (iii) E = ·118R + 1·75, F = ·077R + 1·75.

3. (i) $n = 2\cdot02 \log N - 4\cdot14$; (ii) $n = 2\cdot33 \log N - 4\cdot79$; (iii) $n = 2\cdot32$ $\log N - 4\cdot47$.　　　　**4.** 795·8 lbs. per hour.

5. $B = \cdot0208A + 6\cdot3$, ·84%.　　　　**6.** $M = 1\cdot42 + 4\cdot66N$.

7. (i) $a = \cdot041$, $b = \cdot173$; (ii) $a = 119$, $b = 45\cdot7$, error 2·6%.

8. $L = 1\cdot49\,T^2 + \cdot537$.　　　　**9.** $a = 8\cdot8$, $b = -14$.

10. $a = 2500$, $b = 26$, $W = 2500 + 26P$, $W = 4320$, $W \div P = 76$, 51, 61·7.

11. (i) $d = \cdot75t + \cdot48''$; (ii) $d = 1\cdot2\sqrt{t}$; (iii) ·67, ·79, ·9.

12. $1\cdot7d + \cdot23$ in., $A = \cdot6d^2$, $A = (\cdot593d^2 - \cdot3)$ sq. in.

13. $C = \cdot344$, $n = 1\cdot79$.

Exercises XXXIII., p. 198.

1. $38d.$, $59d.$　　**2.** $-39°\cdot2$, $1\cdot47''$, $2\cdot25''$.　　**3.** 13·77·

5. 22·1, 38·2, 63·3, 27·8, 0·5 million per annum.

6. $4''$ is $68\cdot3s.$, $5''$ is $91\cdot65s.$; $63\frac{1}{2}s.$

7. 2·23, 3·22.　　**8.** 20·06, −1·86.　　**9.** 4·3, −1·376.　　**10.** 1, 2, 4.

11. 2·18.　　**12.** 2·11.　　**13.** 2·22.　　**14.** $n = 1\cdot08$.

15. (i) ·3594; (ii) 1·4435.　　**16.** 1·2953.　　**17.** ·225.

18. −3, −1, 4.　　**19.** −4, −13, 17.　　**20.** −2, 5·898, −3·898.

Exercises XXXIV., p. 214.

1. $5 + 4\cdot2t$, 26.　　**2.** $S = 255$ when t is 5, aver. vel., 82·007, actual 82.

3. 81·8, 80·02, 80·0002, actual speed 80.　　**4.** 3·15, when $r = 0\cdot5$.

5. Aver. force $= 3535$ lbs.; work $= 3535 \times 70 = 247450$ ft. lbs.

6. 10·4, 10·004, 10·0004, 10.　　**7.** aver. val. $= 1\cdot924$.

8. 16·32, or −1·376.　　**9.** ·5491.　　**10.** 2·35.

11. 8 and 4.　　**12.** 2·23, 3·22, aver. value 0·57.

13. Rate of increase 2·014, aver. value $= 10\cdot08$.

14. 0·1558, 1·902.　　**15.** 0·30056, 1·785.

16. (i) nax^{n-1}; (ii) $5ax^4 + \frac{1}{3}bx^{-\frac{2}{3}} + pcx^{p-1} - \dfrac{dx^{-\frac{1}{q}-1}}{q}$.

17. 210, 210.　　　　**19.** 5, 15.

Exercises XXXV., p. 219.

1. 10·87 sq. ft.　**2.** $13\frac{1}{7}$ ft.　　**3.** 488·87 ft.　　**6.** 60 ft.

7. 600·3 sq. ft.　　　　**8.** 8 and 6.　　**9.** 21·82 sq. ft.

10. 480.　　**11.** £4. 7s. 6d.　　**12.** 1 ft. 6 in.　　**13.** 2376·9.

Exercises XXXVI., p. 222.

1. 6·186 sq. ft.　**2.** 84 sq. ft.　　**3.** 210 sq. in.　　**5.** 60 sq. yds.

6. 2390.　　**7.** 3000 sq. ft.　　**8.** 150, 200, 250, 45,000 sq. yds.

9. 270 sq. ft.　**10.** ·538 sq. ft.　**11.** 15 ac.

12. 1·155 miles, ·2421 sq. miles.

Exercises XXXVII., p. 223.

1. ·5, ·75, 1·5, 3·75, 5·499, 40·75·
2. 5·498, 7·854, 14·92, 25·13, 95·82, 212·058.   3. 22 ft. 7·434 in.
4. 4967.   5. 26400, 6·365 ft.   6. 63·65, 58·76.   7. 5¾ miles.
8. 180.   9. 1·91 ft., 2·228 ft.   10. 5712 ft.

Exercises XXXVIII., p. 227.

1. 64 in.   2. 3820 sq. in.   3. 4854 sq. in.
4. (i)·944; (ii)·004; (iii)·02; (iv)·2.   5. (i)·003218; (ii)·00933; (iii)8·553.
6. £2. 18s. 10·9d.   7. 140·3 sq. ft.   8. 11385·3 sq. ft.
9. 13·36 sq. in.   10. 7·658 sq. in.   11. 43·5 in.
12. 82·47 sq. ft.   13. 488·9 sq. ft., 64 ft.   14. ·982 sq. ft.
15. 524·8 sq. in.   16. 1472 sq. ft.   17. 56 ft. 8 in.
18. 65·2 ft.   19. 196 ft.   20. 102·09 sq. ft.   21. 2065·03 sq. ft.

Exercises XXXIX., p. 239.

1. $\frac{r}{4} \times 5^2 = 19·635$, Simpson's 19·45, error 0·8 %.   2. 12797 cub. ft.
3. 2720 lbs.   4. 80·2 sq. yds.   5. 236 lbs.
6. 49988 sq. in., 320·4 in.   7. 58·2, 58·92.   8. 375·2 sq. ft.
9. 674·08 sq. ft.   10. 2853·9 sq. ft.
11. 2794·7 sq. ft.   12. 923·3 sq. ft., 15·39 ft.

Exercises XL., p. 244.

1. 2½ ft., 13s. 1½d.   2. 64 cub. ft., 398·72.
3. 13 cub. ft., 81, 810 lbs.   4. 6·191.   5. 3000 kilos.
6. 2359.   7. 11·51 cub. ft., 9·245 cub. ft.
8. 3·984 cub. ft., 230·9 lbs.   9. 151·4 lbs.   10. 3·578 tons.
11. 33·48 tons.   12. 9·048 in.   13. 3·329 feet.
14. 5·556 cub. ft., 10·44 cub. ft.   15. 1500 kilos.

Exercises XLI., p. 248.

1. (i) 2·82, 106·4 sq. in.; (ii) 3·538″, 66·68 sq. in.; (iii) 502·62 cub. in., 251·3 sq. in.
2. 16 ft.   3. 8 ft.   4. ·06186 in., 2·64 in.
5. 23·58 cub. in., 66·16 sq. in.   6. 2222 lbs.
7. 238·8 sq. ft.; 4352 lbs.   8. 39·5 in.
9. 6·443 in.   10. 9563 yds.

Exercises XLII., p. 250.

1. (i) 4·887″; (ii) 5·3″; (iii) 301·6 cub. in., 188·6 sq. in.
2 10·47 ft.   3. 50·264 cub. in., 13·068 lbs.
4. 47·13 cub. ft., 54·95 sq. ft.   5. 7″.

6. 33 cub. ft.   **7.** 132 cub. ft.

8. 92·02 sq. in., 92·5 sq. in., ·26%. The third figure is only approximately correct and hence seven of the ten figures are unnecessary.

Exercises XLIII., p. 253.

1. 8·555″, 452·4 sq. in.   **2.** 655·8 lbs.   **3.** 2267 lbs.

4. (i) 7·442″; (ii) 3·385″.   **5.** 11·62″.   **6.** 4·083″.

7. ·79 lbs.   **8.** 2·33 tons.   **9.** 16 ft. 3 in.

10. 1″.   **11.** 40·62 sq. in.

Exercises XLIV., p. 256.

1. 631·7 sq. in., 631·7 cub. in., 176·9 lbs.   **2.** 7843 lbs.   **3.** $3\frac{1}{2}''$.

4. 9·8″.   **5.** 128·03 cub. ft.   **6.** 6636 cub. in., 1725·36 lbs.   **7.** 3327 lbs.

8. (i) 118·4 cub. in., 236·9 sq. in. ; (ii) 1″, 5·065″; (iii) ·9187.

9. External radius 5·2″, Internal radius 3″.

10. 702·6 cub. in., 182·7 lbs.

Exercises XLV., p. 259.

1. 91·6 cub. ft., 92·6 cub. ft.   **2.** 104 sq. in., 14980 cub. in.

3. 28·9 cub. ft.   **4.** 54·86 cub. ft.   **5.** 133 cub. ft.

6. 3405·7 cub. yds.   **7.** 40421 cub. ft.   **8.** 2544966 cub. ft.

9. 100·7 cub. ft.   **10.** 792000 cub. ft.   **11.** 3·69 ft., 73·4.

Exercises XLVI., p. 273.

1. 7·07.   **2.** 10·5   **3.** 14·15, 15·36, 17·2, 18·22.

4. 44·2″, 48° 17′, 22° 36′, 32° 52′.

5. 3·55″, 22°·7, 40°·5.   **6.** 2·5″, 2·24″, 1·8″, 2·69″.

7. 3·283, cos a = ·4568, cos β = ·7004, cos θ = ·5483.

8. $x = 3·624$, $y = 9·959$, $z = 16·96$.   **9.** 59°·7, 30°·3.

10. (i) 7·071 ; (ii) cos a = ·4242, cos β = ·5657, cos θ = ·7071·

11. $x = 1·747$, $y = 2·083$, $z = 1·268$.   **12.** 849·6 miles per hour.

13. 2439 miles, 15320, 42·55.   **14.** 69·1 miles.

15. $x = 1·293$, $y = 1·477$, $z = 2·298$.

16. (i) 5·643, (ii) ·4429, ·5493, ·7087· 1·0001.

Exercises XLVII., p. 276.

1. 4·188 radians ; 10·47 ft. per sec.

2. 2·2 radians, 13·2 ft. per sec.   **3.** 2·62 radians.

4. 4·4 ; 61·58.   **5.** 1 radian ; 38·2.

6. 9·425 ; 56·55 ft. per sec.   **7.** 12·56 ; 21·99.

Exercises XLVIII., p. 288.

1. 145·5, 20 N. of E.   2. 16, 44°·5·   3. 6·75 knots, 21° S. of E.

4. (i) 3·08, 42·5 N. of E.; (ii) ·94, 35° ·4 W. of S.; (iii) 3·79, 8·3 E. of N.

5. 2·035, 7°·8 W. of S.; 5·77, 25° E. of N.; 6·5, 11°·5 W. of S.
   $A \cdot B = 2·472$, $A \cdot C = 2·863$.

6. (i) 50·6, 26°; (ii) 42·5, − 6°·7.   7. 31·3, 52° 50′.

8. 6000 ft.-lbs. per sec.; (ii) 2652 ft.-lbs. per sec.; (iii) 0; (iv) − 1060.

9. (i) 25·07, 44° 26′; (ii) 25·07, 44° 26′; (iii) 23·68, 2° 46′; (iv) 23·68, 2° 46′.

10. 7·36 miles per hour, 28·5 W. of N.

11. $\theta = 60°$, 51° 46′, 60°, $A + B + C = 144·8$, $a = 38°$ 42′, $\beta = 101°$ 6′, $\theta = 53°$ 36′.

12. $A = 4·368$, $a = 76°$ 40′, $B = 1·6$, $\beta = 67°$ 24′.

Exercises XLIX., p. 292.

1. $3x^2 - 7x - 2$. 2. $2x^2 + 3x - 5$. 3. $x^2 + 2ax - a^2$.

4. $x^3 - 11x + 17$. 5. $5x^4 - 6x^2 - 7$. 6. $a^2 - a + 4$.

7. $5x^2 - 3x + 4$. 8. $x^3 + 4x - 21$. 9. $4x^2 - 5x + 8$.

10. $\dfrac{2x}{y} - \dfrac{2y}{x}$. 11. $3x^2 - 5xy + y^2$. 14. $3x^2 - 16x + 5$.

Exercises L., p. 298.

1. 10, − 14. 2. 8, − 40. 3. 10, 2. 4. 3, − 1.

5. 7, − 4. 6. $\dfrac{1}{4}$, $-\dfrac{163}{250}$. 7. $\dfrac{3}{2}$, $\dfrac{2}{3}$. 8. a, b.

9. 2, − 1. 10. 4, $-3\frac{5}{15}$. 11. $\dfrac{1}{2}$, $-\dfrac{3}{2}$. 12. 8, $-\dfrac{7}{3}$.

13. 1, $\dfrac{1 \pm \sqrt{17}}{2}$. 14. 1, 3, 5, − 1. 15. $\pm \dfrac{1}{\sqrt{5}}$, $\pm \sqrt{\dfrac{1}{17}}$.

16. $(1 + \sqrt{2}) \pm \sqrt{(2 + 2\sqrt{2})}$. 17. $\pm \sqrt{\dfrac{5}{2}}$. 18. 4, $\dfrac{1}{4}$, $2\frac{1}{2}$, $\dfrac{2}{5}$.

19. $\dfrac{3}{2}$, $\dfrac{2}{3}$, -2, $-\dfrac{1}{2}$. 20. $\dfrac{a + 3b}{4}$, $\dfrac{a + 5b}{6}$. 21. $x = 4, 2$; $y = 2, 4$.

22. $x = 6\frac{1}{3}$, 3; $y = -2\frac{5}{6}$, $\dfrac{1}{2}$. 23. $x = 6·8$, 4; $y = -5·4$, 3.

24. $a + b$, $a - b$; $x^2 - 2(a^2 + b^2)x + (a^2 - b^2)^2 = 0$. 25. 4·2426, − 14·142,

   $x^2 - \dfrac{a^2 + 1}{a}x + 1 = 0$.

Exercises LI., p. 300.

1. (i) 9·5 ft. (ii) 32. 2. (i) 10. (ii) $\dfrac{1}{2}$. 3. 8 in., 18 in.

4. 6000 sq. yds. 5. 60 miles. 6. £10, £15.

7. 4 miles per hour.

8. £900 for 8 months, £600 for 10 months ; rate 6 per cent.

10. 4, 6, 480.   **11.** 5 or $\dfrac{5}{7}$.

Exercises LII., p. 303.

1. $52\frac{1}{2}$.   **2.** $52\frac{1}{2}$.   **3.** 4890.   **4.** 80.   **5.** -120.

6. $-99\frac{3}{4}$.   **7.** $-133\frac{1}{5}$.   **8.** 4864.   **9.** $n(11-2n)$.   **10.** 25.

11. 10.   **12.** 630.   **13.** $369\frac{7}{16}$.   **14.** 5.   **15.** 5, 7, 9, 9, 7, 5.

16. n^2-n+1.   **17.** 25.   **18.** $\dfrac{n(n+1)}{2}$.   **19.** n^2.

Exercises LIII., p. 307.

1. $5\cdot327$.   **2.** $-3\cdot6$.   **3.** $7\cdot556$.   **4.** 765, -255.

5. (i) $-34\cdot18$.   (ii) -103.   **6.** (i) $-12\cdot65$ (ii) $-21\cdot57\cdot$

8. 27, 3.   **9.** $57\cdot6$.   **10.** (i) $\dfrac{32}{99}$.   (ii) $\dfrac{7}{9}$.   (iii) $-\dfrac{211}{8}(\sqrt{3}-\sqrt{2})$.

11. 3, 6, 9....   **12.** ±2.   **13.** A.P. $-\dfrac{1}{2\sqrt{2}+3}$.   **15.** 2, 6, 18.

Exercises LIV., p. 309.

1. $2\frac{2}{5}$, 3, 4, 6.   **2.** 5, 4, $3\cdot2$.   **3.** 7.   **4.** 5.

5. $\dfrac{1}{2}$, $\dfrac{1}{3}$, $\dfrac{1}{4}$.   **6.** 4, 16.   **7.** 1, $\dfrac{6}{5}$, $\dfrac{3}{2}$.

8. 24.   **9.** $2\frac{2}{11}$, $2\frac{2}{5}$, $2\frac{2}{3}$.

10. $\dfrac{13}{4}$, 3, $\dfrac{36}{13}$; 2, $\dfrac{13}{4}$, $\dfrac{9}{2}$; 2, 3, $\dfrac{9}{2}$; 2, $\dfrac{36}{13}$, $\dfrac{9}{2}$

EXAMINATION PAPER, 1901.

1. $7\cdot446$; $0\cdot01254$; $5\cdot68$; 1546.   **2.** (a) $1\cdot691$; (b) $0\cdot515$.

3. $d=3\cdot43''$; 1552 sq. in.   **4.** $4\cdot009$; $6\cdot662$; $53°$; $10\cdot66$ sq. in.

5. £1100000.   **6.** 260.   **7.** 1350 ; 1700.

8. (a) $\dfrac{\sqrt[3]{a^2(a^2+b^2)}+3}{a\times\sqrt{b}}$; (b) $\dfrac{1}{x-4}-\dfrac{1}{x-3}$; (c) 2 miles per hour.

9. *Hint.* Let x be the number, then $x+\dfrac{2\cdot25}{x}=y$;

$$\therefore \frac{dy}{dx}=1-\frac{22\cdot5}{x^2}=0 \text{ or } x=\sqrt{2\cdot25}=1\cdot5.$$

10. $x=3\pm\sqrt{3}.=4\cdot732$; $1\cdot268$.   **11.** 230.   **12.** $v=150-10t$.

13. (1) $5\cdot643$; (2) $\cos\alpha=\cdot4429$; $\alpha=63°\cdot7$; $\cos\beta=\cdot5493$; $\beta=56°\cdot7$; $\cos\theta=\cdot7087$; $\theta=44°\cdot9$; (3) 1.0001.

EXAMINATION PAPER, 1902.

1. $3\cdot123$, 1704, $1\cdot722$, $\cdot0198$.
2. (a) 14407, 16604, 18557, 18815; (b) 55 ft.
3. (a) $(x^3 - xy^2)^{\frac{3}{2}}_{x+y+z}$; (b) $\dfrac{2}{x+3} - \dfrac{1}{x-5}$; (c) $45\cdot6$, $30\cdot4$.
4. $r=3\cdot5\cdot$ 5. $40\cdot1$ ft. per sec., $40\cdot01$ ft. per sec., 40 ft. per sec.
6. $a=2\cdot2$, $b=0\cdot11$, $z=4\cdot4$.   7. $53\cdot56$ sq. in.
8. $AD=2\cdot6624$ ft., $6\cdot2167$ sq. ft., $BD=1\cdot864$, $DC=2\cdot806$, $AC=3\cdot868$.
9. Value of v is about 9.
10. Converted marks are: $118\cdot9$, $160\cdot8$, and $213\cdot3$.
11. 13550, 14350, 14740.   12. $x=2\cdot012$.

EXAMINATION PAPER, 1903.

1. $284\cdot7$, 2817, $3\cdot339$, $1\cdot93$, 1768000, $11\cdot03$.
2. (a) $40°$; (b) $t=1\cdot5$ sec., $s=2685$ ft.; (c) 96.
3. (a) $\left\{ \dfrac{2x^{\frac{3}{2}} + y^2 z^{\frac{1}{3}}}{x+y^{\frac{1}{2}}} + 4 \right\}^{\frac{1}{2}}$; (b) $\dfrac{2}{x-4} + \dfrac{1}{x+1}$; (c) $3\cdot229$, $1\cdot753$;
   (d) $55° 55'$.
4. 304 sq. in , $232\cdot2$ sq. in., 33669 cub. in.
5. Average rates, $A=2\cdot8$, $B=2\cdot4$; A's$=11\frac{1}{2}$, $4\cdot1$, $1\cdot5$.   7. $1\cdot645$.
8. (a) Thickness radially, $7\cdot124$ in.; thickness the other way, $11\cdot4$ in.; inside radius, $99\cdot7$ in.
   (b) $R=\dfrac{16k}{\pi^2 \rho} \times \dfrac{1}{d^4}$, where k is a constant.
9. $y=18e^{0\cdot26x}$.   10. $0\cdot84$, $1\cdot65$, $58°$, $145°$, $87°$.   11. 21 horses.
12. 12080, $33\cdot55$, $5\cdot592$ in.   13. $0\cdot046$, March, $0\cdot072$.
14. $T=0\cdot95U + 0\cdot525W = 0\cdot6U + 0\cdot28$; $W=3\cdot28$.

MATRICULATION EXAMINATION, 1902.

1. $720\cdot2$ grams.   2. $\dfrac{pa+qb}{p+q}$; $26\cdot8$.   3. $\dfrac{gt^2}{4\pi^2}$; $0\cdot815$.
4. $\dfrac{mbd}{ac}$; $3\cdot59$ per cent.   6. $\dfrac{1}{8}x^3 - \dfrac{1}{64}x^4$.   7. $(2x+1)(x-1)$; $1, -\dfrac{1}{2}$.
8. $3\cdot302$, $-0\cdot302$.   10. 120 ft., 50 ft.

1. $0\cdot125$, 1.   3. $5005 (2a)^6 (3b)^9$; $1\cdot059$.
4. $M=221\cdot4$, $H=0\cdot1704\cdot$ 6. $246\cdot6$ yds.
8. $B=51°$, $b=84\cdot05$, $c=108\cdot1$; $\Delta=2861$.   9. $117° 19'$.

INDEX.

GLASGOW: PRINTED AT THE UNIVERSITY PRESS BY ROBERT MACLEHOSE AND CO. LTD.

Lightning Source UK Ltd.
Milton Keynes UK
UKHW02f2333160818
327365UK00011B/933/P